# Progress in Probability
Volume 38

Series Editors
Thomas Liggett
Charles Newman
Loren Pitt

# Stochastic Analysis
# and Related Topics V:
## The Silivri Workshop, 1994

H. Körezlioğlu,  B. Øksendal,   A. S. Üstünel
Editors

Birkhäuser

Boston • Basel • Berlin

H. Körezlioğlu
ENST, Dept. Réseaux
46, rue Barrault
75013 Paris, France

B. Øksendal
Department of Mathematics
University of Oslo
N-0316 Oslo 3, Norway

A. S. Üstünel
ENST, Dept. Réseaux
46, rue Barrault
75013 Paris, France

**Library of Congress Cataloging-in-Publication Data**

Stochastic analysis and related topics V : the Silivri workshop, 1994
  / H. Korezlioglu, B. Oksendal, A. S. Ustunel.
      p.   cm.  -- (Progress in probability ; 38)

  3-7643-3887-3 (acid-free paper)
    1. Stochastic analysis--Congresses.  I. Korezlioglu, H. (Hayri)
  II. Øksendal, B. K. (Bernt Karsten), 1945-  . III. Üstünel, A. S.
  (Ali Süleyman)   IV. Series.
  QA274.2S7716   1996                     95-49835
  519.2--dc20                             CIP

Printed on acid-free paper

© 1996  Birkhäuser Boston

*Birkhäuser* ®

ISBN-13: 978-1-4612-7541-1     e-ISBN-13: 978-1-4612-2450-1
DOI: 10.1007/978-1-4612-2450-1
Typeset by the authors

9 8 7 6 5 4 3 2 1

# CONTENTS

# FOREWORD

This volume contains the contributions of the participants to the Oslo-Silivri Workshop on Stochastic Analysis, held in Silivri, from July 18 to July 29, 1994, at the Nazım Terzioğlu Graduate Research Center of Istanbul University. There were three lectures:

- *Mathematical Theory of Communication Networks* by V. Anantharam,

- *State-Space Models of the Term Structure of Interest Rates*, by D. Duffie,

- *Theory of Capacity on the Wiener Space,* by F. Hirsch.

The main lectures are presented at the beginning of the volume. The contributing papers cover different domains varying from random fields to distributions on infinite dimensional spaces.

We would like to thank the following organizations for their financial support:

- VISTA, a research cooperation between the Norwegian Academy of Scineces and Letters and Den Norske Stats Oljeselskap A. S. (Statsoil).

- Ecole Nationale Supérieure des Télécommunications de Paris.

In the summer of 1994 we lost our dear friend and colleague ALBERT BADRIKIAN. We are dedicating this volume to his memory.

H. Körezlioğlu, B. Øksendal, A. S. Üstünel

# MATHEMATICAL THEORY OF COMMUNICATION NETWORKS

VENKAT ANANTHARAM *
EECS DEPARTMENT
UNIVERSITY OF CALIFORNIA
BERKELEY, CA 94720
ananth@vyasa.eecs.berkeley.edu

### Abstract

We describe some recent advances in the mathematical theory of communication networks.

## 1   Introduction

Developments in telecommunications, manufacturing, and transportation, together with mathematical developments in the theories of interacting particle systems, large deviations, Markov processes, and point processes, have stimulated research on stochastic models having the feature that streams of customers (or packets, or calls) arrive at a system of processing stations, where they occupy resources, move between stations, and eventually leave. Such models are of interest to engineers because, if chosen with sufficient care, they are able to predict the behaviour of engineering systems and aid in their design and operation. They are of interest to mathematicians because they raise a number of fascinating questions, some of which are still unresolved, and also provide a source of examples for more general theories. This area has by now a well established identity, going by the name of *stochastic networks*.

In this article we describe some basic results and sketch some recent advances in this area that have been motivated by problems in telecommunications. The exposition closely follows the pattern of a series of nine fifty-minute lectures delivered by the author at the Fifth Workshop on Stochastic Analysis of Oslo-Silivri held in Silivri, Turkey, in July 1994, at the kind invitation of

---

*Research supported in part by NSF grant NCR 88-57731

Professors Hayri Körezlioğlu, Bernt Øksendal, and Süleyman Üstünel. Only a limited range of topics could be covered during the lectures, and the decision of which topics to cover was left to the idiosyncrasies of the author; I would like to apologize in advance to colleagues whose work is not adequately exposed in this article. Some references to other survey articles that are useful in developing a broader perspective on stochastic networks are to be found at the appropriate points in the article. The focus throughout is on stochastic models.

This article has been organized into three units, each of them in rough correspondence to a distinct engineering context. In Section 2 we discuss circuit-switched networks, which are useful models for telephony. In Section 3 we discuss datagram networks, which are useful models for the existing generation of data networks. Similar models are also useful in manufacturing and transportation applications. Finally, in Section 4 we discuss more recent questions raised by the drive to merge the telephone and data networks into integrated broadband networks. The nature of the questions that are of interest here has been greatly influenced by the enormous bandwidth available in fiber-optical links.

# 2  Circuit-switched networks

## 2.1  The Erlang blocking probability formula

Consider a communication link between two large cities A and B. This link facilitates communication between individuals living in A and those living in B. If the cities are large enough one can plausibly argue, based on the well know limit theorems for a large number of rare events, that the process of call requests between A and B forms a Poisson process, say of rate $\nu$. We assume each accepted call is assigned a fixed amount of bandwidth. We say the link has capacity $C$ circuits if the maximum number of simultaneous calls it can support is $C$. We assume that a call request arriving when the link is fully loaded is rejected (blocked). If we also make the simplifying assumption that each accepted call holds its assigned bandwidth for an independent exponential duration of mean 1, then we have a simple Markov description of the process of calls in progress. It is a birth and death process on the finite state space $\{0, 1, \ldots, C\}$ with up rates $\nu$ and downrate $k$ in state $k$. Of particular interest is the stationary probability that a call request will be blocked. This is seen to be

$$E(\nu, C) = (\frac{\nu^C}{C!})(\sum_{k=0}^{C} \frac{\nu^k}{k!})^{-1} .$$

This formula is called the *Erlang formula* for blocking probability, see [32].

Two comments should be made at this point. First, the Erlang blocking probability formula gives the time-stationary probability that a call request is blocked. What is more relevant is the stationary probability that an arriving call request finds itself blocked. Here the two probabilities are the same, as a consequence of a property called PASTA (*Poisson Arrivals See Time Averages*), see [80]. The relation between event and time averages is a recurring theme in stochastic networks. For a review of the literature in this area, see Brémaud et al. [16].

The second point is that the Erlang blocking probability formula continues to give the time-stationary probability that a call request is blocked when the assumption of exponential service times is relaxed (for instance, to independent identically distributed service times of mean 1). To determine the stationary blocking probability, all that is important about the service time distribution is its mean. This is an example of an *insensitivity* property, see [18]. Several such insensitivity results are known for stochastic networks. Some of these are described in the text of Walrand [78].

## 2.2   The circuit switched network model

Consider a graph whose links are numbered $1, \ldots, J$. Link $j$ is assumed to have capacity $C_j$ (a positive integer). In addition there is a finite set of *routes* numbered $1, \ldots, R$. To each route $r$ is associated a $J$-dimensional column vector of nonnegative integers $a_{jr}, 1 \leq j \leq J$ and a Poisson process of rate $\nu_r$. The interpretation is that call requests along route $r$ arrive at the times of this process and will be accepted iff each link $j$ has at least $a_{jr}$ free circuits, in which case the call holds $a_{jr}$ circuits on link $j$ for an exponentially distributed time of mean 1, after which it simultaneously releases all these resources. The holding times of accepted call requests are independent and independent of the arrival processes. Call requests that arrive to find insufficient resources are rejected (blocked). See Figure 1.

Let $n(t) = (n_r(t), 1 \leq r \leq R)$ denote the $R$-dimensional column vector giving the number of calls along each route that are in progress at time $t$. This process is a Markov process whose state space is

$$\{n \in \mathbf{Z}_+^R \ : \ An \leq C\}$$

where $A = [a_{jr}]$ is a $J \times R$ matrix, $C$ is the $J$-dimensional column vector of the $C_j$, and the inequality is interpreted coordinatewise. This process is time-reversible, and its stationary distribution can be written as

$$\pi(n) = Z^{-1} \prod_r \frac{\nu_r^{n_r}}{n_r!} \tag{1}$$

where

$$Z = \sum_{n \,:\, An \leq C} \prod_r \frac{\nu_r^{n_r}}{n_r!} \ .$$

Let $L_r$ denote the time stationary probability that a call request along route $r$ is blocked. By PASTA, this is the same as the stationary probability that an arriving call request along route $r$ will find itself blocked. Of course the stationary distribution (1) gives us exact formulas for the $L_r$. Unfortunately these are not of much use in applications, as their use entails computing the normalizing constant $Z$, which is difficult. In view of this, we seek good approximations for the blocking probabilities.

## 2.3 The Erlang fixed point approximation

We attempt to define a notion of effective overall arrival rate of requests for individual circuits at each link $j$, call it $\rho_j$. Assuming this has been somehow defined, and that the process of overall arrivals is a Poisson process, the Erlang blocking probability formula would say that the stationary probability that link $j$ has all circuits occupied is

$$E_j = E(\rho_j, C_j) \ . \tag{2}$$

Further, the stationary rate at which individual circuits are occupied would be $\rho_j(1 - E_j)$. We now pretend that each circuit request on a link $j$ is rejected independently with probability $E_j$. This allows us to compute the effective overall rate at which circuits are occupied at link $j$ as

$$\rho_j(1 - E_j) = \sum_r a_{jr} \nu_r \prod_i (1 - E_i)^{a_{ir}} \ . \tag{3}$$

Equations (2) and (3) are to be thought of as a set of fixed point equations for the unknown quantities $E_1, \ldots, E_J$ in terms of the parameters of the model (the matrix $A$, the capacities, and the arrival rates of call requests). This technique of writing fixed point equations by making independence assumptions is called the *Erlang fixed point approximation technique*. Kelly [45], has proved that there is a *unique* solution to these fixed point equations. The quantity $\rho_j$ is called the *reduced load* at link $j$. If the assumption that individual circuits block independently approximates reality, one may hope that the approximate equality

$$1 - L_r \approx \prod_j (1 - E_j)^{a_{jr}}$$

is valid in some sense. In [45] it is shown that if one considers a sequence of circuit-switched networks indexed by a paramenter $N$, with the same matrix $A$, and with $\frac{1}{N}\nu_r(N)$ and $\frac{1}{N}C_j(N)$ converging to limits as $N \to \infty$ then we have

$$1 - L_r(N) = \prod_j (1 - E_j(N))^{a_{jr}} + o(1) \ .$$

An excellent survey of these and other results on circuit-switched networks is that of Kelly [46].


## 2.4  Dynamic Routing

The introduction of digital switches and the common-channel signalling system – see for instance [69, Sec. 12-2] for historical perspective – made it possible to consider dynamic routing strategies, where the route of a call between nodes of the network can be chosen adaptively based on traffic conditions. Such strategies are also called *non-hierarchical* routing strategies, in contrast to the hierarchical strategies that were earlier used in the telephone network. Consider the simple network of Figure 2. For each pair of nodes there is Poisson process of call requests of rate $\nu$. If at least one circuit is free on the direct link between the nodes the call is accepted and occupies one circuit for an exponentially distributed time of mean 1. However, if the link is blocked, the call request tries to make the two link connection between the nodes via the remaining node; this is feasible if there is at least one circuit available on each of the other links. If so, the call request is accepted and holds one circuit on each of these links for an exponentially distributed time of mean 1, after which it releases both of them simultaneously.

While this network can be described by a finite state Markov process, the state space requires more detail than just the occupancy numbers of the individual links. We are still interested in the stationary probability that an arriving call request is blocked. The Erlang fixed point approach may be adopted to approximate this. Let $B$ denote the stationary probability that an individual link is blocked. The effective arrival rate of requests for circuits on a link is the sum of the direct arrival rate and the arrival rate of requests on each of the other links that have to attempt alternate routing. Assuming that links block independently, this overall rate is seen to be $\nu + 2\nu B(1 - B)$, leading to the fixed point equation

$$B = E(\nu + 2\nu B(1 - B), C) \ . \tag{4}$$

A sketch of the solutions of this fixed point equation may be found in Figure 1 (i) of Gibbens et al. [37]. Remarkably, for large enough $C$, there is a range of $\nu/C$ where this equation has *multiple solutions*.

The existence of such multiple solutions suggests the possibility of metastable regimes of operation for a circuit-switched network with dynamic routing. In fact, simulations have revealed the existence of hysteresis phenomena in such networks, see [1, 37]. Namely, for certain parameter values, there is more than one qualitatively different regime of operation for the same offered traffic, with the network spending long periods of time in one or the other regime and rapidly moving from one to the other in response to fluctuations in the demand. Intuitively a situation where most calls are using alternate routes is likely to persist for a while because arriving calls will then find the network close to saturation and will be unable to make their direct connections. On the other hand, for the same offered traffic, it might also be the case that if most of the calls in progress are using their direct route, arriving calls will be able to make their direct connection.

This fascinating phenomenon has led to several analytical investigations. A simple model for dynamic routing is considered in [37]. There are $n$ links, each link comprised of $C$ circuits. At each link calls arrive as a Poisson process of rate $\nu$. If the link is not saturated then the call occupies one circuit. If the link is saturated the call chooses two distinct links at random from the $n-1$ remaining links. If neither one is saturated the call occupies one circuit from each of these two links. Otherwise the call is lost. All circuit holding times are exponentially distributed with unit mean, independent of one another and of the arrival times. Further, a call holding circuits from two links is assumed to release them *independently*.

Let $u_k^n(t)$, $0 \le k \le C$ be the fraction of the $n$ links that have $k$ occupied circuits at time t. Then $u^n(t) = (u_0^n(t), u_1^n(t), ..., u_C^n(t))$ evolves on a $C$-dimensional simplex. It is shown in [37] that as $n \to \infty$, if the initial condition $u^n(0)$ converges weakly to a limit $u(0)$ then the process $(u^n(t), t \in [0, \infty))$ converges weakly to a deterministic process $(u(t), t \in [0, \infty))$ satisfying the following set of differential equations.

$$
\begin{aligned}
\dot{u}_0 &= u_1 - (\nu + 2\nu u_C(1 - u_C))u_0, \\
\dot{u}_k &= (k+1)u_{k+1} + (\nu + 2\nu u_C(1 - u_C))u_{k-1} \\
&\quad - (k + \nu + 2\nu u_C(1 - u_C))u_k, \qquad 0 < k < C \\
\dot{u}_C &= -Cu_C + (\nu + 2\nu u_C(1 - u_C))u_{C-1}.
\end{aligned}
\tag{5}
$$

The set of fixed points of this set of equations can be seen to be in one to one correspondence with the solutions of the Erlang fixed point equation (4).

A spatio-temporal version of this model was considered by Anantharam

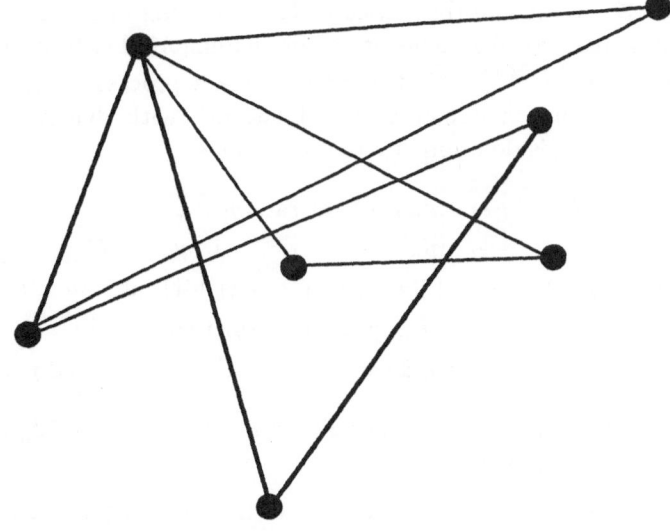

**Each link may have a different capacity**

**A 3-link route is highlighted**

Figure 1: A circuit-switched network with 9 links

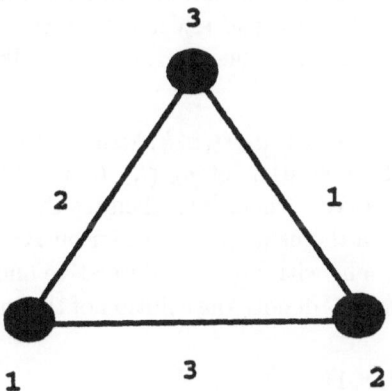

**Each link has C circuits**

Figure 2: A simple network with dynamic routing

[2]. Let $\mathbf{Z}^d/M$ denote the lattice in $\mathbf{R}^d$ consisting of points all of whose co-ordinates are rational with denominator dividing $M$. Let $W$ denote $\{0, 1, \ldots, C\}$. Let $M^*$ denote $\binom{(2M+1)^d-1}{2}$. Consider a Markov process $(\eta_t^M, t \geq 0)$ on $W^{\mathbf{Z}^d/M}$ which caricatures a circuit switched network with dynamic routing. The Markov process is described by the transitions

$$\eta(x) \longrightarrow \eta(x) - 1 \text{ at rate } \eta(x) \ ,$$
$$\eta(x) \longrightarrow \eta(x) + 1 \text{ at rate } \nu \text{ if } \eta(x) \neq C \ ,$$
$$(\eta(x), \eta(y), \eta(z)) \longrightarrow (\eta(x), \eta(y) + 1, \eta(z) + 1) \text{ at rate } \nu/M^*$$
$$\text{if } x, y, z \text{ are distinct sites with}$$
$$\eta(x) = C, \eta(y) < C, \eta(z) < C \text{ and } y, z \in x + [-1, 1]^d \ .$$

Standard techniques, for instance Theorem 3.9 of Liggett [56], ensure that the process is well defined.

In the model each site of the lattice is thought of as representing a portion of a link consisting of $C$ circuits. The value at a site represents the number of occupied circuits in the corresponding link. Thus occupied circuits become free at rate 1 and at each link there is a Poisson process of call requests with rate $\nu$. The dynamic routing is captured by the way call requests are handled : Each call request occupies one circuit on its link if available; if the link is saturated the call randomly picks two other links which are in its $[-1, 1]^d$ neighbourhood, and uses one circuit from each of these links if possible. Otherwise the call is blocked and rejected from the system. Note that because we have a compressed lattice, the interaction actually has range $M$ on the scale of links.

For $x \in \mathbf{Z}^d/M$, let $u_M(t, x, k)$ denote $P(\eta_t^M(x) = k)$, $0 \leq k \leq C$. We may extend the definition of $u_M(t, \cdot, k)$ to $\mathbf{R}^d$ by setting $u_M(t, x, k) = u_M(t, [x]_M, k)$ for $x \in \mathbf{R}^d$, where $[x]_M$ denotes the minimum element in $\mathbf{Z}^d/M$ which dominates $x$ in the usual partial order on $\mathbf{R}^d$. Let $u(0, x, k), 0 \leq k \leq C$, be continuous functions with bounded derivative and with $\sum_{k=0}^{C} u(0, x, k) = 1$. Let $u(t, x, k), 0 \leq k \leq C$ denote the solution of the integrodifferential equations

$$\frac{\partial u(t, x, 0)}{\partial t} = u(t, x, 1)$$
$$-\nu(1 + 2^{1-2d} \iint_{q, r \in [-1, 1]^d} u(t, x + q, C)(1 - u(t, x + q + r, C))dqdr)u(t, x, 0) \ ,$$

$$\frac{\partial u(t, x, k)}{\partial t} = (k + 1)u(t, x, k + 1)$$
$$+\nu(1 + 2^{1-2d} \iint_{q, r \in [-1, 1]^d} u(t, x + q, C)(1 - u(t, x + q + r, C))dqdr)u(t, x, k - 1)$$

$$-(k + \nu(1 + 2^{1-2d} \iint_{q,r\in[-1,1]^d} u(t, x + q, C)(1 - u(t, x + q + r, C))dqdr)u(t, x, k) ,$$

for $0 < k < C$ ,

$$\frac{\partial u(t, x, C)}{\partial t} = -Cu(t, x, C)$$

$$+\nu(1 + 2^{1-2d} \iint_{q,r\in[-1,1]^d} u(t, x + q, C)(1 - u(t, x + q + r, C))dqdr)u(t, x, C - 1) .$$

Then we have the following result, see [2].

**Theorem 2.1** *Fix $T < \infty$. Suppose that we start $(\eta_t^M)_t$ with initial config-uration the product measure having $P(\eta_0^M(x) = k) = u_M(0, x, k)$, $0 \le k \le C$. If $u_M(0, x, k) \to u(0, x, k)$ uniformly on compact sets, $0 \le k \le C$, then $u_M(t, x, k) \to u(t, x, k)$ for all $t \in [0, T]$, $x \in \mathbf{R}^d$ and $0 \le k \le C$.*

Theorem 2.1 is a statement about pointwise convergence of probabili-ties. There is a corresponding functional limit theorem. This functional limit theorem allows us to describe the limit behaviour of an arbitrary choice of spatial integrals $\phi^{(1)}, \ldots, \phi^{(n)}$ at times $t_1, \ldots, t_n \in [0, T]$ as long as the $\phi^{(i)}$ decrease sufficiently rapidly. This allows, for example, to describe the evolu-tion in time of spatial averages of the state over compact regions of the lattice (which caricatures compact regions of our network with dynamic routing). For details, see [2]. When we look for spatially homogeneous solutions of eqns. (6) which are time invariant, we are led to the same equations as that in the model of [37], so that these are in one to one correspondence with the solutions of the Erlang fixed point equation (4). Thus we see that for large enough $C$, there is a range of $\nu$ over which eqns. (6) admit multiple spatially homogeneous solutions. These may be loosely thought of as different phases associated to the network.

Dynamic routing schemes with symmetry on complete networks were analyzed by Marbukh, [59, 60]. Exploiting the symmetry and making the as-sumption that in the limit as the number of nodes in the network grow to infinity the evolution of any fixed finite collection of them becomes asymp-totically independent (so called *propagation of chaos* assumption; see below) limiting differential equations analogous to (5) were derived for the empirical fraction of links having various occupation numbers. These differential equa-tions also have multiple fixed points for certain ranges of the parameters; the intuitive explanation for this phenomenon is the same as that above. This propagation of chaos assumption was later proved by Graham and Méléard

[38]. Graham and Méléard [39] have also proved a fluctuation limit theorem around the law of large numbers that follows from the propagation of chaos proved in [38].

## 2.5    Propagation of Chaos

Some version of propagation of chaos is involved in all of the above examples, including the writing of Erlang fixed point approximations. What this means is the following : Consider a Markovian system of $n$ identical interacting particles and focus attention on the first $p$ of them. Suppose that as the number of particles increases to infinity the initial distribution of the first $p$ particles becomes asymptotically independent, and the empirical distribution of particle states approaches a limit. In a model having the propagation of chaos property, one can write down a time inhomogeneous Markov process such that the evolution of any given particle is described by this process, started from the appropriate initial distribution. The rates appearing in this process consist of rates associated with autonomous changes of the state of the particle and rates which represent the aggregate effects of interactions of the tagged particle with the large number of other particles. Further, the distinguished finite collection of particles will evolve independently, each according to this time inhomogeneous process. The intuition is that the probability that the distinguished collection of particles interacts with one another becomes asymptotically negligible, and because the interaction is symmetric the interactions with the other particles can be replaced by empirical rates. The terminology comes because the chaotic initial condition (finite collections of particles have asymptotically independent initial conditions) propagates. Propagation of chaos is a well studied property of interacting Markov process models. A recent survey is that of Sznitman [74].

A simple and useful mean field model in which to investigate the hysteresis phenomenon associated to dynamic routing in circuit-switched networks is a model of interacting Markov chains introduced by Uchiyama [76]. Consider a Markov process $X^n(t) = (X_1^n(t), X_2^n(t)...X_n^n(t))$ on $S^n = S \times S... \times S$ where $S$ is a finite set. $(X^n(t))_t$ is characterized by two sets of nonnegative constants $L = \{L^x(y) : x, y \in S, x \neq y\}$ and $K = \{K^{x,y}(x', y') : x, y, x', y' \in S, (x, y) \neq (x', y')\}$ and an infinitesimal generator given by

$$G_n = \sum_{k=1}^n L_k + \frac{1}{n-1} \sum_{1 \leq k < l \leq n} K_{k,l}$$

with

$$L_k \phi(\vec{x}) = \sum_{x_k' \in S} [\phi(\vec{x}_k') - \phi(\vec{x})] L^{x_k}(x_k')$$

and

$$K_{k,l}\phi(\vec{x}) = \sum_{x_k' \in S} \sum_{x_l' \in S} [\phi(\vec{x}_{k,l}') - \phi(\vec{x})] K^{x_k, x_l}(x_k', x_l')$$

where $\vec{x} = (x_1, ..., x_n) \in S^n$, $\phi$ is a real function on $S^n$, $\vec{x}_k'$ [resp. $\vec{x}_{k,l}'$] is an element of $S^n$ obtained from $\vec{x}$ by replacing $x_k$ [resp. $x_k$ and $x_l$ ] with $x_k'$ [resp. $x_k'$ and $x_l'$], and the sum $\sum_{k<l}$ is taken over all pairs $(k, l)$ such that $1 \le k < l \le n$. Also assume that $K^{x,y}(x', y') = K^{y,x}(y', x')$. (More generally a fixed number of exchangeable multi-particle interactions can be considered, instead of just considering two-particle interactions as above.)

We think of $S$ as the state space of an individual particle. Hence $X_k^n(t)$ is the physical state of the $k$th particle in a system of $n$ identical particles. The process evolves as follows: each particle $k$ evolves autonomously through a Markovian motion according to $L_k$. Pairwise interaction between particles $k$ and $l$ is controlled by $K_{k,l}$: two particles in state $x$ and $y$ change simultaneously to states $x'$ and $y'$ respectively at rate $\frac{K^{x,y}(x',y')}{n-1}$. The factor $n-1$ is there so that the interaction rate per particle is constant as $n \to \infty$. Another way to think of the pairwise interaction is as follows : for each $y, x', y' \in S$, each particle in state $x$ chooses another particle at random from the remaining $n-1$ at rate $\frac{1}{2}K^{x,y}(x', y')$ and if this particle is in state $y$ they change together to states $x'$ and $y'$ respectively. If the chosen particle is not in state $y$ nothing happens.

Now let $u_i^n(t) = \frac{1}{n}\sum_{l=1}^n 1(X_l^n(t) = i)$. Then $u^n(t) = (u_1^n(t), ..., u_{|S|}^n(t))$ is a Markov chain on the $|S|$−dimensional simplex $\Delta$ given by

$$\Delta = \{\vec{u} \in R^{|S|}, \sum_{i=1}^{|S|} u_i = 1\}$$

For $i \ne j, 1 \le i, j \le |S|$, let $T_{ij}$ be an operator defined on $\vec{u} \in \Delta$ by

$$T_{ij}\vec{u} = \vec{u} + \frac{1}{n}(e_j - e_i)$$

where $e_i$ is the unit vector in the $i$th direction. Then the infinitesimal generator $A_n$ of the Markov Process $u^n(t)$ on $\Delta$ is the operator given by

$$A_n\phi(\vec{u}) = \sum_{\substack{i,j \in S \\ i \ne j}} [\phi(T_{ij}\vec{u}) - \phi(\vec{u})] L^i(j) u_i n \tag{6}$$

$$+ \sum_{\substack{i,j,i',j' \in S \\ i \ne i', (i,i') \ne (j,j')}} [\phi(T_{ij}T_{i'j'}\vec{u}) - \phi(\vec{u})] \frac{K^{i,i'}(j,j')}{n-1} \frac{u_i n u_{i'} n}{2} \tag{7}$$

$$+ \sum_{\substack{i,j,j'\in S \\ (i,i)\neq(j,j')}} [\phi(T_{ij}T_{ij'}\vec{u}) - \phi(\vec{u})]\frac{K^{i,i}(j,j')}{n-1}\frac{u_in(u_in-1)}{2} \quad (8)$$

where $\phi$ is a continuous function on $\Delta$.

Let $u(t) \in \Delta$ evolve according to the following equation started with $u(0)$.

$$\dot{u}_i(t) = \sum_{\substack{j\in S \\ j\neq i}} L^j(i)u_j - \sum_{\substack{j\in S \\ j\neq i}} L^i(j)u_i$$

$$+ \sum_{i'\in S}\sum_{j\in S} K_1^{i',j}(i)u_{i'}u_j - \sum_{i'\in S}\sum_{j\in S} K_1^{i,j}(i')u_iu_j \quad (9)$$

for $i \in S$ where $K_1^{x,y}(x') = \sum_{y'\in S} K^{x,y}(x',y')$.

Then the idea of propagation of chaos is captured by the following theorem.

**Theorem 2.2** *For the system of $n$ interacting particles above let $(X_1(t), ..., X_p(t))$ denote the state of the first $p$ particles and $u_x^{p+1,n}(t)$ the fraction of particles $p+1 \leq l \leq n$ that are in state $x$ at time $t$. Let $u^{p+1,n}(t) = (u_x^{p+1,n}(t), x \in S)$. Suppose $(X_1(0), ..., X_p(0), u^{p+1,n}(0))$ converges weakly to a product distribution $\mu^{(1)} \otimes ... \otimes \mu^{(p)} \otimes \delta_{u(0)}$ in $E = S^p \times \Delta$.*

*Let $u(t)$ solve the ODE (9) starting at $u(0)$. For $1 \leq l \leq p$ let $P^{\mu^{(l)}}$ be the probability measure on $D_E[0,\infty)$ corresponding to the time inhomogeneous Markov chain $X(t)$ with state space $S$, with initial distribution $\mu^{(l)}$, and such that the rate of jumping from state $s$ to state $s'$ is*

$$\lambda(u(t), s, s') = L^s(s') + \sum_{\substack{i,j\in S \\ (i,s)\neq(j,s')}} K^{i,s}(j, s')u_i(t)$$

*Then the process $(X_1(t), ..., X_p(t), u^{p+1,n}(t))$ converges weakly to a product distribution $P^{\mu^{(1)}} \otimes ... \otimes P^{\mu^{(p)}} \otimes \delta_{u(t)}$ in $D_E[0,\infty)$.*

This theorem is due to Uchiyama [76]. A proof of this theorem using modern tools from the theory of weak convergence of Markov processes, see e.g. Ethier and Kurtz [33], can be found in Anantharam and Benchekroun [6], together with applications to computing approximations to sojourn times in networks of queues. The model of [37] can be verified to be a model of this type. In this model Theorem 2.2 gives the additional information that the evolution of any finite collection of links is asymptotically independent in the limit as the total number of links approaches infinity, and that each link evolves according to a time-inhomogeneous birth and death process with up and down rates given explicitly in term of the overall time varying empirical occupation distribution, which follows eqn. (5).

## 2.6  Large deviations

It is of particular importance to give simple rules which can predict which of the metastable regimes is likely to dominate for a given set of parameter values. One approach to this problem would be via the theory of large deviations. Intuitively, there is a "potential well" associated to each equilibrium, and the equilibrium that is likely to dominate is the one whose associated well is deepest, in that it takes the longest time to escape from the well via a rare fluctuation. See Wentzell and Freidlin [79] for a rigorous formalization of such intuition.

There does not appear to be a solution yet to this challenging problem. To close the section, we briefly describe a recent result that is able to answer a related, albeit much more special question. Consider the model of [76] (with a fixed number of exchangeable multiple particle interactions) in the special case where $S = \{0, 1\}$, i.e. when each of the interacting chains is a 2-state chain. For each $n$ the process $(X^n(t))_t$ is a finite state Markov process, and therefore admits a unique equilibrium distribution. Let $\alpha_n$ denote the stationary distribution of the empirical distribution of the interacting chains. Since $S = \{0, 1\}$ this is a distribution on the set $\{\frac{k}{n}, 0 \leq k \leq n\}$. We may write

$$\alpha_n(\frac{k}{n}) = \binom{n}{k} \exp(n h_n(\frac{k}{n}))$$

for some function $h_n$. We extend the definition of $h_n$ to $[0, 1]$ by linear interpolation. In Anantharam [5], we prove the following result.

**Theorem 2.3** *The functions $h_n$ converge uniformly to a limit $h$. A consequence is that the distributions $\alpha_n$ obey a large deviations principle with action functional*

$$I(u) = -h(u) + D(u, \frac{1}{2}) + p \quad 0 \leq u \leq 1$$

*where*

$$D(u, \frac{1}{2}) = u \log 2u + (1 - u) \log(2(1 - u))$$

*and*

$$p = \sup_{u \in [0,1]} [h(u) - D(u, \frac{1}{2})] .$$

It is of particular interest that the action functional is in general non-convex. Its local minima correspond to the fixed points of the limiting differential equation (9). The global minima of the action functional may then be considered as corresponding to the dominating regimes for the given parameter values. Arriving at a theorem of this sort for the case of general $S$ in the model of [76], and therefore in the model of [37], would be of great interest.

# 3  Datagram networks

Data networks work by breaking messages into packets, which are routed through the network and reassembled at the destination. The mathematical analysis and design of such networks involves studying networks of queues.

## 3.1  The M/M/1 queue

A basic queueing model is the M/M/1 queue. Packets enter a buffer of infinite capacity at the times of a Poisson process of rate $\lambda$. Each packet brings in an amount of work which is an exponential random variable of mean $\mu^{-1}$, these variables being independent from packet to packet, and independent of the arrival process. There is a server performing work at rate 1, that operates as follows : on finishing serving a packet it picks an arbitrary packet from the buffer, if any, and proceeds to serve that packet till all the work it has brought in is complete. Packets that have had their work completely served immediately depart the buffer. The first M is a mnemonic for the Poisson (memoryless) character of the arrival process, the second M for the exponential (memoryless) character of the work distributions, the 1 reminds us there is a single server at work. (The queueing literature is littered with a taxonomist's paradise of such abbreviations.) The work brought in by a packet may be taken as representative of its length, or more generally of the time required to carry out some processing of it.

It is conventional to think of the packets as being served in the first-come-first-served (FCFS) order, but as long as the identity of the individual packets is of no concern the exponential assumptions ensure that the order in which packets are served is irrelevant.

Let $(X(t), t \geq 0)$ be the process of total of number of packets in the buffer, including the packet being served, if any. Then $(X(t), t \geq 0)$ is a birth and death process on the non-negative integers $\mathbf{Z}_+$ with up-rate $\lambda$ and down-rate $\mu$. It admits a stationary distribution if and only if $\lambda < \mu$, in which case, with $\rho = \lambda\mu^{-1}$ denoting the *traffic intensity*, the stationary distribution is

$$\pi(n) = \rho^n(1 - \rho) \ , \ n \geq 0 \ .$$

## 3.2   The Jackson network model

Let us make three observations. The first is *Burke's theorem*, also called the *output theorem* for the the M/M/1 queue, see [17]. Consider an M/M/1 queue with $\lambda < \mu$ in stationarity as a process defined for all $t \in \mathbf{R}$. Let $(A_t, t \in \mathbf{R})$ denote the (Poisson) arrival process, and $(D_t, t \in \mathbf{R})$ the departure process of packets leaving the queue. Burke's theorem states that $(D_t, t \in \mathbf{R})$ is a Poisson process of rate $\lambda$ whose past at any time $t \in \mathbf{R}$ is independent of the present state of the buffer, i.e.,

$$(D_s, s \leq t) \amalg X_t \ , \ t \in \mathbf{R} \ .$$

This is a direct consequence of the time reversibility of a stationary birth and death process. Indeed, the departure process of the forward-time process is precisely the arrival process of the reverse-time process. Nevertheless it is a striking, and even counterintuitive, result as naive intuition would suggest, for instance, that an enormous number of departures just prior to a given time $t$ would result in an increased likelihood of the buffer being empty at time $t$.

The second observation is that Bernoulli sampling of a Poisson process results in independent Poisson processes. More precisely, if at each time of a Poisson process $(N_t, t \in \mathbf{R})$ of rate $\lambda$ we draw an independent sample of a Bernoulli random variable whose probability of being 1 is $p$, and split the process into two streams $(N_t^1, t \in \mathbf{R})$ and $(N_t^0, t \in \mathbf{R})$ consisting respectively of those points for which the value of the Bernoulli random variable was 1 or 0, then these are independent Poisson streams of rates $\lambda p$ and $\lambda(1 - p)$ respectively.

The third observation is that the sum of independent Poisson processes is a Poisson process.

We now describe a model of interconnected queues due to Jackson [44] by which to describe a system into which packets enter, move between servers, and eventually leave. A Jackson network has $J$ infinite buffers. Packets arrive from the external world at the times of a Poisson process of rate $\gamma$. The outside world is conventionally indexed by 0. An arriving packet is routed to buffer $j$ with probability $r_{0j}$, with $\sum_{j=1}^J r_{0j} = 1$, where it queues in FCFS order. The work required by a packet at node $j$ is an exponential random variable of mean $\mu_j^{-1}$. Each buffer is served by its own server, that works on the leading packet in the buffer at rate 1 till it completes the work required by this packet, after which it immediately begins work on the next packet in the buffer, if any. With probability $r_{jk}$ a packet completing service at node $j$ is routed to buffer $k$ where it queues in FCFS order, and it leaves the system with probability $r_{j0}$, where $\sum_{k=0}^{J} r_{jk} = 1$. Service times are i.i.d and independent of the arrival process. Routing is Bernoulli, independent of the arrival process and the service times.

We assume that the Jackson network is irreducible, i.e. it is possible for an exogenous arrival to visit any queue before leaving the system. We also assume that it is stable, namely that the solutions of the flow balance equations:

$$\lambda_i = \gamma r_{0i} + \sum_{j=1}^{J} \lambda_j r_{ji}, \qquad 1 \le i \le J \tag{10}$$

satisfy

$$\lambda_i < \mu_i, \qquad 1 \le i \le J . \tag{11}$$

Let $X_j(t)$ denote the number of packets in buffer $j$ at time $t$. and let $X(t) = (X_1(t), \ldots, X_J(t))$. Then $(X(t))_t$ is a Markov process. Note that as long as the focus is on this Markov process, so that the identities of the individual packets are ignored, it is not important that the service at the individual buffers be FCFS, or even that it be *non-preemptive* (i.e. the sever could leave a partially worked-on packet and move to begin work on another packet). All that is important is that the servers be *work-conserving*, i.e. that they work whenever there is at least one packet in the corresponding buffer. Under the stability and irreducibility conditions the Markov process $(X(t))_t$ admits a unique stationary distribution. The importance of the Jackson network model in applications stems from the simple form of its stationary distribution, which is given by

$$\pi(x_1, \ldots, x_J) = \prod_{i \in [J]} \rho_i^{x_i}(1 - \rho_i) \tag{12}$$

where $\rho_i = \lambda_i \mu_i^{-1}$ is called the load factor at node $i$. Such a stationary distribution is said to be of *product-form*, since the individual stationary queue sizes are independent. As might be imagined, this facilitates the computation of stationary quantities. Note that it is far from the case that the evolutions of the individual buffer sizes are independent in stationarity. An interesting line of research in stochastic networks is to develop more sophisticated network models with product-form stationary distributions. A recent survey of some of this literature is that of Nelson [63].

Walrand [77] has provided a clever explanation for the product-form stationary distribution. Suppose that we introduce a small delay in the feedback of packets after they have finished service at a buffer and before they are routed to another buffer in the system. Schematically, the situation can be represented as in Figure 3.

Let $(X^\Delta(t))_t$ be the process of buffer sizes in this modified system, where $X^\Delta(t) = (X_1^\Delta(t), \ldots, X_J^\Delta(t))$. We can guess the stationary distribution of $(X^\Delta(t))_t$ using the intuition gained from the M/M/1 queue. Suppose that $X_j^\Delta(0-)$ are independent geometric random variables with parameter $\rho_j = \lambda_j \mu_j^{-1}$, $1 \leq j \leq J$ (where $\lambda_j$ was defined in eqn. (10)). Also assume that the delay line holds independent pieces of Poisson processes of rate $\sum_i \lambda_i r_{ij}$ respectively destined for buffer $j$, and that these pieces are independent of $X^\Delta(0-)$. This sets up the initial conditions at time 0. We now argue that the situation at time $\Delta$ is exactly the same. Over the time interval $[0, \Delta)$ the individual buffers are being fed by independent Poisson process of rate $\lambda_j$ respectively, since this is the result of summing independent Poisson processes of rates $\gamma r_{0j}$ and $\sum_i \lambda_i r_{ij}$ (see eqn. (10)). This process is independent of the initial condition $X_j^\Delta(0-)$, which is the stationary initial condition of an M/M/1 queue of arrival rate $\lambda_j$ and service time of mean $\mu_j^{-1}$, so it follows that $X^\Delta(\Delta-)$ is also a collection of independent geometric random variables of parameter $\rho_j$ respectively. But Burke's theorem tells us that the departures from the buffers over the interval $[0, \Delta)$ are independent of this vector ! Since Bernoulli sampling of each Poisson process results in independent Poisson processes, and the sums of independent Poisson processes are Poisson, the delay line once again holds pieces of Poisson processes of rates $\sum_i \lambda_i r_{ij}$ respectively destined for buffer $j$, that are independent of $X^\Delta(\Delta-)$. When we let $\Delta \to 0$ we can understand why the stationary distribution of the Jackson network is product-form.

## 3.3   Stability of Jackson-type networks

The stability condition (11) for the Jackson network is easy to understand: the effective rate at which work enters a buffer should be strictly less the service rate. Of course, discussing "effective service rate" presupposes that the network is stable. One of the main concerns in stochastic networks over recent years has been to come to grips with the question of necessary and sufficient stability conditions in more general queueing network models.

Consider the following generalization of the Jackson network model: We retain the Bernoulli routing feature, but generalize the process of packet arrival times to a general renewal process of rate $\gamma$, and the service times of packets at the individual buffers to general independent identically distributed service times with mean service time $\mu_j^{-1}$ at buffer $j$. We also insist that service at the individual buffers is FCFS, non-preemptive, and work-conserving.

Intuition would suggest that if the flow balance equations (10) satisfy (11), the resulting process is stable in some sense.

This problem turned out to be very difficult to resolve. Early works on this problem are due to Borovkov [12], and Sigman [73]. Relatively satisfactory solutions to the problem have only been found quite recently, see Foss [34, 35], Meyn and Down [61], Chang et al. [20], Baccelli and Foss [10], and Dai [25].

We now give a glimpse of the elegant solution of this problem in [10]. Here the problem of stability of datagram networks is approached from a very general point of view. Consider first a broad generalization of the single server M/M/1 queue. On a sample space $(\Omega, \mathcal{F}, \mathcal{P})$ admitting a shift $\theta$ under which $P$ is ergodic, we are given nonnegative random variables $(\sigma_0, \tau_0)$ satisfying $E[\sigma_0] = \mu^{-1} < \infty$ and $E[\tau_0] = \lambda^{-1} < \infty$. Let $(\sigma_n, \tau_n) = (\sigma_0 \circ \theta^n, \tau_0 \circ \theta^n)$. Thus $\{\sigma_n, \tau_n\}_n$ is a stationary ergodic sequence. $\sigma_n$ has the interpretation of the work brought in by the $n$th customer to a server and $\tau_n$ of the interarrival time between the arrival of the $n$th customer and the $n + 1$st customer. The server works at rate 1 if there is work in the system. Let $W_n$ denote the workload in the system seen by the $n$th customer. Then, starting from some initial condition, the workload evolves according to the equation

$$W_{n+1} = (W_n + \sigma_n - \tau_n)^+ \tag{13}$$

This equation is called the *Lindley equation*. We ask for what parameter values this recursion admits a stationary solution, i.e., a proper random variable $W_0$ such that, with $W_n = W_0 \circ \theta^n$, we have $(W_n)_n$ satisfying (13). The M/M/1 queue is the special case where the sequence $\{\sigma_n, \tau_n\}_n$ is i.i.d. with $\sigma_0$ and $\tau_0$ independent exponential random variables. For the M/M/1 queue, we know that the condition for existence of a stationary solution to (13) is precisely $\lambda < \mu$. The following remarkable result is due to Loynes [57].

**Theorem 3.1** *If $\lambda < \mu$ the Lindley equation admits a unique stationary solution. If $\lambda > \mu$ there is no stationary solution to the Lindley equation.*

**Proof:**
The proof is representative of a large number of results of this type, so it is worthwhile to sketch the ideas. Fix $m \geq 0$. We define random variables $(W_n^m, n \geq -m)$ with $W_{-m}^m = 0$ and obeying Lindley's equation

$$W_{n+1}^m = (W_n^m + \sigma_n - \tau_n)^+ . \tag{14}$$

We observe that $W_{n+1}$ is a monotone increasing function of $W_n$ in (13). Thus $(W_n^m, m \geq 0)$ is increasing in $m$ for fixed $n$, and has a limit $W_n^\infty$ which obeys

$$W_{n+1}^\infty = (W_n^\infty + \sigma_n - \tau_n)^+ . \tag{15}$$

Since $W_{n+1}^m = W_n^{m+1} \circ \theta$, we have $W_{n+1}^\infty = W_n^\infty \circ \theta$. It remains to consider when $W_n^\infty$ is proper. We may write

$$W_n^{m+1} \circ \theta = W_{n+1}^m = W_n^m - W_n^m \wedge (\tau_n - \sigma_n) . \tag{16}$$

Taking expectations, we get

$$E[W_n^m \wedge (\tau_n - \sigma_n)] \leq 0 \tag{17}$$

and so

$$E[W_n^\infty \wedge (\tau_n - \sigma_n)] \leq 0 . \tag{18}$$

Since $(W_n^\infty)_n$ is a stationary ergodic sequence, $P(W_n^\infty = \infty)$ is either 0 or 1. From (18) we see that $P(W_n^\infty = \infty) = 1$ implies that $E[\tau_n] \leq E[\sigma_n]$, i.e., $\lambda \geq \mu$. This proves that if $\lambda < \mu$ then a stationary solution to the Lindley equation exists. For the situation when $\lambda > \mu$, first note that $(W_n^\infty)_n$ constructed above is the minimal $\theta$-invariant solution of (13), and that

$$W_0^\infty = (\sup_n \sum_{k=1}^n (\sigma_{-k} - \tau_{-k}))^+ = \infty \tag{19}$$

when $\lambda > \mu$. For more details regarding the solutions of (13) see Baccelli and Brémaud [9, Sec. 2.2]. □

The stability question for datagram networks can be studied in the stationary ergodic framework. A fundamental distinction that emerges is between a station-centered and a customer-centered point of view, see [10]. In the station-centered model, called the *Jackson-type network*, the service times and routing variables are associated to the stations, and handed out to the packets by the server as they are picked up for service. One can visualize identical, featureless packets moving around the network and queuing up in FCFS fashion at the individual nodes. When it is the turn of such a packet to receive service at a node it picks up its service time from a list maintained at the node; when this service is complete it then picks up a routing variable from a list maintained at the node to decide which node to move to next. In contrast to this, in the customer-centered model, called the *Kelly-type network*, the service and routing variables are associated to the individual packets. One can visualize the packets arriving with marks giving their entire route through the network, together with the service times the packet will require at each visit to each node along the route. As before the service at each node is in FCFS order of arrival, but the mark is carried by the packet throughout its route. We discuss each of these models in turn.

The generalized Jackson network with renewal arrivals, i.i.d. service, and Bernoulli routing, can be considered as either a Jackson-type network or a

Kelly-type network. Baccelli and Foss [10] have given a very satisfactory solution to the stability problem for Jackson-type networks. Consider a network consisting of a fixed finite number of stations, each of which has infinite waiting room and a single server that works at unit rate. Consider a stationary ergodic marked point process of arrivals, each of which brings with it a route through the nodes of the network and a service variable for each visit to each node along the route. This process can be described on a sample space $(\Omega, \mathcal{F}, \mathcal{P})$ admitting a shift $\theta$ under which $P$ is ergodic, by giving a pair $(\xi_0, \tau_0)$ : the variable $\xi_0$ describes the route through the nodes of the network and the corresponding service variables for each visit to each node along the route, and $\tau_0$ gives the time to the next arrival. We now visualize the entire mark peeling off the arriving customer at the time of its arrival and joining lists of service and routing variables maintained at the individual *nodes*, in sequence, at the tail of such lists. The featureless packets now move around the network picking up service and routing variables that are handed to them from these lists by the *nodes*.

A key observation is that this mechanism of peeling off the marks of the arriving packets and attaching them to lists at the appropriate nodes ensures that service and routing variables are always available at a node when needed. What is meant by this statement is the following : Consider an initially empty network (also with empty service time and routing lists at each node) and suppose that $n$ packets arrive at the network at times $t(1) < t(2) < \ldots < t(n)$. Each packet brings a mark (a route through the network and a service time requirement at each visit to each node along the route). This mark is peeled off immediately on arrival and distributed among the appropriate lists at the appropriate nodes. The now featureless packets move through the network, queueing at the nodes in FCFS order and picking up service time and routing variables from the lists. Then we will *not* have a situation where a packet looking for a service time variable to pick up to enter service at a node or a routing variable to pick up on finishing service at a node finds the list empty. This statement needs a proof; indeed, with the mechanism of peeling off marks being used, it is possible for a packet to use a service time or routing variable brought in by a packet that arrives *after* it does.

A second key observation is a monotonicity property proved by Foss [34], and Shantikumar and Yao [71]; see also [20, Prop. 4.1] and [10], Theorem 10. This says that, with the mechanism of peeling off marks on arrival that was just described, when we consider the situation with $n$ customers entering into an empty network with empty lists, then delaying the arrival times or increasing any of the service requirements delays the times at which service completions take place; in fact, for any nodes $k$ and $l$ and any $j \geq 1$, the $j$th service at node $k$ that sends a packet to node $l$ will be delayed. Further the time by which these services are delayed can be bounded in terms of the time by which the arrivals are delayed and the increases in the individual service times at the nodes, see [10, Corollary 3].

These observations now allow us to set into motion a machinery to use the Loynes's scheme and exploit Kingman's subadditive theorem by considering the effect of dilations and time shifts of the arrival process. The essential features of this process have been abstracted by Baccelli and Foss [11], and results in a *saturation rule* for stability. Namely if we consider a limiting system in which the network is started empty and immediately receives an infinite number of arrivals, and compute the rate at which packets are released from this scenario then the original network will admit a stationary regime for all arrival rates strictly less than this saturated evacuation rate, and will not admit a stationary regime if the arrival rate strictly exceeds this saturated evacuation rate. In [10] it is shown that this characterization of stability is precisely that given by the requirement (11) on the $\lambda_i$ satisfying (10). The construction of the stationary regime for the network when (11) holds follows the lines of Loynes scheme, because the monotonicity property proved above allows the variables of interest in the network to be described in terms of a monotone stochastic recursion.

## 3.4  Weak Solutions of Stochastic Recursions

The recursion (13) is an example of a *stochastic recursion*. Let $(\Omega, \mathcal{F}, \mathcal{P})$ be a probability space admitting a shift $\theta$ under which $P$ is stationary and ergodic. Let $(E, \mathcal{E})$ be a Polish space and $\varphi_0$ a random variable defined on $(\Omega, \mathcal{F}, \mathcal{P})$ that takes values in the space of measurable maps from $(E, \mathcal{E})$ into itself. Let $\varphi_n(\omega) = \varphi_0(\theta^n \omega)$. Then, under $P$, $\{\varphi_n, n \geq 0\}$ is a stationary ergodic sequence of random maps from $(E, \mathcal{E})$ into itself. In many applications the stability question can be framed as one of finding the conditions under which recursions of the form

$$x_{n+1} = \varphi_n(x_n) \tag{20}$$

admit a stationary solution. For instance (13) is an equation of this type with $(E, \mathcal{E}) = (\mathbf{R}, \mathcal{B})$ and $\phi_0(x) = (x + \sigma_0 - \tau_0)^+$. Note that $\phi_0$ is monotone. By and large, all such recursions that have been successfully studied in the literature exploit some kind of monotonicy of $\phi_0$, and are handled by proceeding path-wise using a version of the Loynes' scheme outlined in the proof of Theorem 3.1. Recently Anantharam and Konstantopoulos [8], have developed another approach to solving such stochastic recursions, which does not require monotonicity, but is based on the weakening of the solution concept, along lines also proposed in [15]. We briefly sketch this approach; for details, see [8].

A *weak solution* of the recursion (20) is a measure $Q$ on a measurable space $(\tilde{\Omega}, \tilde{\mathcal{F}})$ admitting a measurable shift $\tilde{\theta}$ such that $Q$ is $\tilde{\theta}$-invariant, and a pair of random variables $X_0$ and $\Phi_0$ on $(\tilde{\Omega}, \tilde{\mathcal{F}})$ taking values respectively in $E$ and in the space of measurable maps from $(E, \mathcal{E})$ into itself, such that

$$X_0 \circ \tilde{\theta} = \Phi_0(X_0) . \tag{21}$$

In [8] we observe that it is often possible to construct such a weak solution along the lines of the skew product construction in ergodic theory (see, e.g. Krengel [50]), even in the absence of any kind of monotonicity of $\phi_0$. Assume that $(\Omega, \mathcal{F})$ is a Polish space. Consider the product space $\Omega \times E$, endowed with the product $\sigma$-field $\mathcal{F} \otimes \mathcal{E}$ and the new measurable shift $\Theta(\omega, x) : \Omega \times E \to \Omega \times E$ defined by

$$\Theta(\omega, x) = (\theta\omega, \ \varphi_0(\omega)[x]). \tag{22}$$

Note the following composition rule.

$$
\begin{aligned}
\Theta^n(\omega, x) &= (\theta^n\omega, \ \varphi_0(\theta^{n-1}\omega)\varphi_0(\theta^{n-2}\omega)\ldots\varphi_0(\omega)[x]) \\
&= (\theta^n\omega, \ \varphi_{n-1}(\omega)\varphi_{n-2}(\omega)\ldots\varphi_0(\omega)[x]) , \\
\Theta^{n+m}(\omega, x) &= \Theta^n(\Theta^m(\omega, x)) .
\end{aligned}
$$

In [8], the following result is proved.

**Theorem 3.2** *Let $Q_0$ be a probability distribution on $\Omega \times E$ whose $\Omega$ marginal is $P$. Let $Q_n$ denote the probability distribution $Q_0 \circ \Theta^{-n}$ on $\Omega \times E$. Suppose that the sequence $\{Q_n, n \geq 0\}$ is tight. Let $Q$ be any subsequential weak limit of $\{Q_n, n \geq 0\}$. Then $Q$ on $(\Omega \times E, \mathcal{F} \otimes \mathcal{E})$ with the shift $\Theta$ and with $X_0(\omega, x) = x$ and $\Phi_0(\omega, x) = \phi_0(\omega)$ is a weak solution of (20).*

Examples of non-monotone recursions that can be handled using this approach are discussed in [8]. The following uniqueness theorem is also useful, see [8]. Let $C_b(\Omega \times E)$ denote the space of bounded continuous functions on $\Omega \times E$ .

**Theorem 3.3** *Suppose that for every $\Theta$-invariant probability distribution $Q$ on $\Omega \times E$ having $\Omega$ marginal $P$, and all $f \in C_b(\Omega \times E)$*

$$\lim_{n \to \infty} \frac{1}{n} \sum_{j=0}^{n-1} f(\Theta^j(\omega, x)) = \mathcal{A}(f) \tag{23}$$

*exists $Q$-a.s. and is a constant for $Q$-a.a. $(\omega, x)$. Then if there is a $\Theta$-invariant probability distribution $Q$ on $\Omega \times E$ having $\Omega$ marginal $P$, it is unique. Conversely, if there is a unique $\Theta$-invariant probability distribution $Q$ on $\Omega \times E$ having $\Omega$ marginal $P$, then for all $f \in C_b(\Omega \times E)$ the limit in (23) exists $Q$-a.s.. and is constant for $Q$-a.a. $(\omega, x)$.*

## 3.5 Kelly-type networks

We now turn to a discussion of Kelly-type networks, where the service time and routing variables are carried by the packets as they move around the network and queue in FCFS fashion at the nodes. The stability question for Kelly-type networks is very poorly understood. Indeed, already in special cases, examples are known where the rate conditions (11) do not guarantee stability. We now briefly sketch the genesis of these examples. Note that the first two examples outlined below are *not* Kelly-type networks.

The example of Figure 4 is described by Lu and Kumar [58], and attributed to Seidman; see also Kumar and Seidman [54]. There is a single server that can serve either of the buffers 1 and 4, and another server that can serve either of the buffers 2 and 3; each server works at rate 1. There are priority rules for which buffer the server can serve, as indicated in the figure : buffer 4 has priority over buffer 1, and buffer 2 has priority over buffer 3. Assume that there is an arrival at every integer time, and that its service requirements are 2/3 at each of the buffers 2 and 4, and that the service requirement is 0 at each of the buffers 1 and 3 : this means that the server just has to "kiss" the packet before allowing it to leave. Also assume that kissing at buffer 3 takes place just before kissing at buffer 1. Consider the initial condition (at time 0-) when there are $M$ packets in buffer 1 and the other buffers are empty. All $M$ inital packets are immediately kissed and go to buffer 2. Further, if we define the time $\tau$ by

$$\frac{2}{3}(\tau + M) = \tau \qquad (24)$$

so that $\tau = 2M$, we see that, in view of the priority rules, at time $\tau-$ there will be 3M packets in buffer 3, and all the other buffers will be empty. These then get kissed at buffer 3, leaving 3M packets at buffer 4 with all the other buffers empty. Finally, in view of the priority rules, at time 4M- we are in a situation where there are 4M packets in buffer 1, with all the other buffers empty. The system has returned to a scaled version of the initial condition, with scaling bigger than 1.

What is remarkable about this example is that the total work that needs to be done per packet by each server is $\frac{2}{3} < 1$. Clearly rate conditions do not suffice to determine the stability region in this system with service priority rules.

The system above involves a fluid model, and instantaneous events. Motivated by a similar example in [54], Rybko and Stolyar [68], considered the model of Figure 5. There are two classes of packets. Assume that the respective arrival processes are independent Poisson processes of rate 1. The service time required by a packet in buffer $ij$ is an exponential random variable of mean $m_{ij}$; all these service times are independent, and independent of the arrival

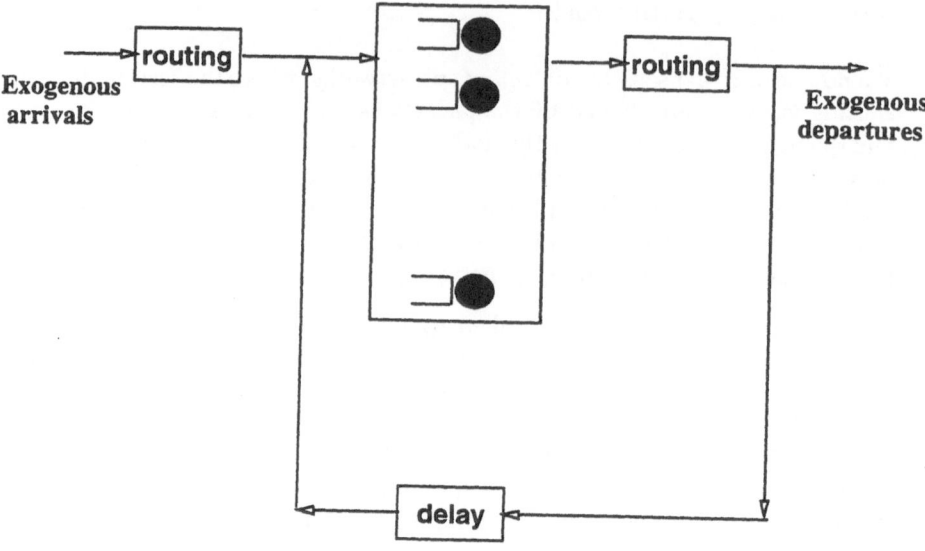

Figure 3: Explaining the product form stationary distribution

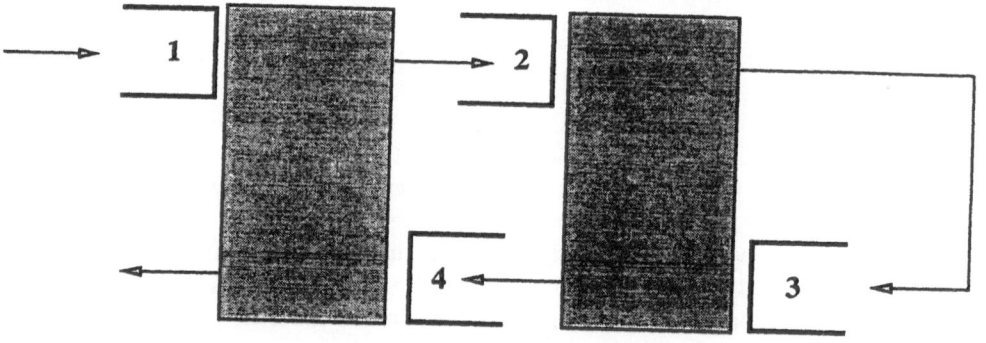

4 has priority over 1     2 has priority over 3

Figure 4: The Lu-Kumar Example

processes. There is a server that works at rate 1 serving buffers 11 and 22, giving preemptive priority to buffer 22, and another server that works at rate 1 serving buffers 12 and 21, giving preemptive priority to buffer 12. The rate based condition for stability is then

$$m_{11} + m_{22} < 1$$
$$m_{12} + m_{21} < 1 .$$

In [68] the situation with $m_{12} = m_{22} = m_2 > \frac{1}{2}$ and $m_{11} = m_{21} = m_1 > 0$, and with $m_1 + m_2 < 1$ is considered. It is shown that the Markov process describing the system is *not* positive recurrent. Thus rate conditions do not suffice to determine stability in this example.

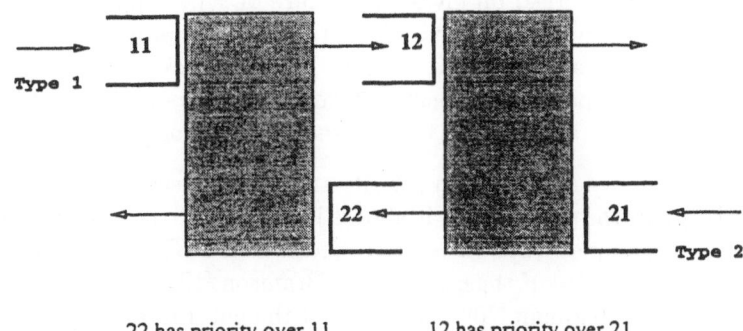

22 has priority over 11        12 has priority over 21

Figure 5: The Rybko-Stolyar Example

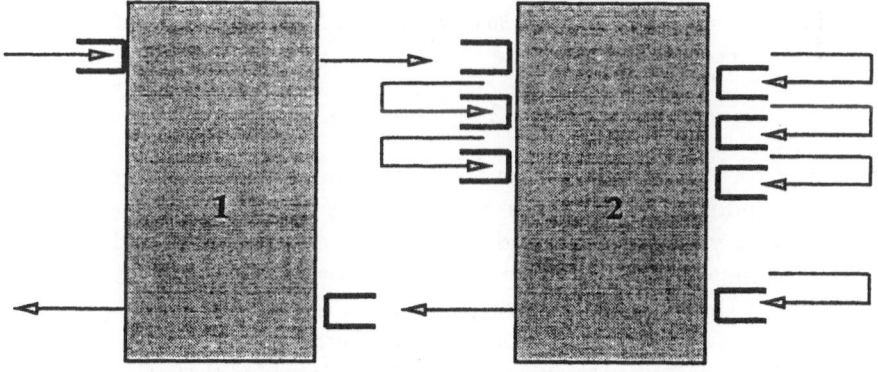

**There is an enormous number of quick visits to node 2**

Figure 6: Bramson's first example

A sketch of the intuition behind this result is the following : Let $Q_{ij}(t)$ denote the queue size at buffer $ij$ at time $t$, including the packet in service, if any. Then the network can described by a Markov process with state $(Q_{ij}(t), 1 \leq i, j \leq 2)$. To begin with, we may restrict attention to the subset of the state space such that $Q_{12}(t)Q_{22}(t) = 0$, since from any initial condition we reach such a state and then never leave this subset of states. This means that if $Q_{12}(t) > 0$, then $Q_{22}(t) = 0$, so that type 1 calls can be thought of as entering a tandem system of two rate 1 servers with infinite buffers in front of them. This picture is no longer valid when buffer 12 empties and a packet at buffer 11 is served and enters buffer 22 before a packet from buffer 11 is served and enters buffer 12. Now arriving type 1 packets build up till we have a situation where buffer 22 empties and a packet served at buffer 11 reaches buffer 12 before a packet served at buffer 21 enters buffer 22. Compressing the intervals where type 2 packets are being worked on gives the picture of type 1 packets entering a tandem system of two rate 1 servers but with an arrival process that includes large bursts of arrivals corresponding to the built up packets of type 1 during a cycle when type 2 packets are being worked on. It is now shown that these bursts destabilize the system. For details, see Section 6 of [68].

Both examples above are based on priority rules for the servers. Nevertheless, Bramson [13] had the insight that it is possible to mimic the phenomenon underlying these examples when the service discipline is FCFS, i.e. in Kelly-type networks. The invention of the examples of Bramson [13, 14], and fluid-model based examples by Seidman [70]. showing that the natural rate conditions (11) do not suffice to guarantee stability in Kelly-type networks is one of the most striking recent developments in the area of stochastic networks.

The example of [13] is illustrated in Figure 6. There is a single stream of customers entering the network according to a Poisson process of rate 1 and proceeding according to the route

$$\rightarrow 1 \rightarrow 2 \rightarrow 2 \rightarrow \ldots \rightarrow 2 \rightarrow 2 \rightarrow 1 \rightarrow \qquad (25)$$

The service requirements at each visit along the route are exponentially distributed random variables, with mean $c$ at the first visit to node 2 and the last (second) visit to node 1 and mean $\delta$ at the first visit to node 1 and all except the first visit to node 2. It is demonstrated in [13] that for $c$ sufficiently close to 1, if the number of visits, $J$, to node 2 is sufficiently large (depending on $c$) and $\delta$ is sufficiently small (depending on $c$ and $J$) the Markov process describing the system is transient. Note that this means that the natural rate conditions

$$c + \delta < 1$$
$$c + (J-1)\delta < 1$$

do not suffice to guarantee stability. For details, see [13].

The second example of Bramson, presented in [14], demonstrates that for any given $\rho < 1$, however small, it is possible to have Kelly-type networks where the load factor at each node is less than $\rho$, but the network is nevertheless unstable. However, note that the smaller the prescribed $\rho$, the larger the number of nodes needed in order to construct an example of this sort.

Understanding the stability question for Kelly-type networks is one of the important and exciting challenges thrown up by the area of stochastic networks. Progress towards this problem in a Markovian framework is reported by Dai [25], Dai and Meyn [26], and Kumar and Meyn [53].

# 4 Integrated Broadband Networks

Network design engineers would like to have in place a network that can offer a wide variety of services, such as audio, video, data, etc. using a common protocol suite. Such a network is conceived of as the facilitator of the multimedia revolution that is booted about in the popular press. Progress in fibre optics and switching technology has made this goal appear within reach. From the viewpoint of stochastic analysis, a number of new questions have been brought into prominence by this drive. In this section we briefly introduce a couple of recent developments in research, aimed at addressing these questions.

## 4.1 Effective Bandwidths

The theory of effective bandwidths, see Hui [43] and Kelly [47], is currently a topic of considerable research, The rationale is to try to carry over some of the intuition available from the design of circuit-switched networks to the design of networks that handle bursty traffic. This is made possible by results in large deviations theory.

To convey the basic idea, we first recall the Gärtner-Ellis theorem, see [30, 36]. For more details, see Dembo and Zeitouni [27, Sec. 2.3]. Let $(X_n, n \geq 1)$ be a sequence of random variables and $S_n = \sum_{i=1}^{n} X_i$. Suppose that the limit

$$\Lambda(\theta) = \lim_{n \to \infty} \frac{1}{n} \log E[\exp(\theta S_n)] \tag{26}$$

exists (possibly infinite) for all $\theta \in \mathbf{R}$ and is lower semicontinuous. Define the effective domain of $\Lambda(\cdot)$ as $\{\theta : \Lambda(\theta) < \infty\}$. Suppose that $\theta = 0$ is in the interior of the effective domain, that $\Lambda(\cdot)$ is differentiable throughout the interior of its effective domain, and that the derivative approaches $\infty$ in absolute value for any sequence approaching a boundary point of the effective domain. Then $(\frac{S_n}{n}, n \geq 1)$ obeys a large deviations principle with convex good rate function $\Lambda^*(\cdot)$ given by the convex dual of $\Lambda$ :

$$\Lambda^*(x) = \sup_{\theta}[\theta x - \Lambda(\theta)] , \qquad (27)$$

i.e. for any closed set $F \subseteq \mathbf{R}$ we have

$$\limsup_{n \to \infty} \frac{1}{n} \log P(\frac{S_n}{n} \in F) \leq - \inf_{x \in F} \Lambda^*(x) \qquad (28)$$

and for any open set $G \subseteq \mathbf{R}$ we have

$$\liminf_{n \to \infty} \frac{1}{n} \log P(\frac{S_n}{n} \in G) \geq - \inf_{x \in G} \Lambda^*(x) . \qquad (29)$$

Here a rate function is said to be *good* if it has compact level sets.

Consider now the Lindley equation (13) with $X_n = \sigma_n - \tau_n$, which we reproduce here :

$$W_{n+1} = (W_n + X_n)^+ . \qquad (30)$$

Then we may prove the following result, see [29].

**Theorem 4.1** *Let $(X_n)_n$ be a stationary ergodic process with $E[X_n] < 0$, and satisfy the conditions of the Gärtner-Ellis theorem. Then there is a unique stationary solution to (30) which satisfies*

$$\Lambda(\theta) \leq 0 \equiv \lim_{B \to \infty} \frac{1}{B} \log P(W_n > B) \leq -\theta . \qquad (31)$$

The probability that the stationary queue size exceeds some level is of particular interest in applications. Indeed, in practice buffer sizes are finite, and this probability can be taken as representative of the probability of buffer overflow. Since these probabilities are very small in well designed systems, an exponent of the form of the limit on the right hand side of (31) is of considerable interest. Theorem 4.1 gives a broad general connection between this exponent and the large deviations behaviour of the driving process $(X_n)_n$.

Consider next a collection of independent sources of different types $1, \ldots, J$. We think of time as divided into slots of identical length and assume that the work brought in by a traffic stream of type $j$ in successive slots has the distribution of $(A_n^j)_n$, where $(A_n^j)_n$ satisfies the conditions of the Gärtner-Ellis theorem, with

$$\Lambda_j(\theta) = \lim_{n \to \infty} \frac{1}{n} \log E[\exp(\theta \sum_{i=1}^{n} A_n^j)] \ . \tag{32}$$

There are $n_j$ traffic streams of type $j$, $1 \leq j \leq J$. These streams all share a buffer which is served by a work-conserving server that can serve at most $c$ units of work during a slot. Assuming that the work brought in by a traffic stream arrives at the beginning of the slot, and letting $W_n$ denote the total work in the buffer at the end of slot $n-1$, we see that $(W_n)_n$ obeys a Lindley equation of the form (30) with $X_n$ being the total work brought in during slot $n$, less $c$.

For $\theta > 0$, let $\alpha_j(\theta) = \frac{\Lambda_j(\theta)}{\theta}$. The function $\alpha_j(\cdot)$ is called the *effective bandwidth* function of sources of type $j$. The reason behind this is the following theorem, obtained by Kesidis et al. [48], which follows from Theorem 4.1, see [29] for a proof.

**Theorem 4.2** *For the buffer shared by $n_j$ arrivals of each type $j$, $1 \leq j \leq J$, and served by a work conserving server that can serve at most $c$ units of work during a slot, as above, we have*

$$\sum_{j} n_j \alpha_j(\theta) \leq c \equiv \lim_{B \to \infty} \frac{1}{B} \log P(W_n > B) \leq -\theta \ . \tag{33}$$

Thus the idea of effective bandwidth appears to convert problems of designing for small buffer overflow probability into problems similar to fitting calls on a link in a circuit switched network. Much of the current work in this area is aimed at trying to carry over this analogy to networks of queues. There has also been considerable interest in computing formulas for the effective bandwidths of different source types. For a better perspective on this rapidly evolving area see [19, 31, 64].

## 4.2 Virtual backlog

Before a traffic stream is admitted to the network it is regulated to prevent its potential burstiness from adversely affecting the ability of the network to handle other traffic streams sharing the network. This process is called *flow control*. One of the features of broadband networks is the difficulty of effectively using feedback for flow control. This is because a large number of packets would have already entered the network before feedback has a chance to come into play. For this reason the flow control schemes currently being proposed are largely open loop in nature. Of these, variants of the *leaky bucket* flow control scheme, see [75], are by far the most popular.

Broadband networks are likely to use the ATM (Asynchronous Transfer Mode) protocol suite, see [28, 40]. In an ATM network, traffic is broken up into packets of fixed length, called *cells*. The basic idea of the leaky bucket scheme is to regulate the admission of cells into the network by means of a stream of tokens generated at a constant rate with constant interarrival times. The tokens collect in a token buffer, and the cells collect in a cell buffer. A cell is released only when there is a token available, in which case it consumes one token. Tokens arriving when the token buffer is full are lost, as are cells arriving when the cell buffer is full. This scheme has been found to be very effective in reducing the burstiness of offered traffic in practice, and this burstiness reducing property has also been analytically justified, see [7, 51]. For instance, Anantharam and Konstantopoulos [7] prove the following result : Consider a stationary point process as bringing one unit of work with each point. Define a stationary point process to be less bursty than another such process if the steady state queue length in any single server queue working at rate strictly bigger than the total rate of arrival of work in either process is stochastically smaller than that for the other process. Then for *any* stationary arrival process of offered traffic into a leaky bucket flow controller, if the token arrival rate is at least as big as the rate of offered traffic, the burstiness of the stationary departure process is monotonically increasing in the size of the token buffer. Here we assume that the cell buffer has infinite capacity. Note that the degenerate case of infinite token buffer corresponds to not regulating the arrival process at all, so this result shows that the departure process from the leaky bucket (the traffic admitted to the network) is less bursty than the process of arrivals into the leaky bucket (the offered traffic from the source).

We observe that the leaky bucket scheme regulates offered traffic so that it satisfies *burstiness constraints* of the form that the amount of traffic over a time interval is bounded by an affine function of the length of the interval. Indeed, if the size of the token buffer is $C$, and the rate of token generation is $\rho$, the total number of cells that can be released by the bucket over a time interval $[a, b]$ can be no more than $C+1+\rho(b-a)$. The class of traffic flows that satisfy such burstiness constraints is therefore an interesting one to consider. There has been a considerable amount of work on such models of traffic flow, beginning with Cruz [22, 23]; see also Chang [19], and Parekh and Gallager [65, 66].

A key observation is that it is possible to characterize the past of a burstiness constrained flow in terms of a simple recursively updatable statistic which we call the *virtual backlog*. To make this precise let us assume, for convenience, that time is discrete, although similar ideas carry over to continuous time.

We will use the term *message flow* to denote a sequence of nonnegative real numbers $(a_n, n \geq 0)$. We model traffic on the links of a network by a message flow. A message flow $(a_n, n \geq 0)$ is said to be $(\sigma, \rho)$ constrained, if for all $0 \leq n_0 \leq n_1 < \infty$, we have

$$\sum_{n_0}^{n_1} a_k \leq \sigma + \rho(n_1 - n_0 + 1) . \tag{34}$$

Let $(a_n, n \geq 0)$ be a $(\sigma, \rho)$ constrained flow. We say that it has initial *virtual backlog* $\sigma_0$, if in addition to (34) the flow obeys the constraints

$$\sum_{0}^{n} a_k \leq \sigma_0 + \rho(n + 1) \tag{35}$$

for all $n \geq 0$. Then we may easily prove the following result, see [4] for instance.

**Lemma 1** *Let $(a_n, n \geq 0)$ be a $(\sigma, \rho)$ constrained flow with initial virtual backlog $\sigma_0$. Suppose that $a_0$ is revealed. Then the information gained about $(a_n, n \geq 1)$ is completely summarized by the statement that $(a_n, n \geq 1)$ is a $(\sigma, \rho)$ constrained flow with initial virtual backlog $\sigma_1$, where*

$$\sigma_1 = \min(\sigma_0 + \rho - a_0, \sigma) \tag{36}$$

This observation allows one to adopt a system theoretic viewpoint to the design of resource allocation and control strategies in networks handling burstiness constrained flows. Indeed, the virtual backlog is a *recursively updatable state* that *completely summarizes* the past of a burstiness constrained flow. We have been able to use this intuition to approach the problem of designing optimal open loop flow control strategies for traffic in broadband networks from a prescriptive point of view. For instance, Konstantopoulos and Anantharam [49] pose and solve the problem of regulating an arbitrary traffic stream to create a $(\sigma, \rho)$ constrained traffic stream, while subjecting it to minimal delay. The permissible schemes are allowed in principle to use the *entire* past history of the offered traffic. The possibility of additional constraints such as a limit on the amount of traffic that can be buffered or a limit on the acceptable delay of traffic is also considered. The optimal control schemes that we find are very simple in nature (in fact they are greedy schemes), and are based on the virtual backlog of the admitted traffic stream. They can therefore be implemented with very little intelligence at the flow controller. For the problem with a constraint on the amount of traffic that can be buffered the optimal solution in fact turns out be a leaky bucket flow controller.

We now describe an abstract point of view for the control problems faced by network elements inside a network handling burstiness constrained traffic, relying on the idea of virtual backlog, see [4]. This idea was first discussed in [3].

We visualize the network element as being fed by a number of burstiness constrained flows and as implementing actions based on past information. Since we are working in discrete time throughout, we will assume that the *state* of the network element at time $n$, $n = 0, 1, \ldots$ is given by an element $\xi_n \in \Xi$. At time $n$ the network element is to choose a *control action* $u_n \in U$. At a switch, for instance, $U$ might represent a choice of matching between input ports and output ports. The evolution of the state of the network element occurs in response to the incoming flows at the current time and the choice of control action, resulting in the abstract evolution equation

$$\xi_{n+1} = f(\xi_n, u_n, \underline{a}_n) \tag{37}$$

Note that when there are $K$ driving message flows with the the $i$ th flow being $(\sigma^i, \rho^i)$ constrained, $1 \leq i \leq K$, the domain of $f$ is $\Xi \times U \times \prod_{i=1}^{K}[0, \sigma^i + \rho^i]$. A *randomized adapted control strategy* at the network element is a choice of a probability distribution on $U$ at each time $n$ as a function of $\xi_{[0,n]}$, $u_{[0,n-1]}$, $\underline{\sigma}_0$, and $\underline{a}_{[0,n-1]}$. Here $\underline{\sigma}_0$ is the vector of initial virtual backlogs of the flows.

The problem of designing a good control strategy can be posed via the theory of zero-sum stochastic games. For instance, we may postulate a function $c(\xi, u, \underline{a})$ representing the cost associated to taking the action $u$ when the current element state is $\xi$ and the current message flows are given by $\underline{a}$. We think of this as a cost *paid by the controller to the sources*. Let $0 < \beta < 1$ be a discount factor. We may then formulate the problem of the controller as one of minimizing the total infinite horizon discounted cost

$$\sum_{n=0}^{\infty} \beta^n c(\xi_n, u_n, \underline{a}_n) \tag{38}$$

The minimization is to be done over all possible randomized adapted control strategies of the controller and over all possible randomized adapted control strategies of the sources. Since the source action space is convex, this is equivalent to a worst case formulation where the controller attempts to minimize the overall discounted cost over all possible burstiness constrained source sequences.

Using techniques that are fairly standard in the theory of stochastic games, one can prove the following theorem. For the basic ideas behind the Shapley recursion, see Shapley [72].

**Theorem 1** *Suppose $\Xi$ is a complete separable metric space and $U$ is compact. Suppose $f$ is continuous. Suppose there is a continuous nonnegative function* *g defined on $\Xi$ having compact level sets, a polynomial $p(\cdot)$, and a constant $K < \infty$ such that, for all $\xi \in \Xi$, $u \in U$, and $\underline{a} \in \prod_{i=1}^{K}[0, \sigma^i + \rho^i]$*

*(1) $|g(f(\xi, u, \underline{a})) - g(\xi)| \leq K$.*

*(2) $c(\xi, u, \underline{a}) \leq p(g(\xi))$.*

*Then the stochastic game above admits a continuous value function that is the unique fixed point of a Shapley recursion. Further, both the controller and the sources have optimal stationary randomized adapted strategies which depend only on the state of the game. These are respectively given by the outer extremizers in the min-max and the max-min forms of the Shapley recursion.*

By iterating the Shapley recursion, optimal control strategies for the controller within the context of this formulation can be identified. Further, they can *in principle be implemented in real time.* Indeed, the entire history of a burstiness constrained flow can be kept track of by the *recursively updatable* virtual backlog, which is a simple finite dimensional statistic. This therefore appears to be a promising approach to handle resource allocation problems in broadband networks. Analytical results about the qualitative nature of optimal control strategies in specific problems would also be welcome. In practice, of course, one would expect that the basic burstiness parameters should be updated on a slower time scale.

# 5  Concluding remarks

We have given a rather selective sketch of some of the recent research in the area of stochastic networks, organizing them along the lines of distinct applications contexts. Large areas of highly interesting work, both from the viewpoint of mathematics and from the viewpoint of applications, have been left entirely untouched; this includes work on diffusion approximations, see [21, 41, 24, 42]; on the use of self-similar traffic models, which appear to match actual samples of local area network traffic much better than Markovian models, see Leland et al. [55]; on the analysis of different scheduling policies for re-entrant lines, see Kumar [52]; on the stochastic perfomance analysis of interconnection networks that form the switching fabric of high speed networks, see [67]; among many others. The reader whose appetite is whetted by this survey, and who follows the track of some of the references, will no doubt quickly find a wealth of interesting and useful problems to occupy his or her attention.

34

# References

[1] Akinpelu, J. M. (1984). The Overload Performance of Engineered Networks with Nonhierarchical and Hierarchical Routing. *AT & T Bell Laboratories Technical Journal*, Vol. 63, No. 7, pp. 1261-1281.

[2] Anantharam, V. (1991). A Mean Field Limit for a Lattice Caricature of Dynamic Routing in Circuit Switched Networks. *Annals of Applied Probability*, Vol. 1, No. 4, pp. 481-503.

[3] Anantharam, V. (1992). Bandwidth allocation to burstiness constrained flows. Talk delivered at the *IEEE Information Theory Workshop*, Salvador, Brazil, June 1992.

[4] Anantharam, V. (1993). An approach to the design of high speed networks for bursty traffic. *Proceedings of the International Teletraffic Congress sponsored seminar on Teletraffic Methods for Current and Future Telecom Networks*, Indian Institute of Science, Bangalore, India, Nov. 15-19, 1993, pp. 163-168.

[5] Anantharam, V. (1994). Large deviations for a Mean Field Model of Interacting Particles. *In preparation.*

[6] Anantharam, V., and Benchekroun, M. (1993). A Technique for Computing Sojourn Times in Large Networks of Interacting Queues. *Probability in the Engineering and Information Sciences*, Vol. 7, pp. 441-464.

[7] Anantharam, V., and Konstantopoulos, T. (1994). Burst reduction properties of the leaky bucket flow control scheme in ATM networks. *IEEE Transactions on Communications*, Vol. 42, No. 12, pp. 3085-3089.

[8] Anantharam, V., and Konstantopoulos, T. (1994). Stationary Solutions of Stochastic Recursions describing Discrete Event Systems : General results. *Technical Report*, No. SCC-94-03, ECE Department, University of Texas, Austin.

[9] Baccelli, F., and Brémaud, P. (1994). *Elements of Queueing Theory : Palm-Martingale Calculus and Stochastic Recurrences.* Springer-Verlag, Berlin.

[10] Baccelli, F., and Foss, S. (1994). Ergodicity of Jackson-type queueing networks. *Queueing Systems : Theory and Applications*, Vol. 17, pp. 5-72.

[11] Baccelli, F., and Foss, S. (1994). On the Saturation rule for the Stability of Queues. *Preprint. To appear in Journal of Applied Probability.*

[12] Borovkov, A. A. (1986). Limit Theorems for Queueing Networks. *Theory of Probability and its Applications*, Vol. 31, pp. 413-427.

[13] Bramson, M. (1994). Instability of FIFO queueing networks. *Annals of Applied Probability*, Vol. 4, No. 2, pp. 414-431. Correction published in Vol. 4, No.3, pg. 952.

[14] Bramson, M. (1994). Instability of FIFO queueing networks with quick service times. *Annals of Applied Probability*, Vol. 4, No. 3, pp. 693-718.

[15] Brandt, A., Franken, P., and Lisek, B. (1990). *Stationary stochastic models*. John Wiley and Sons, New York.

[16] Brémaud, P., Kannurpatti, R., and Mazumdar, R. (1992). Event and Time Averages : A review. *Annals of Applied Probability*, Vol. 24, pp. 377-411.

[17] Burke, P. J. (1956). The output of a queueing system. *Operations Research*, Vol. 4, pp. 699-704.

[18] Burman, D. Y., Lehoczy, J. P., and Lim, Y. (1984). Insensitivity of Blocking Probabilities in a Circuit-switching Network. *Journal of Applied Probability*, Vol. 21, pp. 853-859.

[19] Chang, C.-S. (1994). Stability, Queue Length, and Delay of Deterministic and Stochastic Queueing Networks. *IEEE Transactions on Automatic Control*, Vol. 39, No. 5, pp. 913-931.

[20] Chang, C.-S., Thomas, J. A., and Kiang, S. H. (1994). On the stability of open networks : a unified approach by stochastic dominance. *Queueing Systems : Theory and Applications*, Vol. 15, pp. 239-260.

[21] Chung, K.L. and Williams, R.J. (1983). *Introduction to Stochastic Integration*. Birkhäuser, Boston.

[22] Cruz, R.L. (1991). A calculus for network delay, part I: Network elements in isolation. *IEEE Transactions on Information Theory*, Vol. 37, 114-131.

[23] Cruz, R.L. (1991). A calculus for network delay, part II: Network analysis. *IEEE Transactions on Information Theory*, Vol. 37, 132-141.

[24] Dai, J. G., and Wang, Y. (1993). Nonexistence of Brownian models for certain multiclass queueing networks. *Queueing Systems : Theory and Applications*, Vol. 13, pp. 41-46.

[25] Dai, J. G. (1994). On Positive Harris Recurrence for Multiclass Queueing Networks : A Unified Approach via Fluid Limit Models. *Preprint*.

[26] Dai, J., and Meyn, S. (1994). Stability and Convergence of Moments for Multiclass Queueing Networks via Fluid Limit Models. *Preprint.*

[27] Dembo, A., and Zeitouni, O. (1993). *Large Deviations : Techniques and Applications.* Jones and Bartlett Publishers, London.

[28] de Prycker, M. (1993). *Asynchronous Transfer Mode : Solution for Broadband ISDN.* Ellis Horwood, New York.

[29] de Veciana, G., and Walrand, J. (1993). Effective bandwidths : call admission, traffic policing, and filtering for ATM networks. *Preprint. To appear in the IEEE/ACM Transactions on Networking.*

[30] Ellis, R. S., (1984). Large deviations for a general class of random vectors. *Annals of Probability,* Vol. 12, pp. 1-12.

[31] Elwalid, A. I., and Mitra, D. (1993). Effective bandwidth of general Markovian traffic sources and admission control of high-speed networks. *IEEE/ACM Transactions on Networking,* Vol. 1, pp. 329-343.

[32] Erlang, A. K. (1917). Solution of some problems in the theory of probabilities of significance in automatic telephone exchanges. *Post Office Electrical Engineers Journal,* Vol. 10, pp. 189-197.

[33] Ethier, S. N., and Kurtz, T. G. (1986). *Markov Processes: Characterization and Convergence* John Wiley and Sons.

[34] Foss, S. G. (1989). Some properties of open queueing networks. *Problems of Information Transmission,* Vol. 25, pp. 241-246.

[35] Foss, S. G. (1991). Ergodicity of Queuing Networks. *Siberian Mathematical Journal,* Vol. 31, pp. 184-203.

[36] Gärtner, J. (1977). On large deviations from the invariant measure. *Theory of Probability and its Applications,* Vol. 22, pp. 24-39.

[37] Gibbens, R. J., Hunt, P. J., and Kelly F. P. (1990). Bistability in Communications Networks. In *Disorder in Physical Systems,* Grimmett, G. R., and Welsh, D. J. A., eds., Oxford University Press, pp. 113-128.

[38] Graham, C., and Méléard, S. (1993). Propagation of Chaos for a fully connected loss network with alternate routing. *Stochastic Processes and their Applications,* Vol. 44, pp. 159-180.

[39] Graham, C., and Méléard, S. (1993). Fluctuations for a fully connected loss network with alternate routing. *Preprint.*

[40] Händel, R., Huber, M. N., and Schröder, S. (1994). *ATM Networks : Concepts, Protocols, Applications.* Addison-Wesley, Reading, Massachusetts.

[41] Harrison, J.M. (1985). *Brownian Motion and Stochastic Flow Systems.* Wiley, New York.

[42] Harrison, J.M., and Nguyen, V. (1993). Brownian models of multiclass queueing networks : Current status and open problems. *Queuing Systems : Theory and Applications*, Vol. 13, pp. 5-40.

[43] Hui, J. (1988). Resource allocation for broadband networks. *IEEE Transactions on Selected Areas in Communications*, Vol. 6, pp. 1598-1608.

[44] Jackson, J. R. (1963). Jobshop-like queueing systems. *Management Science*, Vol. 10, pp. 131-142.

[45] Kelly, F. (1986). Blocking probabilities in large circuit-switched networks. *Advances in Applied Probability*, Vol. 18, pp. 473-505.

[46] Kelly, F. (1991). Loss Networks. Special Invited paper in the *Annals of Applied Probability*, Vol. 1, No. 3, pp. 319-378.

[47] Kelly, F. (1991). Effective Bandwidths at multi-class queues. *Queueing Systems : Theory and Applications.* Vol. 9, Nos. 1-2, pp. 5-16.

[48] Kesidis, G., Walrand, J., and Chang, C.-S. (1993). Effective Bandwidths for Multiclass Markov Fluids and Other ATM Sources. *IEEE/ACM Transactions on Networking*, Vol. 1, No.4, pp. 424-427.

[49] Konstantopoulos, T., and Anantharam, V. (1993). Optimal Flow Control Schemes that Regulate the Burstiness of Traffic. *SCC Technical Report SCC-93-15*, ECE Department, University of Texas, Austin. (submitted to *IEEE Transactions on Communications*).

[50] Krengel, U. (1985). *Ergodic Theorems.* de Gruyter Studies in Mathematics, No. 6, Walter de Gruyter, Berlin.

[51] Kuang, L. (1994) On the variance reduction property of the buffered leaky bucket. *IEEE Transactions on Communications*, Vol. 42, No. 9, pp. 2670-2671.

[52] Kumar, P. R. (1993). Re-entrant lines. Queueing Systems : Theory and Applications, Vol. 13, Nos. 1-3, pp. 87-110.

[53] Kumar, P. R., and Meyn, S. (1993). Duality and Linear Programs for Stability and Performance Analysis of Queueing networks and Scheduling Policies. *Technical Report*, University of Illinois, Urbana. To appear in *IEEE Transactions on Automatic Control.*

[54] Kumar, P. R., and Seidman, T. (1990). Dynamic Instabilities and Stabilization Methods in Distributed Real-Time Scheduling of Manufacturing Systems. *IEEE Transactions on Automatic Control*, Vol. 35, pp. 289-298.

[55] Leland, W. E., Taqqu, M. S., Willinger, W., and Wilson, D. V. (1994). On the Self-Similar Nature of Ethernet Traffic (Extended Version). *IEEE/ACM Transactions on Networking*, Vol. 2, No. 1, pp. 1-15.

[56] Liggett, T. (1985). *Interacting Particle Systems*. Grund. der Math. Wiss., Vol. 276, Springer-Verlag, Berlin.

[57] Loynes, R. M. (1962). The stability of queues with non independent interarrival and service times. *Proceedings of the Cambridge Philosophical Society*, Vol. 58, pp. 497-520.

[58] Lu, S. H., and Kumar, P. R. (1991). Distributed scheduling based on duedates and buffer priorities. *IEEE Transactions on Automatic Control*, Vol. 36, pp. 1406-1416.

[59] Marbukh, V. V. (1985). Asymptotic Investigation of a Complete Communications Network with a large number of points and bypass routes. *Problems of Information Transmission*, pp. 212-216.

[60] Marbukh, V. V. (1985). Fully Connected Message-Switching Network with a Large number of Nodes, Bypass routes and a limited number of Waiting places at the nodes. *Problems of Information Transmission*, pp. 154-161.

[61] Meyn, S. P., and Down, D. (1994). Stability of generalized Jackson networks. *Annals of Applied Probability*, Vol. 4, No. 1, pp. 124-148.

[62] Meyn, S. P., and Tweedie, R. L. (1993). *Markov chains and Stochastic Stability*. Springer-Verlag, London.

[63] Nelson, R. D. (1993). The Mathematics of Product-Form Queuing Networks. *Computing Surveys*, Vol. 25, No. 3, pp. 339-369.

[64] O'Connell, N. (1994). Large deviations in Queueing Networks. *Preprint*.

[65] Parekh, A. K., and Gallager, R. G. (1993). A Generalized Processor Sharing Approach to Flow Control in Integrated Services Networks : The Single Node Case. *IEEE/ACM Transactions on Networking*, Vol. 1, pp. 344-357.

[66] Parekh, A. K., and Gallager, R. G. (1994). A Generalized Processor Sharing Approach to Flow Control in Integrated Services Networks : The Multiple Node Case. *IEEE/ACM Transactions on Networking*, Vol. 2, No. 2, pp. 137-150.

[67] Robertazzi, T. G., (Editor) (1993). *Performance Evaluation of High Speed Switching Fabrics and Networks : ATM, Broadband ISDN, and MAN Technology.* IEEE Press, New York.

[68] Rybko, A., and Stolyar, A. L. (1992). Ergodicity of stochastic processes that describe the functioning of of open queueing networks. *Problems of Information Transmission*, Vol. 28, pp. 3-26.

[69] Schwartz, M. (1987). *Telecommunication Networks : Protocols, Modeling and Analysis.* Addison-Wesley, Reading, Massachusetts.

[70] Seidman, T. (1994). First-come first-served is unstable. *Preprint.*

[71] Shantikumar, J. G., and Yao, D. (1989). Stochastic monotonicity in general queueing networks. *Journal of Applied Probability*, Vol. 26, pp. 413-417.

[72] Shapley, L. (1953). Stochastic Games. *Proceedings of the National Academy of Sciences*, Vol. 39, pp. 1095-1100.

[73] Sigman, K. (1990). The stability of open queueing networks. *Stochastic Processes and Their Applications*, Vol. 35, pp. 11-25.

[74] Sznitman, A. (1989). Propagation of chaos. In *Ecole d'Eté de Probabilités de Saint-Flour*, Lecture Notes in Mathematics, Vol. 1464, Springer-Verlag, Berlin.

[75] Turner, J. (1986). New Directions in Communications (or Which Way in the Information Age). *IEEE Communications Magazine*, Vol 24, pp. 8-15.

[76] Uchiyama, K. (1985). Fluctuation in Population Dynamics. In *Lecture Notes in Biomathematics*, Vol. 70, pp. 222-229.

[77] Walrand, J. (1983). A Probabilistic Look at Networks of Quasireversible Queues. *IEEE Transactions on Information Theory*, Vol. IT-29, pp. 825-831.

[78] Walrand, J. (1988). *An Introduction to Queueing Networks.* Prentice Hall, Englewood Cliffs, N.J.

[79] Wentzell, A.D., and Freidlin, M. I. (1984). *Random perturbations of dynamical systems.* Grundlehren der mathematischen Wissenschaften, Vol. 260, Springer-Verlag, Berlin.

[80] Wolff, R. W. (1982). Poisson arrivals see time averages. *Operations Research*, Vol. 30, pp. 223-231.

# State-Space Models of the Term Structure of Interest Rates

**Darrell Duffie**

Graduate School of Business

May 28, 1995

**Abstract:** This is a survey of models of the term structure of interest rates, concentrating on models in which the term structure has a Markov state-space representation.

This survey is based on presentations at the Fifth Oslo-Silivri Workshop on Stochastic Analysis, which took place in Silivri, Turkey in July, 1994. I am grateful to the organizers, H. Körezlioğlu, B. Økesendal, and A.S. Üstünel. Sections 1 through 5 extend and update work presented at The Royal Society, London, November 10-11, 1993, and published in *Philosophical Transactions of the Royal Society*, Series A, Volume 347 (1994), pp. 577-586, co-authored with Rui Kan. Please address correspondence to Duffie at The Graduate School of Business, Stanford University, Stanford California 94305-5015 USA. Fax 415-725-7979; email duffie@baht.stanford.edu.

## 1. Introduction

Stochastic models for random fluctutations of the term structure of interest rates are commonly used in the finance industry for at least the following purposes:

- Pricing fixed-income derivative securities, such as options and mortgage-backed securities.
- Analyzing the risk of fixed-income portfolio strategies.
- Managing the interest-rate risk of fixed-income positions.

By "fixed income," we mean assets whose payoffs depend on the term structure itself. In a wide sense, this can include bonds; bond derivatives such as options, swaps, or caps; defaultable bonds; and even foreign bonds or derivatives based in sometimes complicated ways on domestic and foreign interest rates. There are many other reasons for understanding the process by which interest rates are determined and change over time, but our focus will be on models that are particualry useful for the above 3 purposes.

While various classes of stochastic models are used, the most common language of term-structure modelers in industry and universities is that of continuous-time stochastic calculus, which reached popularity following the impact of the Black and Scholes (1973) option pricing formula and the associated modeling ideas developed by Merton (1973) and others. Even in the early work of Pye (1966), however, one finds discrete-time models that capture much of the mechanics and principles of term structure models in a Markov state-space setting, where we will focus our efforts.

We will review how such models are constructed and applied, with particular emphasis on Markov diffusions, or jump diffusions, that represent the current term structure. Within this class, one can make reasonable tradeoffs between economic realism and computational tractability, bearing in mind that no tractable model can fully capture the complexity of unexpected changes in interest rates.

Section 2 gives the basic definitions and pricing relationships implied by the absence of arbitrage. Section 3 introduces Markov models in a one-dimensional diffusion setting, with examples. Section 4 moves to a multi-dimensional diffusion setting, emphasizing the tractability offered by a special "affine" class. Section 5 is devoted to derivative pricing and hedging. Sections 6 and 7 introduce jumps and default, respectively. Section 8 describes the Heath-Jarrow-Morton (HJM) model, while Section 9 places HJM models in a Markov state-space setting. Section 10 considers foreign term-structure derivative valuation in an HJM framework. Section 11 adds comments on new research directions.

## 2. Setup

We begin with a probability space $(\Omega, \mathcal{F}, P)$ and filtration $\{\mathcal{F}_t : t \in [0, \infty)\}$ of sub-$\sigma$-algebras of $\mathcal{F}$ satisfying the usual conditions. (For technical details, see, for example, Protter (1990), or other standard references.) Given is a progressively measurable "short rate" process $r$ that is integrable, in the sense that $\int_0^T |r_t|\, dt < \infty$ almost surely for all $T > 0$. We may think of $r_t$ as the interest rate at time $t$ on loans of infinitesimal maturity. More properly, a short rate proess $r$ implies the ability to invest one unit of account at any time $t$, and, with continual re-investment, to receive at any time $s \geq t$ the payoff $\exp\left(\int_t^s r_u\, du\right)$.

For our purposes here, a financial security is a contract that promises to pay, at some time $T$, possibly a stopping time, some $\mathcal{F}_T$-measurable random variable $Z$. Each such claim $(Z, T)$ is assigned a price process given by a semimartingale $S^{Z,T}$. An "arbitrage" is, loosely speaking, a strategy for trading various claims, at no initial investment cost, so as to generate only positive (and non-zero) cash flows. According to a formalization of this conceived by Harrison and Kreps (1979), and subsequently developed by many,[1] under technical conditions there is no arbitrage if and only if there is a probability measure $Q$, equivalent to $P$, under which, for any such triple $(Z, T, S^Z)$,

$$S_t^{Z,T} = E\left[\exp\left(-\int_t^T r_s\, ds\right) Z \,\middle|\, \mathcal{F}_t\right], \quad t \leq T, \tag{1}$$

where "$E$" denotes expectation under $Q$. We fix such a measure $Q$ throughout.

A simple example of a claim, a zero-coupon bond, is defined by letting $T$ be determinisitic and $Z = 1$. For such a claim $(1, T)$, the price $S_t^Z$ is denoted $p_{t,T}$. The continuously-compounding yield of a zero-coupon bond of maturity $m \in [0, \infty)$ is then defined as

$$y_{t,m} = -\frac{1}{m}\, \log\, p_{t,t+m}, \quad (t, m) \in \mathbb{R}_+^2. \tag{2}$$

At time $t$, the "term structure of interest rates" is the function mapping the maturity $m$ to the yield $y_{t,m}$ at that maturity.

---

[1] See, for example, references cited in Duffie (1996).

For practical applications, the basic issue is how to model the probabilistic behavior of the short-rate process $r$ under $Q$. One wants a model for $r$ that is sufficiently rich to capture the essential nature of actual interest rates, while at the same time sufficiently tractable for purposes of econometric estimation and for computation of the prices of contingent claims by (1), for a range of commonly traded claims. There are also many theoretically interesting questions regarding the equilibrium determination of the short rate process $r$ and the equivalent "martingale" measure $Q$. It is known that, under weak technical conditions, any short rate process $r$ can be supported in a simple and standard model of market equilibrium, with easily risk preferences. [See, for example, Heston (1991) and Duffie (1996, Exercise 10.3).]

We will be focus here only on practical issues, and disregard other aspects of the market equilibrium problem. We will review some basic classes of models for the behavior of the short rate process $r$ under a given equivalent martingale measure $Q$. We will concentrate on Markovian state-space models, beginning with "single-factor" models, moving to finite-dimensional multi-factor models, and finally describe infinite-dimensional state-space models, in the framework of Heath, Jarrow, and Morton (1992), using the approach of stochastic partial differential equations. Along the way, we shall review such applications as the valuation and hedging of derivative securities.

For many applications, it would be useful to model the distribution of interest rate processes under the original probability measure $P$. Conversion from $P$ to $Q$ and back will not be dealt with here, but is an important issue, particulary from the point of view of statistical fitting of models as well as the measurement of risk.

## 3. Single-Factor Models

The simplest class of models that we consider takes the short rate process $r$ to be the solution of a stochastic differential equation of the form

$$dr_t = \mu(r_t)\, dt + \sigma(r_t)\, dW_t, \tag{3}$$

where $W$ is a Standard Brownian Motion that is a martingale under $Q$, and where $\mu : I\!R \to I\!R$ and $\sigma : I\!R \to I\!R$ have enough regularity to ensure the existence of a unique solution to (3). [See, for example, Ikeda and Watanabe (1981).] Since $r$ is a strong Markov process under

$Q$, we have $p_{t,T} = F(r_t, t)$, for some measurable function $F : \mathbb{R} \times [0, T] \to \mathbb{R}$, and we can therefore view the entire yield curve $y_t = \{y_{t,m} : m \geq 0\}$ defined by (2) as measurable with respect to $r_t$. Hence the label "single-factor model" applies, since a single state variable, in this case the short rate $r_t$, determines all yields and is a sufficient statistic (under the equivalent martingale measure $Q$) for all future yield curves.

While simple and, as it turns out, quite tractable, the single-factor class of models given by (3) is (like any theoretical model) at variance with reality. Consequently, on a given day, the yield curve associated with the model differs from that observed in the marketplace. If significant, a discrepancy may suggest the development of a new theoretical model. In the finance industry, however, one needs to use some particular model, even if imperfect. In practice, a discrepancy between the actual and theoretical yield curves is eliminated by introducing, at each current time $t$, time-dependence in the functions $\mu$ and $\sigma$, to arrive at a "calibrated model," $\mu^t : \mathbb{R} \times [0, \infty) \to \mathbb{R}$ and $\sigma^t : \mathbb{R} \times [0, \infty) \to \mathbb{R}$, of the form

$$dr_s = \mu^t(r_s, s)\, ds + \sigma^t(r_s, s)\, dW_s, \quad s \geq t, \tag{4}$$

under technical conditions on $(\mu^t, \sigma^t)$. The *calibrated model* $(\mu^t, \sigma^t)$ is computed with numerical algorithms, examples of which are described in Black, Derman, and Toy (1989) and Black and Karasinski (1992). With calibration, an exact match can be achieved between the actual and modeled yield curves.

It is common to calibrate not only with the yield curve, but also with certain "volatility-related" information available in the market through the prices of option-related securities such as caps. A cap is a portfolio $\{(Z_i, T_i) : i \in \{1, \ldots, N\}\}$ of claims, of which $(Z_i, T_i)$ is the claim defined by letting $T_i - T_{i-1}$ be a constant, say 0.5 years, independent of $i$, and by letting $Z_i = \max(Y_i - \overline{Y}, 0)$, where $\overline{Y}$ is a constant and $Y_i$ is the simple interest rate that applies at time $T_{i-1}$ for bonds maturing at time $T_i$, that is, $(1 + Y_i)^{-1} = p_{T_{i-1}, T_i}$. A cap is often purchased as a hedge against a loan whose interest payments are periodically re-set to current market rates. A cap is but one example of a term-structure derivative, a claim (or portfolio of claims) derived from the term structure of interest rates. Through the basic pricing relationship (1), one can search for a calibrated model $(\mu^t, \sigma^t)$, within a given class of models, that not only matches the term structure of yields observed in the market with those determined by the model, but also matches the market prices of caps with those determined by the model.

In general, calibration is designed to capture the pricing implications of certain "bench-mark" financial securities, usually those with a high level of trading activity, in order to estimate the prices of less common claims through the basic relationship (1).

Suppose a term-structure model has been calibrated at time $t$. At the next time period $t + 1$, given the likley mis-specification of the model, it is common in practice to recalibrate so as to achieve a new model $(\mu^{t+1}, \sigma^{t+1})$ that is once again consistent with market prices of certain benchmark claims. Since the necessity for re-calibration was not considered when using the previous version of the model for pricing purposes, this suggests a theoretical inconsistency in the application of the model. The compromise involved seems reasonable under the circumstances.

Most, if not all, of the parametric single-factor models appearing in the literature or in industry practice, are of the form

$$dr_t = [\alpha_1(t) + \alpha_2(t)\, r_t + \alpha_3(t)\, r_t \log r_t]\; dt + [\beta_1(t) + \beta_2(t)\, r_t]^\gamma\; dW_t, \tag{5}$$

for time-dependent deterministic coefficients $\alpha_1$, $\alpha_2$, $\alpha_3$, $\beta_1$, and $\beta_2$, and for some exponent $\gamma \geq 0.5$. (For existence and uniqueness of solutions, additional coefficient restrictions apply.) Table 1 lists the origins[2] of various special cases of this parametric class, indicating with "•" the coefficients that are non-zero (sometimes constant) for each special case, and indicating the choice of exponent $\gamma$.

## 4. Multi-Factor Models

While single-factor models offer tractability, there is compelling reason to believe that a single state variable, such as the short rate $r_t$, is insufficient to capture reasonably well the distribution of future yield curve changes. The econometric evidence in favor of this view includes the work of Chen and Scott (1992b, 1993), Duffie and Singleton (1995), Litterman and Scheinkman (1988), Pearson and Sun (1990), and Stambaugh (1988). [For empirical comparisons among most of the single-factor models considered in Table 1, see Chan, Karolyi, Longstaff, and Sanders (1992).]

---

[2] By offering extensions with time-varying coefficients, Ho and Lee (1986) and Hull and White (1990) have popularized the constant coefficients models of Merton (1973) and Vasicek (1977), respectively.

Table 1.   Parametric Single-Factor Models

| $dr_t = [\alpha_1(t) + \alpha_2(t)r_t + \alpha_3(t)r_t \ \log \ r_t] \ dt + [\beta_1(t) + \beta_2(t)r_t]^\gamma \ dB_t$ | | | | | | |
|---|---|---|---|---|---|---|
| | $\alpha_1$ | $\alpha_2$ | $\alpha_3$ | $\beta_1$ | $\beta_2$ | $\gamma$ |
| Cox-Ingersoll-Ross | • | • | | | • | 0.5 |
| Dothan | | | | | • | 1.0 |
| Brennan-Schwartz | • | • | | | • | 1.0 |
| Merton | • | | | • | | 1.0 |
| Vasicek | • | • | | • | | 1.0 |
| Pearson-Sun | • | • | | • | • | 0.5 |
| Black-Derman-Toy | | • | • | | • | 1.0 |
| Constantinides-Ingersoll | | | | | • | 1.5 |

In principle, of course, the yield curve sits in an infinite-dimensional space of functions, and there is no reason to believe that the direction of its movements will be restricted to some finite-dimensional manifold. An infinite-dimensional state space model is outlined in Section 9. For practical purposes, however, tractability might call for a finite number of state variables, and it is an empirical issue as to how many might be sufficient to offer reasonable empirical properties. Some of the empirical studies mentioned above suggest that 2 or 3 state variables might suffice for many practical purposes.

In any case, we will consider a state process $X$ in some open subset $D$ of $I\!R^n$, defined as the solution of the stochastic differential equation (SDE)

$$dX_t = \mu(X_t) \, dt + \sigma(X_t) \, dW_t, \qquad (6)$$

where $W$ is a Standard Brownian Motion in $I\!R^n$ under $Q$, and where $\mu : D \rightarrow I\!R^n$ and $\sigma : D \rightarrow I\!R^{n \times n}$ satisfy sufficient regularity for existence and uniqueness of solutions. In what follows, we could add time dependence to $\mu$ and $\sigma$ without changing the major ideas.

We also suppose that the short rate process $r$ is given by $r_t = R(X_t)$, for some $R : D^n \rightarrow I\!R$. Thus the zero-coupon bond maturing at $T$ has a price at time $t \leq T$ given from

(1) by

$$F(X_t, t) = E\left[\exp\left(-\int_t^T R(X_s)\,ds\right) \Big| X_t\right]. \tag{7}$$

It is frequently convenient to exploit the Markov setting and solve (7) for the market prices of bonds via the "Feynman-Kac" connection between (7) and the associated partial differential equation (PDE) for $F$. Under technical regularity (see, for example, Friedman (1975)), $F$ satisfies (7) for all $t$ if and only if $F$ solves the PDE

$$F_t + F_x\mu + \frac{1}{2}\text{trace}\left[F_{xx}\sigma\sigma^\mathsf{T}\right] + RF = 0, \tag{8}$$

with boundary condition $F(x, T) = 1$, for $x \in D$.

We may wish to exploit special structure in order to obtain numerical and econometric tractability. For example, it turns out to be convenient for some applications to take $\mu$, $\sigma\sigma^\mathsf{T}$, and $R$ to be affine functions on $D$ into their respective ranges. (An affine function is a constant plus a linear function.) In this case, we say that the primitive model $(\mu, \sigma\sigma^\mathsf{T}, R)$ is *affine*. Likewise, we say that the term structure is itself *affine* if there are $C^1$ functions $c : [0, \infty) \to I\!\!R$ and $C : [0, \infty) \to I\!\!R^n$ such that

$$p_{t,s} = e^{c(s-t)+C(s-t)\cdot X(t)}, \quad t \geq 0, \quad s \geq t,$$

so that yields are affine in the state variables. Conditions for an affine term structure can be deduced from (8) and the required form of solution

$$F(x, t) = e^{c(T-t)+C(T-t)\cdot x}. \tag{9}$$

In order for (9) to apply, we can then see from (8) that the coefficient functions $c$ and $C$ must solve an ordinary differential equation of the Ricatti form

$$C_i'(m) = k_i + K_i \cdot C(m) + C(m)^\mathsf{T} Q_i C(m), \quad i \in \{1, \ldots, n\}, \tag{10}$$

$$c'(m) = k_0 + K_0 \cdot C(m) + C(m)^\mathsf{T} Q_0 C(m), \tag{11}$$

with boundary conditions

$$c(0) = C_i(0) = 0, \quad i \in \{1, \ldots, n\}, \tag{12}$$

where $\{k_0, \ldots, k_n\} \subset \mathbb{R}$, $\{K_0, \ldots, K_n\} \subset \mathbb{R}^n$, and $\{Q_0, \ldots, Q_n\} \subset \mathbb{R}^{n \times n}$ are constant coefficients given in terms of the coefficients defining the affine functions $\mu$, $\sigma\sigma^\top$, and $R$. The Ricatti equation (10)-(11)-(12) can easily be solved numerically, for example by a Runge-Kutta method.

Indeed, Duffie and Kan (1992) show that, under technical conditions, the basic model $(\mu, \sigma\sigma^\top, R)$ is affine if and only if the term structure is affine. This extends the same result for $n = 1$ given by Brown and Schaefer (1991), and earlier hinted at by Cox, Ingersoll and Ross (1985), whose model allows for explict solutions of the coefficient functions $c$ and $C$.

If $R$, $\mu$, and $\sigma\sigma^\top$ are affine with time-dependent coefficients designed for calibration to given data, then the coefficients $k_i$, $K_i$, and $Q_i$ in (10)-(11)-(12) will have time dependencies, but the ODE itself remains numerically tractable.

In general, one could imagine that the state vector $X_t$ might include various economic indices that would affect interest rates such as economic activity, monetary supply variables, central bank policy objectives, and so on. In order to facilitate the pricing and hedging of fixed-income derivatives, however, it is convenient to assume that one can find a change of variables under which we may view $X_t$ as yield-related variables. Given the solution $(c, C)$ of (10)-(11)-(12), relation (8) provides an affine change of variables under which the state may be taken to be an $n$-dimensional "yield-factor" process $Y$, where for some fixed maturities $\tau(1), \ldots \tau(n)$, we take

$$Y_{ti} = y_{t, \tau(i)} = -\frac{1}{\tau(i)} \left[ c(\tau(i)) + C(\tau(i)) \cdot X_t \right], \quad i \in \{1, \ldots, n\}.$$

We need only ensure that the "basis maturities" $\tau(1), \ldots, \tau(n)$ are chosen so that the matrix $K$ in $\mathbb{R}^{n \times n}$, defined by $K_{ij} = -C_j(\tau(i))/\tau(i)$, is non-singular. In that case, we have $Y_t = k + K X_t$, where $k_i = -c(\tau(i))/\tau(i)$, and the new state dynamics are given by

$$dY_t = \mu^*(Y_t)\, dt + \sigma^*(Y_t)\, dW_t, \tag{13}$$

where

$$\mu^*(y) = k\mu(K^{-1}y - k)$$

$$\sigma^*(y) = K\sigma(K^{-1}y - k),$$

for $y \in D^* = \{Kx + k : x \in D\}$.

If $\sigma$ is constant, $X$ and $Y$ are Gauss-Markov processes of the Ornstein-Uhlenbeck form. For abstract factors, this Gaussian model was developed by Langetieg (1980) and

Jamshidian (1990, 1991). A Gauss-Markov yield-factor model was developed by El Karoui and Lacoste (1992) in the forward-rate setting of Heath, Jarrow, and Morton (1992), and in the current state-space setting, was developed as a special case of stochastic volatility models by Duffie and Kan (1992).

A simple example of an affine model that is not Gaussian (that is, for which $\sigma$ depends on the state) is the multivariate Cox-Ingersoll-Ross model:

$$dX_{it} = (a_i - b_i X_{it})\, dt + c_i \sqrt{X_{it}} \ dW_{it}, \quad i \in \{1, \ldots, n\}, \tag{14}$$

for positive constants $a_i$, $b_i$, $c_i$, with $R(x) = x_1 + \cdots + x_n$. This model was developed by Cox, Ingersoll, and Ross (1985), exploiting results by Feller (1951), and extended by Richard (1978), Heston (1991), Longstaff and Schwartz (1992), Pearson and Sun (1994), and Chen and Scott (1992a). Restrictions apply: For all $i$, we want

$$a_i > c_i^2/2. \tag{15}$$

As shown by Ikeda and Watanabe (1981), (15) is necessary and sufficient to ensure that $X$ will remain in the obvious open state space $D = \text{int}\,(I\!R_+^n)$. Another special case of an affine model is the 3-dimensional state-space model studied by Chen (1994). Duffie and Kan (1992) study the general affine case, under which one can, without significant loss of generality after a linear change of variables, take

$$\mu(x) = ax + b; \quad \sigma_{ij}(x) = \gamma_{ij}\sqrt{\alpha_{ij} + \beta_{ij} \cdot x}, \tag{16}$$

for some $\gamma_{ij} \in I\!R$, $\alpha_{ij} \in I\!R$, $\beta_{ij} \in I\!R^n$, $a \in I\!R^{n \times n}$, and $b \in I\!R^n$. In this case, the natural state space is

$$D = \left\{ x \in I\!R^n : \alpha_{ij} + \beta_{ij} \cdot x > 0, \quad i, j \in \{1, \ldots, n\} \right\}. \tag{17}$$

Restrictions on the coefficients $(a, b, \gamma, \alpha, \beta)$, analogous to (15) but more complicated, are shown by Duffie and Kan (1992) to imply the affine form and to guarantee the existence and uniqueness of solutions to (6) in $D$ for (16)-(17).

Aside from the affine case, multivariate term-structure models appear in Brennan and Schwartz (1979), Beaglehole and Tenney (1991), Chan (1992), Constantinides (1992), Duffie, Ma, and Yong (1993), El Karoui, Myneni, and Viswanathan (1991), and Jamshidian (1993).

Most of these non-affine multifactor models do not allow direct observation of the state from the yield curve.

If one does not observe the state-vector directly, one can in principle filter the state-variable from yield-curve data. See, for example, Chen and Scott (1993) and Kennedy (1995). There are debates concerning how much this limited observation property detracts from the practical application of the models. It can be said, for example, that we do not observe the yield curve in any case, but merely the prices of coupon bonds, from which one infers statistically (and with noise) the zero-coupon curve by some curve-fitting method such as splines or non-linear least squares. In any case, it seems to be of at least some value to have state variables that can be observed in terms of the yield curve, as in the affine models described above.

## 5. Derivative Pricing and Hedging

A major application of term structure models is the pricing of derivative securities. Given a Markov term structure model, as defined in the previous section, the price of a claim $(Z, T)$ is given at any time $t \leq T$ by

$$E\left[\exp\left(-\int_t^T R(X_s)\,ds\right) Z \,\bigg|\, \mathcal{F}_t\right].$$

If $Z$ is measurable with respect to the yield curve at time $T$, as are bond options and other "path-independent" derivatives, we may take $Z = g(X_T)$ for some $g : D \to \mathbb{R}$, since the yield curve $y_T$ is itself $X_T$-measurable. In this case, the Markov property of $X$ implies that we can write the derivative price as

$$F(X_t, t) = E\left[\exp\left(-\int_t^T R(X_s)\,ds\right) g(X_t) \,\bigg|\, X_t\right], \tag{18}$$

for some $F : D \times [0, T] \to \mathbb{R}$. Under technical regularity we also know that $F$ is the unique solution in $C^{2,1}(D \times [0, T))$, under technical growth conditions, to the parabolic partial differential equation (8) with the boundary condition

$$F(x, T) = g(x), \quad x \in D. \tag{19}$$

One can then solve for the derivative price function $F$ via a numerical solution of the PDE (8)-(19), say by a finite-difference algorithm. (See, for example, Ames (1977).) Fully worked examples are given by Duffie and Kan (1992) for the case $n = 2$. For large $n$, say more than 3, currently available algorithms and hardware may not be up to the task, and Monte-Carlo simulation may be applied. (See, for example, Duffie and Glynn (1992) and Kloeden and Platen (1992).) For the "path-dependent" case, in which $Z$ depends non-trivially on the path $\{X_t : 0 \le t \le T\}$ of the state process, it may also be advisable to resort to Monte Carlo simulaton. There are only rare cases, such as Jamshidian's (1991) solution for bond options in the Gaussian setting, for which one can obtain explicit solutions for derivative prices. [See, also, El Karoui and Rochet (1989).]

The term *hedging* has different meanings in different contexts. For our narrow purposes here, we will take it to mean the construction of a trading strategy involving certain claims that replicates the value of another "target" claim. By virtue of such a replication, one can offset the risk of the target security completely by selling the replicating strategy. In practice, of course, this is unrealistic for it usually assumes both that the underlying valuation model is correct and that the necessary hedging strategy can be executed precisely and without transactions costs. This replication approach to hedging has nevertheless been shown to be useful in applications.

To take a simple case, consider the single-factor model (3), and suppose that the target claim to be hedged is defined by the payoff $Z = g(r_T)$ at time $T$, for some $g : \mathbb{R} \to \mathbb{R}$. As asserted earlier, the price process $V$ for this claim is, under technical regularity, given by $V_t = F(r_t, t)$, for some function differentiable $F$ solving the PDE (8) with boundary condition (19).

For hedging purposes, consider another claim, similarly defined, and with a market value at time $t$ given by $U_t = \Phi(r_t, t)$, for some $\Phi \in C^{2,1}(\mathbb{R} \times [0, T))$, also satisfying (8). We wish to deduce a predictable process $\{\theta_t : 0 \le t \le T\}$, defining a strategy for trading the hedging security, so that the gains or losses associated with the hedging security offset those associated with the target security. The number of units held at time $t$ in state $\omega$ is $\theta(\omega, t)$. Thinking of $\theta_t$ for the moment as a fixed position $b$, and fixing the short rate at time $t$ at some level $x$, the total market value $F(x_t, t) + b\Phi(x, t)$ of the hedged position has a derivative (or "sensitivity") with respect to the short rate of $F_x(x, t) + b\Phi_x(x, t)$. Since the only source of risk in this single-factor setting is the short rate $r$, an intuitive choice for $\theta_t$ is one that equates this derivative to zero. If we do so state by state, time by time, we have

$$\theta_t = -\frac{F_x(r_t, t)}{\Phi_x(r_t, t)}, \quad t \in [0, T],$$

(20)

assuming that $\Phi_x$ is everywhere non-zero.

In order to show that this intuitive position is in fact appropriate, we augment the position $\theta_t$ in the hedging claim with deposits of $J_t$ at the short rate at time $t$. We also require technical conditions for $\theta$ and $J$ so that $\int \theta_t \, dU_t$ and $\int J_t r_t \, dt$ make sense as stochastic integrals. The total market value process $Y$ would then satisfy

$$dY_t = dV_t + \theta_t \, dU_t + J_t r_t \, dt.$$

(21)

From Ito's Lemma,

$$\begin{aligned}
dY_t = {} & \mathcal{D}F(r_t, t) \, dt + F_x(r_t, t) \, dr_t \\
& + \theta_t [\mathcal{D}\Phi(r_t, t) \, dt + \Phi_x(r_t, t) \, dr_t] + J_t r_t \, dt,
\end{aligned}$$

(22)

where $\mathcal{D}$ is the infinitesimal generator associated, as usual, with $(\mu, \sigma)$. From (8), assuming zero dividend rates for simplicity, we have $\mathcal{D}F(x, t) = xF(x, t)$ and $\mathcal{D}\Phi(x, t) = x\Phi(x, t)$ for all $x$ and $t$. Thus, once one applies the specified hedge position $\theta_t = -F_x(r_t, t)/\Phi_x(r_t, t)$, we have

$$dY_t = r_t [F(r_t, t) + \theta_t \Phi(r_t, t) + J_t] \, dt.$$

(23)

In order for the market value $Y$ of the hedged position to remain constant, we can therefore let

$$J_t = -F(r_t, t) - \theta_t \Phi(r_t, t).$$

(24)

The total initial cost of the hedge is then

$$\theta_0 \Phi(r_0, 0) + Z_0 = -F(r_0, 0),$$

(25)

as one would expect from the fact that the trading strategy $(\theta, J)$ in the hedging asset and short-term deposits merely replicates $-F(r_T, T)$. The restrictiveness of a single-factor model is apparent from its implication, just shown, that essentially any derivative (or bond) can be used to perfectly hedge any other. With a state-space of dimension $n$, one can see that $n + 1$ positions would typically be both necessary and sufficient to achieve hedging, in the above sense.

## 6.  Jumps

The models in Sections 3 and 4 imply continuous sample paths for any bond yield. For realism, one may introduce jumps and keep essentially all of the pricing tractability of the Markov diffusion setting, although perfect hedging, as described in the previos section, becomes problematic. For example, consider an extension of the single-factor model (3) for the short rate $r$ with state space $D \subset {\rm I\!R}$ satisfying

$$dr_t = \mu(r_t)\,dt + \sigma(r_t)\,dW_t + dU(t, r_t), \qquad (26)$$

where $U$ is a pure jump process with jumps at inaccessible stopping times, with jump arrival intensity process $\{\lambda(r_t) : t \geq 0\}$, where $\lambda : D \to {\rm I\!R}_+$ is a measurable function, and with jump distribution process $\{\nu(r_t) : t \geq 0\}$, where $\nu : D \to \mathcal{P}({\rm I\!R})$ is a measurable function mapping each interest rate level $x$ to a probability measure $\nu(x)$. For any jump time $\tau$, the jump $\Delta U_\tau = U_\tau - U_{\tau-}$ has $\mathcal{F}_{\tau-}$-conditional distribution $\nu(r_{\tau-})$. This is a loose description of a jump-diffusion, a class of processes treated by Gihman and Skorohod (1972).

In order to characterize the yield curve in terms of the state variable $(r)$, we can once again compute the the zero-coupon bond price function $F$ defined by $F(r_t, t) = p_{t,T}$. From Ito's lemma for semimartingales, we can deduce under technical regularity the integro-differential equation for $F$ that extends (8), given by

$$F_t + F_x\mu + \frac{1}{2}F_{xx}\sigma^2 + xF + \lambda(x)\left[\int_{-\infty}^{\infty} F(x+u)\,\nu(x, du) - F(x, t)\right] = 0, \qquad (27)$$

with boundary condition $F(x, T) = 1$.

One can solve (27) numerically by an extension of a finite-difference algorithm for PDEs. Likewise, in order to calculate the price of a claim of the form $(g(r_T), T)$, a finite-difference method can be used to solve (27) with the boundary condition (19).

For computational tractability, one can deduce restrictions on $(\mu, \sigma^2, \nu, \lambda)$ that imply the affine yield model (9). For example, we can see from (27) that it is sufficient that $(\mu, \sigma^2, \lambda)$ is affine and $\nu$ is a constant distribution (that is, $\nu(x)$ does not depend on $x$). The special Gaussian case has been explored by Das (1994). For the case of $\sigma(x) = C\sqrt{x}$, for $x \geq 0$, we would want the jump distribution to support the non-negative real line only, in order to allow for the existence of solutions to (26).

For an affine jump-duffusion model, in this univariate setting or in the obvious multivariate extension, an ordinary differential equation of the form of (10)-(11)-(12) characterizes the term structure.

With jumps, the hedging calculations in Section 5 do not apply, and in general it is impossible to perfectly hedge a given claim with fewer positions in other claims than the cardinality of the support of the jump distribution.

## 7. Default

In practice, any contingent claim $(Z, T)$ may default. That is, the individual, corporation, or institution responsible for payment of the claim $Z$ at $T$ may not do so. One usually ignores default risk for certain government obligations, but default risk for corporations and some sovereign states cannot be ignored if one is to explain the fact that the prices of identical claims against different issuers have different market prices. For example, the yields of corporate bonds of low credit quality are typically higher than those of government bonds of the same maturity.

A defaultable claim is defined by a pair $((Z, T), (Z', T'))$ of claims. The first claim $(Z, T)$ of the pair represents the obligation of the issuer to pay $Z$ at $T$. The second claim of the pair is the actual payment of $Z'$ in the event that the issuer defaults at some time $T'$ before $T$. A defaultable claim $((Z, T), (Z', T'))$ thus defines a claim $(\widehat{Z}, \widehat{T})$, in the usual sense, by

$$\widehat{Z} = Z1_{T < T'} + Z'1_{T \geq T'}, \qquad \widehat{T} = \min(T, T').$$

Two basic approaches have been applied to the modeling of defaultable claims.

The first approach, which we call "structural," begins with Black and Scholes (1973) and Merton (1974), and is based on a stochastic model of the values of the assets and liabilities of the issuer. In this framework, a contingent claim $(Z, T)$ issued by a corporation will result in actual payment of $Z$ at $T$ in the event that the market value of the corporation's assets remains larger than that of its liabilities through time $T$, and otherwise will result in payment of some claim $(Z', T')$. Typically, with this structural approach, $T'$ is the first time at which the market value of assets is exceeded by the market value of liabilities. Indeed, the Black-Scholes option pricing formula was actually derived, in part, in order to value the debt of a corporation in this manner. Nielsen and Saa-Requejo (1992) is a recent example of the literature extending this structural approach.

The second approach starts from a "reduced-form" model in which the default time $T'$ is a given stopping time that may or may not be directly tied to the solvency of the issuer of the claim $(Z, T)$ in question. This approach has been developed by Artzner and Delbaen (1992), Duffie, Schroder, and Skiadas (1994), Duffie and Singleton (1994), Jarrow and Turnbull (1992), Jarrow, Lando, and Turnbull (1993), Lando (1994, 1995), Madan and Unal (1993), and others. The stopping time is usually assumed to have a "default hazard rate" process, a progressively measurable integrable process $h$ defined by the property that $1_{t>T'} = \Gamma_t$, where

$$d\Gamma_t = 1_{t \leq T'} h_t \, dt + dM_t, \tag{29}$$

where $M$ is a $Q$-martingale. In particular, $T'$ is inaccessible, as opposed to the usual case of the structural models described above, for which $T'$ is the accessible stopping time defined as the first time at which the market value of the assets of the issuer, modeled as a diffusion process, is exceeded by the market value of the liabilities of the issuer, also modeled as a diffusion.

In the reduced-form approach, various models have been proposed for the default recovery value $Z'$. In some cases, the recovery value is a fraction of the market value of a default-free version of the same claim, that is, $Z' = Z_f(T')$, where

$$Z_f(t) = f(t)E\left[\exp\left(-\int_t^T r_s \, ds\right) Z \,\middle|\, \mathcal{F}_t\right], \tag{30}$$

where $f$ is a bounded progressively measurable integrable process. (For practical purposes, we may wish to assume that $0 \leq f(t) \leq 1$ for all $t$.)

In other reduced-form models, such as Duffie and Singleton (1994), default generates a given fractional loss of value, that is, the price process $S$ of the defaultable claim in question has the property that

$$Z' = S(T') = \varphi(T')S(T'-), \tag{31}$$

where $\varphi$ is a bounded progressively measurable fractional recovery process. (Again, we may wish to assume that $0 \leq \varphi(t) \leq 1$ for all $t$.) One can combine the two approaches, as in Duffie and Singleton (1994) and Duffie, Schroder, and Skiadas (1994), by allowing the fractional recovery process $\varphi$ to be determined endogenously, for example by taking $\varphi(\omega, t) = \Phi(\omega, t, S(\omega, t-))$, for some well-behaved $\Phi : \Omega \times [0, T] \times \mathbb{R} \to \mathbb{R}$. In order to recover (30) as a special case, we would have

$$\Phi(\omega, t, v) = \frac{Z_f(\omega, t)}{v}, \quad v \neq 0. \tag{32}$$

One may also allow the hazard-rate process $h$ to be determined in terms of the price process $S$ itself. That is, we may have $h(\omega, t) = H(\omega, t, S(\omega, t-))$, where $H$ is defined as was $\Phi$. Ultimately, of course, the price process $S$ of the defaultable claim must satisfy the pricing relationship (1) in terms of the effective claim $(\widehat{Z}, \widehat{T})$. Allowing $h$ and $\varphi$ to be determined endogenously through such functions $\Phi$ and $H$, respectively, presents no essential difficulties. Under regularity conditions on $(r, Z, T, H, \Phi)$ found in Duffie, Schroder, and Skiadas (1994), a price process $S$ for a given defaultable claim is uniquely well defined as the solution of a recursive stochastic integral equation.

Duffie and Huang (1994) provide an extension of the notion of a defaultable claim in which either of two counterparties to the same contract may default. A natural example is a swap or forward contract, for which the underlying claim $(Z, T)$ calls for both positive and negative outcomes of $Z$. For this case, one allows the default characteristics $h$ and $\varphi$ to depend on the price process $S$ through functions $H$ and $\Phi$, as above. Indeed, this is essential in order to capture the impact on pricing of differences in credit quality between the two counterparties.

If we take the simple case in which $h$ and $\varphi$ are given processes, the defaultable valuation model can be simplified dramatically. One can show in this case the simple pricing relationship

$$S_t = E\left[\exp\left(-\int_t^T \rho_s \, ds\right) Z \,\middle|\, \mathcal{F}_t\right], \quad t < T', \tag{33}$$

where $\rho_t = r_t + h_t(1 - \varphi_t)$. That is, prior to default, we may treat the price of a defaultable claim $((Z, T), (Z', T'))$ with default characteristics $(h, \varphi)$ as identical to that of a default-free claim to $(Z, T)$, with the sole exception that the default-free short interest rate process $r$ is replaced in (1) with a default-adjusted short rate process $\rho$ that captures the pricing effect of the probability distribution of the default time $T'$ as well as the effect of the recovered value $Z' = Z_\varphi$ at default.

In applications, one can take advantage of a state-space representation for $(r, \rho)$. For example, Duffie and Singleton (1994) consider cases in which $r_t = a \cdot X_t$ and $\rho_t = A \cdot X_t$, for a Markov process $X$ in $\mathbb{R}^n$ satisfying an SDE of the form (6), and for $a$ and $A$ in $\mathbb{R}^n$. In order to guarantee that $\rho \geq r$, one can impose restrictions under which $X$ is non-negative and take $A \geq a$.

## 8. Heath-Jarrow-Morton Models

In modeling the term structure, we have so far taken as the primitive a model of the short rate process $r$ of the form $r_t = R(X_t, t)$, where (under some equivalent martingale measure) $X$ solves a given stochastic differential equation. In the one-factor case, one usually takes $r_t = X_t$. This approach has the advantage of a finite-dimensional "state-space." For example, with this state-space approach one can compute certain derivative prices or hedges by solving PDEs.

An alternative approach is to take the entire yield curve as a state variable. This can be done by exploiting the model of forward rates introduced by Heath, Jarrow, and Morton (1992). This section is a summary of the basic elements of the Heath-Jarrow-Morton (HJM) model. The next section moves the HJM model into a state-space setting.

The forward price at time $t$ of a zero-coupon bond for delivery at time $\tau \geq t$ with maturity at time $s \geq \tau$ is given by $p_{t,s}/p_{t,\tau}$, the ratio of zero-coupon bond prices at maturity and delivery, respectively. (For our purposes, one may take this as a definition. In fact, this relationship is a consequence of assuming the absence of arbitrage and of a more basic definition of a forward contract.) The associated *forward rate* is defined by

$$\Phi_{t,\tau,s} \equiv \frac{\log(p_{t,\tau}) - \log(p_{t,s})}{s - \tau}, \tag{34}$$

which can be viewed as the continuously compounding yield of the bond bought forward. The *instantaneous forward rate*, when it exists, is defined for each time $t$ and forward delivery date $\tau \geq t$, by

$$f(t, \tau) = \lim_{s \downarrow \tau} \Phi_{t,\tau,s}. \tag{35}$$

Thus, the instantaneous forward rate process $f$ exists if and only if, for all $t$, the mapping $s \mapsto p_{t,s}$ is differentiable.

A convenient fact is that the price at time $t$ of a zero-coupon bond maturing at $s$ can be computed as

$$p_{t,s} = \exp\left(-\int_t^s f(t,u)\,du\right),\qquad(36)$$

so the term structure can be recovered from the instantaneous forward rates, and *vice versa*.

Given a stochastic model $f$ of forward rates, we will assume that $r_t = f(t,t)$ defines the short rate process $r$. This means that we will treat $r_t$ as the limit of bond yields as maturity goes to zero. Justification of this assumption can be given under technical conditions. See, for example, Carverhill (1995).

The *HJM model of forward rates*, for each fixed maturity $s$, is given by

$$f(t,s) = f(0,s) + \int_0^t \mu(u,s)\,du + \int_0^t \sigma(u,s)\,dW_u,\qquad t \le s,\qquad(37)$$

where $\{\mu(t,s) : 0 \le t \le s\}$ and $\{\sigma(t,s) : 0 \le t \le s\}$ are adapted processes valued in $\mathbb{R}$ and $\mathbb{R}^n$ respectively such that (37) is well-defined. We may think of $\mu$ and $\sigma$ as measurable functions on $\mathcal{T} \times \Omega$, where $\mathcal{T} = \{(t,s) \in \mathbb{R}_+^2 : t \le s\}$.

It turns out that there is an important consistency relationship between $\mu$ and $\sigma$ implied by (1). Specifically, one can show under purely technical conditions that if (1) is to apply to *every* claim of the form $(1,T)$, that is, for all bond prices, then it must be the case that

$$\mu(t,s) = \sigma(t,s)\int_t^s \sigma(t,u)^{\mathsf{T}}\,du.\qquad(38)$$

This basic "drift restriction" is found by an application of Fubini's Theorem for stochastic integrals. See Carverhill (1994) for details.

Given (38), we can use the definition $r_t = f(t,t)$ of the short rate to obtain

$$r_t = f(0,t) + \int_0^t \sigma(v,t)\int_v^t \sigma(v,u)^{\mathsf{T}}\,du\,dv + \int_0^t \sigma(v,t)\,d\widehat{B}_v.\qquad(39)$$

Knowledge of the forward-rate "volatility" process $\sigma$ and the initial forward rate curve $\{f(0,s), s \ge 0\}$ is therefore enough for the computation of all prices.

From (39), one can evaluate the price any claim from the basic formula (1). Aside from Gaussian or log-Gaussian special cases, for example those described by Jamshidian (1989) and by Miltersen, Sandmann, and Sondermann (1995), most valuation work in the HJM setting is done numerically.

## 9. State-Space HJM Models

Except for rather restrictive special cases [as in Cheyette (1992), El Karoui and Lacoste (1992), Ritchken and Sankarasubrmaniam (1993)], there is no finite-dimensional state space for the HJM model, so PDE-based (finite-difference) numerical computational methods cannot be used. Instead, one can build an analogous model in discrete-time discrete-state, and compute prices from "first principles." For the discrete model, the expectation analogous to (1) is obtained by constructing all sample paths for $r$ from the discretization of (39), and by computing the probability (under $Q$) of each.

for a general state-space representation of HJM models, one can take the forward rate curve itself as the state variable. For example, Musiela (1994) has taken the direct approach of allowing the state space $D$ to be the set of $C^1$ functions on $[0, \infty)$ into $I\!\!R$. The current state $X(t)$ is the function mapping the maturity $m$ to the current forward rate $f(t, t + m)$ of that maturity. With the goal of viewing $X$ as a Markov process solving a stochastic partial differential equation, Musiela has shown that one can use (37) and (38) to derive the relationship

$$dX_t(m) = \frac{\partial}{\partial m}\left(X_t(m) + \frac{1}{2}\left\|\int_0^m \tau(t, u)\, du\right\|^2\right) dt + \tau(t, m)\, dW_t, \qquad (40)$$

where $\tau(t, m) = \sigma(t, t + m)$. Then, taking $\tau(t, m) = \Psi(X_t, m)$, for some function $\Psi$ : $D \times [0, \infty) \to I\!\!R^n$, we can view (40) as a stochastic PDE, in the sense of DaPrato and Zabczyk (1992) or Walsh (1984). General conditions on $\Psi$ for existence and uniqueness of Markovian solutions to (4) are yet to be deduced. Musiela (1994) Musiela and Sondermann (1994) have illustrated some special cases, for example in which $X$ is a Gaussian process.

It has sometimes been said that with an HJM model one can avoid the compromise between theory and practice that arises with calibration, since essentially any initial forward rate curve is consistent with a given HJM model. In fact, the HJM model admits movements in the yield curve generated only by a finite-dimensional Brownian motion, and therefore limits the sorts of movements of the yield curve that can be considered without calibration. Recent work by Kennedy (1993, 1995), however, extends the HJM model to allow for an infinite-dimensional Brownian motion (in the framework of stochastic flows). That is, Kennedy (1993, 1995) has extended the basic HJM model (37) to allow the $I\!\!R^n$-valued standard Brownian Motion $W$ to be replaced with a Brownian sheet, for the case of deterministic volatility process $\sigma$. At this point, it seems logical and natural to also view the stochastic PDE (39) as one driven by an infinite-dimensional Brownian motion. There are as yet no results in this vein, however, aside from the special Gussian cases examined by Kennedy. Moreover, there is a need for special cases that are amenable to statistical estimation. Current estimation methods have been brought to bear only on finite-dimensional Markov special cases. See, for example, Ait-Sahalia (1992), Chen and Scott (1992b, 1993), Duffie and Singleton (1994), Pearson and Sun (1994), Gibbons and Ramaswamy (1993).

## 10. Foreign Term-Structure Models

Suppose, as a typical application involving the valuation of term-structure securities in an international setting, that one wishes to price a foreign bond option, say an option granting its owner the right, but not the obligation, to buy a given foreign bond at some time $\tau$ before its maturity, at a pre-agreed price $K$. The underlying zero-coupon bond pays one unit of foreign currency at some maturity date $T$, and has a domestic (say, "dollar") price process of $S$. The bond option therefore defines a claim $((S_\tau - K)^+, \tau)$, in dollars. Our job is to obtain the bond option price process, say $C$.

The foreign currency price process, say $U$, is assumed to satisfy

$$dU_t = \alpha_t U_t \, dt + U_t \beta_t \, dW_t, \tag{41}$$

where $\alpha$ is a real-valued progressively measurable process, and $\beta$, an $\mathbb{R}^n$-valued progressively measurable process, are both bounded.

Suppose foreign interest rates are given by a forward-rate process $F$, as in the Heath-Jarrow-Morton (HJM) setting. That is, the price of the given zero-coupon foreign bond at time $t$, in units of foreign currency, is $\exp\left(\int_t^T -F(t, u) \, du\right)$. It follows that the bond price process, in dollars, is given by

$$S_t = U_t \exp\left(\int_t^T -F(t, u) \, du\right),$$

and that the price of the bond option, in dollars, is

$$C_t = E\left[\exp\left(\int_t^\tau -r_u \, du\right)(S_\tau - K)^+ \,\bigg|\, \mathcal{F}_t\right], \tag{1}$$

where $r$ is the dollar short rate process, which may be specified in terms of the HJM model (37). It is assumed that the foreign forward rates are also given by an HJM model. That is, for each $t$ and $s \geq t$,

$$F(t, s) = F(0, s) + \int_0^t a(u, s) \, du + \int_0^t b(u, s) \, dW_u,$$

where the $s$-dependent drift process $a(\, \cdot \, , s) : \Omega \times [0, T] \to I\!R$ and the $s$-dependent diffusion process $b(\, \cdot \, , s) : \Omega \times [0, T] \to I\!R^n$ are assumed to satisfy sufficient regularity conditions for (1) to apply to foreign bonds of any maturity. Since the probability measure $Q$ was chosen so that (1) applies when all prices and claims are denominated in *dollars*, it is *not* generally true that the drift restriction

$$a(t, s) = b(t, s) \cdot \int_t^s b(u, s) \, du.$$

Instead, by methods analogous to those used to derive the drift restriction (38) on the dollar forward rate curve, one can show that the drift restriction on the foreign forward rate process is given by

$$a(t, s) = b(t, s) \cdot \left[ \int_t^s b(u, s) \, du - \beta_t \right]. \tag{45}$$

The basic pricing relationship (1) can now be applied to value the foreign bond option. Explicit solutions are known for determinsitic $\sigma$, $b$, and $\beta$. For applications and a related formulation, see Amin and Bodurtha (1995).

## 11.  Where Next?

A great deal of work remains to be done, particularly on the topic of statistical estimation of term structure models. In a multi-dimensional state space setting, ecconometric models of the term structure have stayed within a relatively narrow framework. Recent work by Pearson and Sun (1994), Chen and Scott (1992b, 1993), and Duffie and Singleton (1994) stays extremely to the multi-factor Cox-Ingersoll-Ross model (14)-(15). In a univariate setting, Ait-Sahalia (1992) has allowed for non-parametric $\sigma$ for the special case of (3). For the constant-volatility Gauss-Markov (affine) case, Frachot, Janci, and Lacoste (1992) together with Frachot and Lesne (1993) have done some empirical work in the HJM setting. Much remains to be done in integrating the use of statistical models within the practical applications of term structure models mentioned in the introduction.

Judging from the literature on term-structure modeling, much also remains to be done in the development and application of numerical methods, such as finite-difference or finite-element algorithms for solving multi-dimensional Cauchy problems such as (19)-(21), especially for the cases that arise in term-structure models.

## References

Y. Ait Sahalia (1992) "Nonparametric Pricing of Interest Rate Derivative Securities," Graduate School of Business, University of Chicago.

W. Ames (1977) *Numerical Methods For Partial Differential Equations, 2nd edition*, New York: Academic Press.

K. Amin and J. Bodurtha (1995) "Discrete-Time Valuation of American Options with Stochastic Interest Rates," *Review of Financial Studies* **8**: 193-234.

P. Artzner and F. Delbaen (1992) "Credit Risk and Prepayment Option," *ASTIN Bulletin* **22**: 81-96.

D. Beaglehole and M. Tenney (1991) "General Soltutions of Some Interest Rate Contingent Claim Pricing Equations," *Journal of Fixed Income* **1**: 69-83.

F. Black, E. Derman, and W. Toy (1990) "A One-Factor Model of Interest Rates and Its Application to Treasury Bond Options," *Financial Analysts Journal* : 33-39.

F. Black and P. Karasinski (1992) "Bonds and Option Pricing when Short Rates are Lognormal," *Financial Analysts Journal* **1991**: 52-59.

F. Black and M. Scholes (1973) "The Pricing of Options and Corporate Liabilities," *Journal of Political Economy* **81**: 637-654.

M. Brennan and E. Schwartz (1979) "A Continuous Time Approach to the Pricing of Bonds," *Journal of Banking and Finance* **3**: 133-155.

R. Brown and S. Schaefer (1993) "Interest Rate Volatility and the Shape of the Term Structure," *Philosophical Transactions of the Royal Society: Physical Sciences and Engineering* **347**: 449-598.

A. Carverhill (1994) "A Simplified Exposition of the Heath, Jarrow, and Morton Model," Department of Finance, Hong Kong University of Science and Technology.

K.-C. Chan, G. Karolyi, F. Longstaff, and A. Sanders (1992) "An Empirical Comparison of Alternative Models of the Short-Term Interest Rate," *Journal of Finance* **47**: 1209-1227.

Y.-K. Chan (1992) "Term Structure as a Second Order Dynamical System and Pricing of Derivative Securities," Bear Stearns and Company.

L. Chen (1994) "Stochastic Mean and Stochastic Volatility: A Three-Factor Model of the Term Structure of Interest Rates and its Application to the Pricing of Interest Rate Derivatives: Part I," School of Business, Harvard University.

R.-R. Chen and L. Scott (1992a) "Pricing Interest Rate Options in a Two-Factor Cox-Ingersoll-Ross Model of the Term Structure," *Review of Financial Studies* **5**: 613-636.

R.-R. Chen and L. Scott (1992b) "Maximum Likelihood Estimation for a Multi-Factor Equilibrium Model of the Term Structure of Interest Rates," Working Paper, Rutgers University and University of Georgia.

R.-R. Chen and L. Scott (1993) "Multi-Factor Cox-Ingersoll-Ross Models of the Term Structure: Estimates and Tests from a State-Space Model Using a Kalman Filter," Working Paper, Rutgers University and University of Georgia.

O. Cheyette (1992) "Markov Representation of the Heath-Jarrow-Morton Model," Capital Management Sciences.

G. Constantinides (1992) "A Theory of the Nominal Structure of Interest Rates," *Review of Financial Studies* **5**: 531-552.

G. Constantinides and J. Ingersoll (1984) "Optimal Bond Trading with Personal Taxes," *Journal of Financial Economics* **13**: 299-335.

J. Cox, J. Ingersoll, and S. Ross (1985) "A Theory of The Term Structure of Interest Rates," *Econometrica* **53**: 385-408.

G. Da Prato and J. Zabczyk (1992) *Stochastic Equations inInfinite Dimensions,* Cambridge University Press.

S. Das (1993) "Jump-Diffusion Processes and the Bond Markets," Harvard Business School.

M. Dothan (1978) "On the Term Structure of Interest Rates," *Journal of Financial Economics* **7**: 229-264.

D. Duffie (1996) *Dynamic Asset Pricing Theory, Second Edition,* in press, Princeton University Press.

D. Duffie and P. Glynn (1991) "Efficient Monte-Carlo Estimation of Security Prices," Graduate School of Business and Department of Operations Research, Stanford University, Stanford California, forthcoming: *Annals of Applied Probability.*

D. Duffie and M. Huang (1994) "Swap Rates and Credit Quality," Graduate School of Business, Stanford University.

D. Duffie and R. Kan (1992) "A Yield-Factor Model of Interest Rates," Graduate School of Business, Stanford University, Stanford California.

D. Duffie, J. Ma, and J. Yong (1993) "Black's Console Rate Conjecture," Graduate School of Business, Stanford University, Stanford California, forthcoming: *Annals of Applied Probability.*

D. Duffie, M. Schroder, and C. Skiadas (1994) "Recursive Valuation of Defaultable Securities and the Timing of Resolution of Uncertainty," Graduate School of Business, Stanford University, Stanford California.

D. Duffie and K. Singleton (1994) "Econometric Models of the Term Structure of Defaultable Bonds," Graduate School of Business, Stanford University, Stanford California.

N. El Karoui and V. Lacoste (1992) "Multifactor Models of the Term Strucutre of Interest Rates," Working Paper, June, University of Paris VI.

N. El Karoui, R. Myneni, and R. Viswanathan (1992) "Arbitrage Pricing and Hedging of Interest Rate Claims with State Variables: I Theory," Working Paper, January, University of Paris VI.

N. El Karoui and J.-C. Rochet (1989) "A Pricing Formula for Options on Coupon Bonds," Working Paper, October, University of Paris VI.

Feller (1951) "Two Singular Diffusion Problems," *Annals of Mathematics* **54:** 173-182.

A. Frachot, D. Janci, and V. Lacoste (1992) "Factor Analysis of the Term Structure: A Probabilistic Approach," Banque de France.

A. Frachot and J.-P. Lesne (1993) "Econometrics of Linear Factor Models of Interest Rates," Banque de France.

A. Friedman (1975) *Stochastic Differential Equations and Applications, Volume 1,* New York: Academic Press.

M. Gibbons and K. Ramaswamy (1993) "A Test of The Cox-Ingersoll, and Ross Model of the Term Structure," *Review of Financial Studies* **6:** 619-658.

I. Gihman and A. Skorohod (1972) *Stochastic Differential Equations,* Berlin: Springer-Verlag.

M. Harrison and D. Kreps (1979) "Martingales and Arbitrage in Multiperiod Security Markets," *Journal of Economic Theory* **20:** 381-408.

D. Heath, R. Jarrow, and A. Morton (1992) "Bond Pricing and The Term Structure of Interest Rates: A New Methodology for Contingent Claims Valuation," *Econometrica* **60:** 77-105.

S. Heston (1991) "Testing Continuous-Time Models of the Term Structure of Interest Rates," School of Organization and Management, Yale University.

T. Ho and S. Lee (1986) "Term Structure Movements and Pricing Interest Rate Contingent Claims," *Journal of Finance* **41:** 1011-1029.

C.-F. Huang (1987) "An Intertemporal General Equilibrium Asset Pricing Model: The Case of Diffusion Information," *Econometrica* **55:** 117-142.

J. Hull and A. White (1990) "Pricing Interest Rate Derivative Securities," *Review of Financial Studies* **3:** 573-592.

N. Ikeda and S. Watanabe (1981) *Stochastic Differential Equations and Diffusion Processes,* Amsterdam: North-Holland.

F. Jamshidian (1990) "The Preference-Free Determination of Bond and Option Prices From the Spot Interest Rate," *Advances in Futures and Options Research* **4:** 51-67.

F. Jamshidian (1991) "Bond and Option Evaluation in the Gaussian Interest Rate Model and Implementation," *Research in Finance* **9:** 131-170.

F. Jamshidian (1993) "Bond, Futuresm and Option Evaluation in the Quadratic Interest Rate Model," mimeo, Fuji Bank, London.

R. Jarrow, D. Lando, and S. Turnbull (1993) "A Markov Model for the Term Structure of Credit Risk Spreads," Johnson School of Management, Cornell University.

R. Jarrow and S. Turnbull (1992a) "Pricing Options on Financial Securities Subject to Default Risk," Johnson School of Management, Cornell University.

R. Jarrow and S. Turnbull (1992b) "The Pricing and Hedging of Options on Financial Securities Subject to Credit Risk: The Discrete Time Case," Johnson School of Management, Cornell University.

I. Karatzas and S. Shreve (1988) *Brownian Motion and Stochastic Calculus,* Springer-Verlag.

D. P. Kennedy (1994) "The Term Structure of Interest Rates as a Gaussian Random Field," *Mathematical Finance* 4: 247-258.

D. P. Kennedy (1994) "Characterizing and Filtering Gaussian Models of the Term Structure of Interest Rates," Statistical Laboratory, Cambridge University.

P. Kloeden and E. Platen (1992) *Numerical Solution of Stochastic Differential Equations,* Springer-Verlag.

D. Lando (1993) "A Continuous Time Markov Model of The Term Structure of Credit Spreads," Copenhagen Institute of Statistics.

D. Lando (1994) "On Cox Processes and Credit Risky Bonds," Copenhagen Institute of Statistics.

T. Langetieg (1980) "A Multivariate Model of the Term Structure of Interest Rates," *Journal of Finance* 35: 71-97.

R. Litterman and J. Scheinkman (1988) "Common Factors Affecting Bond Returns," Research Paper, Goldman Sachs Finanical Strategies Group.

F. Longstaff and E. Schwartz (1992) "Interest Rate Volatility and the Term Structure: A Two-Factor General Equilibrium Model," *Journal of Finance* 47: 1259-1282.

D. Madan and H. Unal (1992) "Pricing The Risks of Default," University of Maryland, Dept. of Finance, working paper.

R. Merton (1973) "Theory of Rational Option Pricing," *Bell Journal of Economics and Management Science* 4: 141-183.

R. Merton (1974) "On the Pricing of Corporate Debt: The Risk Structure of Interest Rates," *Journal of Finance* 29: 449-470.

K. Miltersen, K. Sandmann, and D. Sondermann (1994) "Closed Form Solutions for Term Structure Derivatives with Log-Normal Interest Rates," Department of Management, Odense University.

M. Musiela (1994) "Stochastic PDEs and Term Structure Models," School of Mathematics, University of New South Wales.

M. Musiela and D. Sondermann (1994) "Different Dynamical Specifications of the Term Structure of Interest Rates and Their Implications," School of Mathematics, University of New South Wales.

L. T. Nielsen and J. Saá-Requejo (1992) "Exchange Rate and Term Strcuture Dynamics and the Pricing of Derivative Securities," INSEAD working paper.

N. Pearson and T.-S. Sun (1994) "An Empirical Examination of the Cox, Ingersoll, and Ross Model of the Term Structure of Interest Rates Using the Method of Maximum Likelihood," *Journal of Finance* : .

P. Protter (1990) *Stochastic Integration and Differential Equations,* New York: Springer-Verlag.

G. Pye (1966) "A Markov Model of the Term Structure," *Quarterly Journal of Economics* 81: 61-72.

S. Richard (1978) "An Arbitrage Model of the Term Structure of Interest Rates," *Journal of Financial Economics* 6: 33-57.

P. Ritchken and L. Sankarasubramaniam (1993) "On Finite State Markovian Representations of the Term Structure," Department of Finance, University of Southern California.

R. Stambaugh (1988) "The Information in Forward Rates: Implications for Models of the Term Structure," *Journal of Financial Economics* **21**: 41-70.

O. Vasicek (1977) "An Equilibrium Characterization of the Term Structure," *Journal of Financial Economics* **5**: 177-188.

J. Walsh (1984) "An Introduction to Stochastic Partial Differential Equations," In P. Hennequin, *Ecole d'eté de Probailité de Saint Flour XIV, Lecture Notes in Mathematics, Number 1180, pp 265-439*. New York: Springer-Verlag.

# Theory of capacity on the Wiener space

Francis HIRSCH
Equipe d'Analyse et Probabilités
Université d'Evry - Val d'Essonne
Boulevard des Coquibus
F-91025 Evry Cedex

This text consists of four parts.

In the first one, we develop a fairly general potential theory related to a kernel. Such notions as capacity, equilibrium potentials and equilibrium measures, are studied.

In the second part, we consider the particular setting of the Wiener space and we specially study the capacities $c_{r,p}$ appearing in the Malliavin calculus ([23]).

In the third part, we introduce the notion of a symmetric $n$-parameter Markov process and we show that, for such processes, hitting probabilities may be estimated in terms of capacities related to an $L^2$-potential theory. This is applied to give a probabilistic interpretation of capacities $c_{r,2}$ on the Wiener space.

In the last part, we introduce, in the general context described in the first part, "Sobolev spaces" of Banach-valued functions and we use them in the so-called quasi-sure analysis. Here again, the case of Wiener space is specially considered.

# 1   Analytic potential theory

In this first part, we study a general potential theory from an analytic point of view. Such a theory was developped by H. Sugita ([32]) in the framework of abstract Wiener spaces, with specific methods, and then generalized by T. Kazumi and I. Shigekawa ([21]) to the case of an arbitrary separable metric space equipped with a probability measure and a Markovian semi-group (under some additional assumptions). The theory presented here is more general (since we do not assume the existence of a Markovian semi-group), and the methods are different. In fact, in a recent joint paper with S. Song ([18]), we developped an even more general theory where, in particular, the basic measure is only assumed to be $\sigma$-finite. This is important in many examples, even the most classical as the potential theory related to the classical Sobolev spaces in $\mathbb{R}^d$. Here, because of the main example that we shall consider (namely the case of the Wiener space), we shall restrict ourselves to finite measures. The methods are close to that of [15].

## Hypotheses

We assume in what follows the following hypotheses:

$(H_1)$ $E$ is a metric space and $m$ is a Borel probability measure on $E$.

$(H_2)$ $U(x, dy)$ is a Borel kernel on $E$ which satisfies:

$$U1 = 1, \quad mU = m, \quad U(\mathcal{C}_b) \subset \mathcal{C}_b.$$

These hypotheses are far to be minimal (see [18]), but they are simple and often satisfied.

**Notation 1.1** Henceforth, we fix a real $p$, $1 < p < \infty$, and we denote by $q$ the conjugate exponent of $p$. The symbol $\| \ \|_p$ will denote the $L^p(m)$-norm. In what follows, $L^p$ is set in place of $L^p(m)$. Kernel $U$ defines a contraction of $L^p$ denoted by $\mathbb{U}_p$.

We also assume:

$(H_3)$ $\mathbb{U}_p$ is injective on $L^p$.

(Actually, this hypothesis is not necessary, because it is possible to use quotient spaces.)

**Notation 1.2** We define space $H_p$ as the image $\mathbb{U}(L^p)$ equipped with the norm $\|\mathbb{U}_p f\|_{H_p} = \|f\|_p$.

Hence, $H_p$ is isometric to $L^p$ and therefore it is a uniformly convex Banach space. If $p = 2$, $H_2$ is a Hilbert space. Space $H_p$ has to be viewed as a "Sobolev space" defined as a space of "Bessel potentials", which is the classical situation.

## Capacity

**Notation 1.3** Following an old idea due to D. Feyel ([6]), we define a *functional capacity* $\gamma_p$ by:

$\forall u$ l.s.c., $u \geq 0$, $\gamma_p(u) = \inf\{\|v\|_{H_p}; v \in H_p \text{ and } v \geq u \text{ } m\text{-a.s.}\}$
  $(\gamma_p(u) = +\infty$ if the above set is empty)
$\forall u : E \longrightarrow \bar{\mathbb{R}}, \gamma_p(u) = \inf\{\gamma_p(v); v \text{ l.s.c. and } v \geq |u|\}$.

Associated with $\gamma_p$, the *capacity* $c_p$ is defined by

$$\forall A \subset E \quad c_p(A) = \gamma_p(1_A).$$

It is easy to see that $c_p$ can also be defined directly, as usually, by:
$\forall O$ open set, $c_p(O) = \inf\{\|u\|_{H_p}; u \geq 1 \text{ } m\text{-a.s. on } O\}$, and
$\forall A \subset E, c_p(A) = \inf\{c_p(O); O \text{ open and } O \supset A\}$.

Clearly, if $u \in L^p$, then $\gamma_p(u) \geq \|u\|_p$. In particular, if $A$ is a Borel set, $c_p(A) \geq (m(A))^{1/p}$.

*In the rest of this section, we generally omit, for simplicity, p in the notation.*
Let us give some direct consequences of the definitions:

(i) $\forall f \; \gamma(f) = \gamma(|f|)$ and $\gamma(1) = 1$.

(ii) $\forall f, g \geq 0 \; \gamma(f+g) \leq \gamma(f) + \gamma(g)$ and $f \leq g \Longrightarrow \gamma(f) \leq \gamma(g)$,
  $\forall f \; \forall \lambda \in \mathbb{R} \; \gamma(\lambda f) = |\lambda| \gamma(f)$.

(iii) For any non negative l.s.c. function $f$ such that $\gamma(f) < \infty$, there exists a unique $\varphi \in H$ such that $\varphi \geq f$ $m$-a.s. and $\|\varphi\|_H = \gamma(f)$. (This is a consequence of the projection theorem in uniformly convex spaces.)

(iv) Let $(f_n)_{n \geq 0}$ be an increasing sequence of non negative l.s.c. functions, then $\gamma(\lim_{n \to \infty} f_n) = \lim_{n \to \infty} \gamma(f_n)$.

We then easily obtain:

**Proposition 1.1** *For any sequence $(f_n)$ of functions*

$$\gamma(\sum_n |f_n|) \leq \sum_n \gamma(f_n) \quad (\sigma\text{-subadditivity}).$$

*For any sequence $(A_n)$ of subsets of $E$*

$$c(\cup_n A_n) \leq \sum_n c(A_n).$$

Let us now give some basic definitions.

**Definitions 1.1**    • A *polar set* is a set $A$ such that $c(A) = 0$. In particular, a Borel polar set is $m$-negligible and, by $\sigma$-subadditivity, a countable union of polar sets is polar.

• A property is said to hold *quasi-everywhere* (q.e.) if it holds out of a polar set.

• A *nest* is an increasing sequence $(F_k)$ of closed sets in $E$ such that $c(F_k^c) \to 0$ when $k \to \infty$. (Superscript $c$ denotes the complement.)

• A *quasi-continuous function* is a function $f$ on $E$ such that there exists a nest $(F_k)$ so that, for any $k$, $f$ is finite continuous on $F_k$ (Lusin's property with respect to the capacity).

It is easy to obtain from the definitions the following basic facts:

(v) $\forall f \quad \gamma(f) = 0 \Longleftrightarrow f = 0$ q.e.

(vi) $\forall f \quad \gamma(f) < +\infty \Longrightarrow f$ is finite q.e.

(vii) $\forall f, g \quad |f| \leq |g|$ q.e. $\Longrightarrow \gamma(f) \leq \gamma(g)$.

An important property is then the following.

**Proposition 1.2** *Let $f$ be a real quasi-continuous function and let $O$ be an open set. Then,*

$$f \geq 0 \ \ m\text{-}a.s. \ \ on \ O \Longrightarrow f \geq 0 \ \ q.e. \ \ on \ O.$$

*Proof.* Let $N = O \cap \{f < 0\}$ and let $(F_k)$ be a nest such that $f$ is continuous on each $F_k$. The set $O_k = \{f < 0\} \cup F_k^c$ is an open set and we have $O_k \cap O = N \cup (F_k^c \cap O)$. Hence, by the definition of the capacity of open sets, as $N$ is $m$-negligible, $c(O_k \cap O) = c(F_K^c \cap O)$. Finally, $c(N) \leq c(O_k \cap O) \leq c(F_k^c)$, and therefore, $c(N) = 0$. $\qquad\square$

# Space $L^1(\gamma)$

**Notation 1.4** Again following D. Feyel [6], we define:
$$
\begin{aligned}
\mathcal{F}^1(\gamma) &= \{f : E \longrightarrow \overline{\mathbb{R}}; \ f \text{ q.e. defined and } \gamma(f) < +\infty\} \\
\mathcal{L}^1(\gamma) &= \{f \in \mathcal{F}^1(\gamma); \ \exists(\varphi_n) \subset C_b \ \gamma(f - \varphi_n) \to 0\} \\
L^1(\gamma) &= \text{Quotient space of } \mathcal{L}^1(\gamma) \text{ by the relation} \\
&\quad \text{of equality quasi-everywhere.}
\end{aligned}
$$

Then, $L^1(\gamma)$ is a vector space that we equip with the norm $\|f\|_\gamma = \gamma(f)$.

**Proposition 1.3**    *1. Each element of $\mathcal{L}^1(\gamma)$ is quasi-continuous.*

*2. $L^1(\gamma)$ is a Banach lattice.*

The proof is not difficult. The following theorem relates space $H$ with space $L^1(\gamma)$.

**Theorem 1.1**    *1. Each element $h \in H$ admits a quasi-continuous $m$-representative $\tilde{h}$ which is unique up to quasi-everywhere equality.*

*2. $\forall h \in H, \tilde{h} \in \mathcal{L}^1(\gamma)$ and $\gamma(\tilde{h}) \leq \|h\|_H$.*

*3. If $h = \mathbb{U}g \in H$ with $g \in L^p$, then for any Borel $m$-representative $g_0$ of $g$, $Ug_0$ is defined q.e. and $\tilde{h} = Ug_0$ q.e.*

*Proof:* First of all, the uniqueness of the quasi-continuous representative follows from proposition 1.2. Let $g \in L^p$. If $\psi$ is an l.s.c. function such that $\psi \geq |g_0|$,

$$\gamma(U|g_0|) \leq \gamma(U\psi) \leq \|\psi\|_p$$

by definition of $\gamma$ and the fact that, by $(H_2)$, $U\psi$ also is l.s.c. By exterior regularity of $m$ ($E$ is metric), we then have $\gamma(U|g_0|) \leq \|g\|_p$. In particular, $U|g_0|$ is finite q.e. and then $Ug_0$ is defined q.e. and $\gamma(Ug_0) \leq \|g\|_p$. Let $(\varphi_n)$ be a sequence in $C_b$ converging to $g$ in $L^p$. By what precedes, $\gamma(Ug_0 - U\varphi_n) \leq \|g - \varphi_n\|_p$ and therefore, by $(H_2)$, $Ug_0 \in \mathcal{L}^1(\gamma)$. □

**Remark 1.1** By identification of $h$ and the class of $\tilde{h}$ in $L^1(\gamma)$, space $H$ may be considered as a subspace of $L^1(\gamma)$ (with a finer norm). We shall often do this identification. Then, space $L^1(\gamma)$ is a Banach *lattice* which contains space $H$. This is one of the main interests of $L^1(\gamma)$ because, in general, $H$ is not a lattice.

By a similar proof to that of theorem 1.1, we obtain the following result.

**Proposition 1.4** *Let $V$ be a Borel kernel satisfying the same hypothesis $(H_2)$ as $U$. Assume that its extension $\mathbb{V}_p$ to $L^p$ satisfies $\mathbb{V}_p(L^p) \subset H$. Then, for any $g \in L^p$, for any $g_0$ Borel $m$-representative of $g$, $Vg_0$ is defined q.e. and belongs to $\mathcal{L}^1(\gamma)$. In particular, $Vg_0$ is a quasi-continuous representative of $\mathbb{V}_p g$.*

We now have the following useful characterization of elements of $\mathcal{L}^1(\gamma)$.

**Proposition 1.5** *For a function $f$, the following properties are equivalent:*

*i) $f \in \mathcal{L}^1(\gamma)$.*

*ii) $f \in \mathcal{F}^1(\gamma)$ and $f$ is quasi-continuous.*

*iii)* $f$ *is quasi-continuous and* $\exists h \in H$ *such that* $|f| \le h$ *a.s.*

*Proof.* By proposition 1.3, $i) \implies ii)$, and, by definition of $\mathcal{F}^1(\gamma)$, $ii) \implies iii)$. Suppose now that $f$ is a bounded quasi-continuous function. Let $(F_k)$ be a nest such that, for any $k$, $f$ is continuous on $F_k$. By Tietze's theorem, there exists $g_k \in C_b$ such that $g_k|_{F_k} = f|_{F_k}$ and $\|g_k\|_\infty \le \|f\|_\infty$. Then $\gamma(f - g_k) \le 2\|f\|_\infty c(F_k^c)$ which tends to 0 when $k$ tends to infinity. Therefore $f \in \mathcal{L}^1(\gamma)$. Suppose finally that property $iii)$ holds. Let $g_0$ be a Borel representative of $g \in L^p$ such that $h = Ug$. Then, $|f| \le Ug_0$ q.e. (by proposition 1.2 and theorem 1.1). Denoting in what follows sup (resp. inf) by $\vee$ (resp. $\wedge$) and setting $f_n = \frac{nf}{n \vee |f|}$, by what precedes $f_n \in \mathcal{L}^1(\gamma)$ and, on the other hand,

$$|f - f_n| \le \tilde{h} - \tilde{h} \wedge n \le U(g_0 - g_0 \wedge n) \quad \text{q.e.}$$

Therefore, by theorem 1.1, $\gamma(f - f_n) \le \|g - g \wedge n\|_p$ which tends to 0 when n tends to infinity. This implies that property $i)$ holds. $\qquad \square$

## Potentials

The following proposition shows that, for any $g \in \mathcal{F}^1(\gamma)$, $\gamma(g)$ may be realized as a minimum.

**Proposition 1.6** *Let* $g \in \mathcal{F}^1(\gamma)$. *Then*

$$\gamma(g) = \min\{\|h\|_H; \quad \tilde{h} \ge |g| \ q.e.\}.$$

*This minimum is achieved by a unique* $\phi_g \in H$ *which is called the* equilibrium potential *of* $g$.

*Proof.* By the projection theorem in uniformly convex spaces and the fact that, by theorem 1.1, $\{h \in H; \quad \tilde{h} \ge |g| \text{ q.e.}\}$ is a closed (convex) subset of $H$, the minimum is achieved by a unique $\phi_g \in H$. We have $\widetilde{\phi_g} \ge |g|$ q.e. and therefore, again by theorem 1.1, $\gamma(g) \le \|\phi_g\|_H$. Let now $\lambda$ satisfy $\gamma(g) < \lambda$. There exists an l.s.c. function $\varphi$ such that $|g| \le \varphi$ and $\gamma(\varphi) < \lambda$. There exists $h \in H$ such that $\|h\|_H < \lambda$ and $h \ge \varphi$ m-a.s. We then have, by proposition 1.2, $\tilde{h} \ge \varphi$ q.e. and therefore $\tilde{h} \ge |g|$ q.e. Hence $\|\phi_g\|_H \le \|h\|_H < \lambda$ and finally $\|\phi_g\|_H \le \gamma(g)$. $\qquad \square$

An important corollary is the following.

**Corollary 1.6.1** *For any increasing sequence* $(f_n)$ *of non negative functions,*

$$\gamma(\lim_n f_n) = \lim_n \gamma(f_n).$$

*In particular, if* $(A_n)$ *is an increasing sequence of subsets of* $E$,

$$c(\cup_n A_n) = \lim_n c(A_n).$$

**Notation 1.5** We denote by $\mathcal{U}$ the set of non negative finite u.s.c. functions on $E$ with compact support.

Concerning decreasing sequences, it is clear that, if $(\varphi_n)$ is a decreasing sequence in $\mathcal{U}$, then $\gamma(\lim_n \varphi_n) = \lim_n \gamma(\varphi_n)$. Likewise, if $(K_n)$ is a decreasing sequence of compact sets, $c(\cap_n K_n) = \lim_n c(K_n)$. As a consequence of these properties of the capacity with respect to monotone sequences, $\gamma$ is a so-called *functional Choquet capacity* and $c$ is a *Choquet capacity*, which implies that Choquet's capacitability theorem is valid.

We now introduce the general definition of a potential.

**Definition 1.2** A *potential* is an element $u \in H$ which satisfies:

$$\forall v \in H \quad v \geq u \Longrightarrow \|v\|_H \geq \|u\|_H.$$

Of course, equilibrium potentials are potentials: Namely, if $v \in H$ and $v \geq \phi_g$, then $\tilde{v} \geq \tilde{\phi_g} \geq |g|$ q.e., and therefore, $\|v\|_H \geq \|\phi_g\|_H$. The converse will follow from the following theorem which shows that any potential is the equilibrium potential of itself. Let us first introduce further notation.

**Notation 1.6**
$\mathcal{P}$: The set of potentials.
$T$: The isometry from $L^p$ onto $H$ given by $\mathrm{U}$.
$S = T^{-1}$.
$T^*$ (resp. $S^*$) denotes the adjoint of $T$ (resp. $S$).
$i$ denotes the canonical embedding from $H$ into $L^p$. Consequently, $\mathrm{U} = i \circ T$. The adjoint of $i$ is denoted by $i^*$.
For any set $F$ of (classes of) functions, $F^+$ denotes the set of non negative elements of $F$. If $F$ is a normed space, we denote by $F'$ (resp. $F^*$) the set of linear continuous functionals (resp. linear functionals) on $F$, and by $F'^+$ (resp. $F^{*+}$) the subspace consisting of those functionals which are non negative on $F^+$. It can be noticed that $T((L^p)^+) \subset H^+$ and therefore $T^*(H'^+) \subset (L^q)^+$.

We have the following characterization which establishes a one-to-one correspondence between $\mathcal{P}$ and $H'^+$.

**Theorem 1.2** *Let $u \in H$. The following statements are equivalent:*

(1) $u \in \mathcal{P}$

(2) $Su \geq 0$ *and* $S^*((Su)^{p/q}) \in H'^+$

(3) $\exists \nu \in H'^+ \quad u = T((T^*\nu)^{q/p})$

(4) $u \geq 0$ *and* $\gamma(\tilde{u}) = \|u\|_H$

(5) $u = \phi_u$

If $u \in \mathcal{P}$, then $\nu$ given by (3) is unique. Potential $u$ is then called the potential generated by $\nu$ and it is denoted by $u_\nu$.

*Proof.* Let $u \in \mathcal{P}$. For any $v \in H^+$ and for any $t \geq 0$, $\|u + tv\|_H \geq \|u\|_H$. Then,

$$\left(\frac{d^+}{dt}\|u + tv\|_H^p\right)_{t=0} = p\int |Su|^{p-1}\text{sign}(Su)\, Sv\, dm \geq 0.$$

Consequently, $Su \geq 0$ and $S^*((Su)^{p-1}) \in H'^+$. Thus (1) $\Longrightarrow$ (2).

If (2) holds, by what precedes and an argument of convexity, for any $v \in H^+$ and for any $t \geq 0$, $\|u + tv\|_H \geq \|u\|_H$ and (1) follows.

Clearly, (2) $\Longleftrightarrow$ (3) and $\nu$ is unique, given by $S^*((Su)^{p/q})$.

Suppose $u \in \mathcal{P}$. Then $Su \geq 0$ and hence $u = TSu \geq 0$. By definition of $\mathcal{P}$, $\min\{\|v\|_H; \ \tilde{v} \geq \tilde{u}\}$ is achieved by $u$ and therefore (4) holds.

By uniqueness of the equilibrium potential, (4) $\Longrightarrow$ (5), and, since equilibrium potentials are potentials, (5) $\Longrightarrow$ (1).  $\square$

**Remark 1.2** If $p = 2$, the map $\nu \in H'^+ \longrightarrow u_\nu \in \mathcal{P}$ given in the previous theorem is the restriction to $H'^+$ of the *linear isometry* $TT^*$ from $H'$ onto $H$, which is the Riesz isometry from $H'$ onto $H$. Namely, for any $\varphi \in H$,

$$< \varphi, \nu >_{H,H'} = (S\varphi, Su_\nu)_{L^2} = (\varphi, u_\nu)_H.$$

The following density property is an easy consequence of the Hahn-Banach theorem.

**Proposition 1.7** *The set $i^*((L^q)^+)$ is dense in $H'^+$ (with respect to the metric defined by the norm of $H'$).*

The following corollary could be used for defining $\mathcal{P}$ in the Hilbert case.

**Corollary 1.7.1** *In the case $p = 2$, the set $UU^*((L^2)^+)$ is dense in $\mathcal{P}$ (with respect to the metric defined by the norm of $H$).*

We now study the dual of $L^1(\gamma)$.

**Proposition 1.8**

$$L^1(\gamma)' = L^1(\gamma)^{*+} - L^1(\gamma)^{*+}$$

The proof is standard. In particular, $L^1(\gamma)^{*+} = L^1(\gamma)'^+$.

**Theorem 1.3** *For any $L \in L^1(\gamma)'^+$, denote by $\nu_L$ the "restriction" of $L$ to $H$:*
$h \in H \longrightarrow < \tilde{h}, L >_{L^1(\gamma),L^1(\gamma)'}$. *Then $\nu_L \in H'^+$ and $\|\nu_L\|_{H'} = \|L\|_{L^1(\gamma)'}$.*
*The map $L \in L^1(\gamma)^{*+} \longrightarrow \nu_L \in H'^+$ is surjective and $H^{*+} = H'^+$.*

*Proof.* Let $L \in L^1(\gamma)'^+$. By theorem 1.1, $\nu_L \in H'^+$ and $\|\nu_L\|_{H'} \leq \|L\|_{L^1(\gamma)'}$.
Let $u \in L^1(\gamma)$. Then $|u| \leq \tilde{\phi}_u$ q.e. and hence

$$| < u, L >_{L^1(\gamma),L^1(\gamma)'} | \leq < \widetilde{\phi_u}, L >_{L^1(\gamma),L^1(\gamma)'} =$$

$$< \phi_u, \nu_L >_{H,H'} \leq \|\nu_L\|_{H'} \|\phi_u\|_H = \|\nu_L\|_{H'} \gamma(u).$$

Therefore $\|L\|_{L^1(\gamma)'} \leq \|\nu_L\|_{H'}$. Finally, since for any $u \in \mathcal{L}^1(\gamma)$ there exists
$h \in H$ such that $|u| \leq \tilde{h}$ q.e., by a classical consequence of the Hahn-Banach
theorem, for any $\nu \in H^{*+}$, map $\nu$ extends into $L \in L^1(\gamma)^{*+}$. Then $\nu = \nu_L \in$
$H'^+$. □

**Remark 1.3** If we assume that $H$ is dense in $L^1(\gamma)$, of course the above map
$L \longrightarrow \nu_L$ is a one-to-one correspondence between $L^1(\gamma)'^+$ and $H'^+$, preserving
the norms.

## Finite energy measures

From now on, we assume a fourth hypothesis:

$(H_4)$ (Tightness of capacity $c$) There exists a nest $(K_k)$ consisting of compact
sets.

**Proposition 1.9** *For any non negative Borel function $f$,*

$$\gamma(f) = \sup\{\gamma(\varphi); \ \varphi \in \mathcal{U} \text{ and } \varphi \leq f\}$$

*(where $\mathcal{U}$ was defined in notation 1.5). Likewise, for any Borel subset $A$ of $E$,*

$$c(A) = \sup\{c(K); \ K \text{ compact and } K \subset A\}.$$

This is a consequence of Choquet's capacitability theorem because, thanks
to $(H_4)$, Borel sets are analytic up to a polar set. Another consequence of
hypothesis $(H_4)$ is the following.

**Proposition 1.10** *Space $L^1(\gamma)$ is a separable Banach space.*

*Proof.* Let $D$ be a countable set in $\mathcal{C}_b$, dense for the topology of uniform
convergence on the sets $K_k$, $k \geq 0$. Then $\{\varphi 1_{K_k}; \ \varphi \in D, \ k \geq 0\} = \mathcal{D}$ is
countable and $L^1(\gamma)$ is contained in the closure of $\mathcal{D}$ in $\mathcal{F}^1(\gamma)$. □

**Definition 1.3** The $\sigma$-algebra generated by Borel sets and polar sets will be called the $\sigma$-algebra of *quasi-Borel sets*.

It is clear that a set $A$ is quasi-Borel iff there exist $B_1$ and $B_2$ Borel sets such that $B_1 \subset A \subset B_2$ and $c(B_2 \backslash B_1) = 0$. The $\sigma$-algebra of quasi-Borel sets also is the $\sigma$-algebra generated by $\mathcal{L}^1(\gamma)$. We have then the following representation theorem.

**Theorem 1.4** *For any $L \in L^1(\gamma)^{\prime+}$, there exists a unique measure on the $\sigma$-algebra of quasi-Borel sets, $l$, such that*

$$\mathcal{L}^1(\gamma) \subset \mathcal{L}^1(l) \text{ and } \forall f \in \mathcal{L}^1(\gamma) \quad <f, L>_{L^1(\gamma), L^1(\gamma)'} = \int f \, dl,$$

*where $\mathcal{L}^1(l)$ denotes the set of $l$-integrable functions. Moreover, for any quasi-Borel set $A$,*

$$l(A) \leq \|L\|_{L^1(\gamma)'} c(A)$$

*(and, in particular, $l$ does not charge polar sets).*

*Proof.* By Daniell's theorem, we have essentially to prove that, if $(\varphi_n)$ is a decreasing sequence in $\mathcal{L}^1(\gamma)$ pointwise converging to 0, then $\lim_n \gamma(\varphi_n) = 0$. Let $(\varphi_n)$ be such a sequence. Let $(A_k)$ be a nest such that, for any $n$ and $k$, $\varphi_n$ is continuous on $A_k$ (such a nest exists). By Dini's lemma, for all $k, m$, $\lim_{n \to \infty} \varphi_n 1_{K_m \cap A_k} = 0$ uniformly. Therefore, for any $N > 0$,

$$\limsup_{n \to \infty} \gamma(\varphi_n) \leq \gamma(\varphi_1 1_{(K_m \cap A_k)^c}) \leq N(c(K_m^c) + c(A_k^c)) + \gamma(\varphi_1 - \varphi_1 \wedge N).$$

There exists $h \in H$ such that $\varphi_1 \leq \tilde{h}$ q.e. There exists $g \in L^p$ such that $h = \mathbb{U}g$. By the same argument as in the proof of proposition 1.5, $\gamma(\varphi_1 - \varphi_1 \wedge N) \leq \|g - g \wedge N\|_p$. It then suffices to let $m$, $k$ and $N$ tend to infinity.

In particular, if $\varphi \in \mathcal{C}_b^+$, then $\int \varphi \, dl \leq \|L\|\gamma(\varphi)$. By increasing limit, for any open set $O$, $l(O) \leq \|L\|c(O)$ and then, by definition of the capacity, for any quasi-Borel set $A$, $l(A) \leq \|L\|c(A)$. $\qquad \square$

This leads to the following definition.

**Definition 1.4** The measures $l$ appearing in the previous theorem are called *finite energy measures*.

**Notation 1.7** We shall denote by $\mathcal{E}$ the set of finite energy measures.

Clearly $\nu \in \mathcal{E}$ iff $\nu$ is a measure on the quasi-Borel $\sigma$-algebra such that

$$\exists C \geq 0 \ \forall f \in \mathcal{L}^1(\gamma)^+ \quad \int f \, d\nu \leq C\gamma(f).$$

In particular, a quasi-Borel measure which is dominated by a finite energy measure is a finite energy measure. By theorems 1.3 and 1.4, we obtain the following correspondence between $H'^+$ and $\mathcal{E}$.

**Proposition 1.11** *For any $\nu \in H'^+$, there exists $l \in \mathcal{E}$ such that*

$$\forall h \in H \quad < h, \nu >_{H,H'} = \int \tilde{h} \, dl.$$

*Conversely, any element $l$ in $\mathcal{E}$ thus defines a unique $\nu \in H'^+$.*

As a consequence we obtain:

**Corollary 1.11.1** *Any finite energy measure $l$ defines, according to the previous proposition, an element $\nu$ of $H'^+$ which generates, by theorem 1.2, a potential $u$. We shall also say that $u$ is the potential generated by $l$ and we shall use the notation $u = u_l$. Conversely, any potential $u$ is generated by a finite energy measure $l$. We then have*

$$\forall h \in H \quad \int \tilde{h} \, dl = \int (Sh)(Su_l)^{p/q} \, dm \quad (= (h, u_l)_H \text{ if } p = 2).$$

*In particular, $\int \tilde{u}_l \, dl = \|u_l\|_H^p$.*

**Notation 1.8** We denote by $(H_5)$ the following hypothesis:

$(H_5)$ Space $H$ is dense in $L^1(\gamma)$.

Clearly, if $(H_5)$ is satisfied, the finite energy measure generating a given potential is uniquely determined by this potential and there are bijective correspondences between $(L^1(\gamma))'^+$, $H'^+$, $\mathcal{E}$ and $\mathcal{P}$.

## Equilibrium measures

We assume hypotheses $(H_1)$ to $(H_4)$.

**Definitions 1.5**　　● Let $g \in \mathcal{F}^1(\gamma)$. Any finite energy measure generating the equilibrium potential $\phi_g$ is called an *equilibrium measure* of $g$. (If $(H_5)$ also is satisfied, this measure is uniquely determined.)

● A *quasi-upper semicontinuous* (q.u.s.c.) function is a function $g$ such that there exists a decreasing sequence $(g_n)$ in $\mathcal{L}^1(\gamma)$ such that $g = \lim_n g_n$ q.e.

The main result is the following.

**Theorem 1.5** *Let $g \in \mathcal{F}^1(\gamma)^+$ be a q.u.s.c. function. Then there is an equilibrium measure of $g$, denoted by $\nu_g$, which is carried by $\{g > 0\} \cap \{g = \tilde{\phi}_g\}$. In particular*

$$\gamma(g)^p = \int g \, d\nu_g.$$

*Proof.* The following proof was suggested by D. Feyel.

We identify, by theorem 1.4, $\mathcal{E}$ with $L^1(\gamma)'^+$ and we denote by $\mathcal{K}$ the set $\{\nu \in \mathcal{E}; \; \|\nu\|_{L^1(\gamma)'} \leq 1\}$ equipped with the weak topology $\sigma(L^1(\gamma)', L^1(\gamma))$ for which it is compact. Let $g \in \mathcal{F}^1(\gamma)^+$ be a q.u.s.c. function and let $(g_n)$ be a corresponding decreasing sequence in $\mathcal{L}^1(\gamma)$. By the Hahn-Banach theorem, for any $n$ there exists a linear functional $L_n$ on $L^1(\gamma)$ such that $L_n(g_n) = \gamma(g_n)$ and, for any $\varphi \in L^1(\gamma)$, $L_n(\varphi) \leq \gamma(\varphi^+)$. Clearly $L_n \in L^1(\gamma)'^+$ and $\|L_n\|_{L^1(\gamma)'} \leq 1$. Therefore there exists $\nu_n \in \mathcal{K}$ such that $\int g_n \, d\nu_n = \gamma(g_n)$. Hence $\gamma(g_n) = \max_{\nu \in \mathcal{K}} \int g_n \, d\nu$. By a classical min-max theorem, since $\mathcal{K}$ is compact, we then have

$$\gamma(g) \geq \max_{\nu \in \mathcal{K}} \int g \, d\nu = \max_{\nu \in \mathcal{K}} \inf_n \int g_n \, d\nu = \inf_n \max_{\nu \in \mathcal{K}} \int g_n \, d\nu = \inf_n \gamma(g_n) \geq \gamma(g).$$

Consequently, there exists $\nu \in \mathcal{K}$ such that $\gamma(g) = \int g \, d\nu$. Replacing $\nu$ by $1_{\{g>0\}}\nu$, we may assume that $\nu$ is carried by $\{g > 0\}$. We may also assume that $\gamma(g) \neq 0$. Let us still denote by $\nu$ the element of $H'^+$ associated with $\nu$ by proposition 1.11. By theorem 1.3, $\|\nu\|_{H'} = \|\nu\|_{L^1(\gamma)'} = 1$ and therefore $\gamma(g) = \int g \, d\nu \leq \int \tilde{\phi}_g \, d\nu = \, < \phi_g, \nu >_{H,H'} = \, < S\phi_g, T^*\nu >_{L^q, L^p} \leq \|\phi_g\|_H \|\nu\|_{H'} = \gamma(g)$. It follows that $\nu$ is carried by $\{g = \tilde{\phi}_g\}$ and, by the case of equality in Hölder's inequality, $S\phi_g = \gamma(g)[T^*\nu]^{q/p}$. Therefore we can set $\nu_g = (\gamma(g))^{p/q}\nu$. $\square$

**Corollary 1.5.1** *Let $F$ be a closed set. Denote by $\phi_F$ the equilibrium potential of $1_F$ and by $\nu_F$ a corresponding equilibrium measure as in the previous theorem. Then*

$$\nu_F(F^c) = 0, \quad \widetilde{\phi_F} = 1 \; \nu_F\text{-a.e.}, \quad c(F)^p = \int d\nu_F.$$

As a consequence, we have the following dual characterization of polar sets.

**Corollary 1.5.2** *Let $A$ be a quasi-Borel set. Then $A$ is polar if and only if*

$$\forall \nu \in \mathcal{E} \quad \nu(A) = 0.$$

*Proof.* The necessity has been shown in theorem 1.4. Conversely, if $A$ is not polar, by proposition 1.9 there exists a compact subset $K$ of $A$ which is not polar. Then $\nu_K(A) \geq \nu_K(K) = c(K)^p > 0$. $\square$

# 2 Capacities on Wiener space

We shall now consider the framework of the classical Wiener space (according to [7], we could, more generally, consider the case of a locally convex Lusin space with a centered Gaussian measure). We shall prove that the classical capacities $c_{r,p}$ appearing in Malliavin's calculus are of the type studied in the first section whose hypotheses are satisfied.

**Notation 2.1** In this section, $E$ denotes the classical Wiener space $\mathcal{C}_0(\mathbb{R}_+; \mathbb{R}^d)$ equipped with its usual topology and $m$ denotes the Wiener measure on $E$, considered as a Borel measure. Hence, hypothesis $(H_1)$ is satisfied.

We denote by $(B_t)_{t\geq 0}$ the coordinates process which is, under $m$, the standard Brownian process in $\mathbb{R}^d$ starting from 0. We denote by $B_t^j$, $1 \leq j \leq d$, the components of $B_t$.

For $t \geq 0$, if $f$ is a non negative Borel function on $E$, we set

$$P_t f(x) = \int f(e^{-t/2} x + \sqrt{1 - e^{-t}} y) \, dm(y) \quad (\text{Mehler's formula}).$$

Then $P_t$ is a Borel kernel, $P_t 1 = 1$, $mP_t = m$, $P_t(\mathcal{C}_b) \subset \mathcal{C}_b$ and

$$\forall f, g \geq 0 \quad \int P_t f \, g \, dm = \int f \, P_t g \, dm \quad (\text{symmetry}),$$

$$\forall t, s \geq 0 \quad P_{t+s} = P_t P_s \quad (\text{semi-group property}).$$

The semi-group of Borel kernels $(P_t)_{t\geq 0}$ is called the *Ornstein-Uhlenbeck semigroup*.

We denote, for $1 < p < \infty$, by $(\mathbb{P}_{t,p})_{t\geq 0}$ the $L^p(m)$-extension of $(P_t)_{t\geq 0}$. Then $(\mathbb{P}_{t,p})_{t\geq 0}$ is a strongly continuous sub-Markovian contraction semi-group on $L^p(m)$, and, for $t \geq 0$, $\mathbb{P}_{t,2}$ is symmetric on $L^2(m)$. We denote by $A_p$ the infinitesimal generator of the semi-group $(\mathbb{P}_{t,p})_{t\geq 0}$: Operator $A_p$ is called the *Ornstein-Uhlenbeck operator* in $L^p(m)$.

We set, for any real number $r > 0$,

$$U^r = \frac{1}{\Gamma(r/2)} \int t^{\frac{r}{2}-1} e^{-t} P_t \, dt.$$

Then $U^r$ satisfies the same properties as $P_t$. In particular, kernel $U^r$ satisfies $(H_2)$. We denote by $\mathbb{U}_p^r$ the $L^p(m)$-extension of $U^r$. We then have $\mathbb{U}_p^r = (I - A_p)^{-r/2}$. Consequently, hypothesis $(H_3)$ also is satisfied ($\mathbb{U}_p^r$ is injective). The associated space $H_p = \mathbb{U}_p^r(L^p(m))$ will be denoted by $\mathbb{D}_p^r$. According to the previous section, $\mathbb{D}_p^r$ is equipped with the norm $\|\mathbb{U}_p^r f\|_{r,p} = \|f\|_p$.

The previous potential theory can then be developped for any fixed $r > 0$ and $1 < p < \infty$. The corresponding capacities will be denoted by $\gamma_{r,p}$ and

$c_{r,p}$. We shall also use the terminology $(r,p)$-polar, $(r,p)$-quasi-continuous, ... in place of $c_{r,p}$-polar, $c_{r,p}$-quasi-continuous, ... Space $\mathbb{D}_p^r$ is decreasing with respect to $r$ and $p$, while $\gamma_{r,p}$ and $c_{r,p}$ are increasing with respect to $r$ and $p$. We shall denote by $\mathbb{D}^\infty$ the set $\bigcap_{r>0,p>1} \mathbb{D}_p^r$.

**Definition 2.1** A *slim set* is a set which is $(r,p)$-polar for any $r > 0$ and $1 < p$.

The following proposition comes easily from the definitions.

**Proposition 2.1** *If* $f \in \mathbb{D}^\infty$, *then there exists an m-representative of* $f$, $\tilde{f}$, *which belongs to* $\bigcap_{r>0,p>1} \mathcal{L}^1(\gamma_{r,p})$ *and which is unique up to equality out of a slim set.*

The following result was first proved in [32] by using the differential definition of $\mathbb{D}_p^1$ (Meyer's inequalities). It also is a direct consequence of the holomorphy of $\mathbb{P}_{t,p}$ which follows from the symmetry of $(\mathbb{P}_{t,2})$ (see [31]).

**Proposition 2.2**

$$\forall r > 0 \ \forall p > 1 \ \forall t > 0 \quad \mathbb{P}_{t,p}(L^p(m)) \subset \mathbb{D}_p^r$$

Then, by proposition 1.4, we get:

**Corollary 2.2.1** *For all* $r > 0$, *for all* $p > 1$, *for all Borel function* $f$ *such that* $\int |f|^p \, dm < \infty$, $P_t f$ *is defined* $(r,p)$-q.e. *and* $P_t f \in \mathcal{L}^1(\gamma_{r,p})$.

As a consequence, we have the following improvement of a classical 0-1 law (cf. [7]).

**Proposition 2.3** *Let* $G$ *be a Borel linear subspace of* $E$. *If* $m(G) > 0$, *then* $m(G) = 1$ *and, more precisely, the complement* $G^c$ *is a slim set.*

*Proof.* We have, by Mehler's formula, $P_t 1_G = m(G)$ on $G$. Letting $t$ tend to 0, we obtain $1 = m(G)$ a.s. on $G$. Therefore, if $m(G) > 0$, then $m(G) = 1$. We then have, again by Mehler's formula, $P_t 1_G = 1_G$ for any $t \geq 0$. Hence, by corollary 2.2.1, $1_G \in \mathcal{L}^1(\gamma_{r,p})$. As $1_G = 1$ a.s., $1_G = 1$ $(r,p)$-q.e. and therefore $c_{r,p}(G^c) = 0$. $\square$

**Example** If $0 < \alpha < 1/2$, the set of elements of $E$ which are not locally Hölder continuous of order $\alpha$ is a slim set.

The following result ([7]) has many applications.

**Proposition 2.4** *Let* $q : E \longrightarrow [0,\infty]$ *be a Borel function which is sublinear (i.e.* $\forall x, y \ q(x+y) \leq q(x) + q(y)$, $q(0) = 0$, $\forall \lambda > 0 \ \forall x \ q(\lambda x) = \lambda q(x)$). *If* $q$ *is finite a.s., then* $q \in \bigcap_{r>0,p>1} \mathcal{L}^1(\gamma_{r,p})$.

*Proof.* Let $\hat{q}(x) = q(x) + q(-x)$. Then $\hat{q}$ also is finite a.s. Since $Z = \{\hat{q} < \infty\}$ is a Borel linear subspace of $E$, $Z^c$ is, by proposition 2.3, a slim set. By Fernique's theorem, $\hat{q} \in \cap_{p>1} L^p(m)$. Then, by corollary 2.2.1, $P_t q \in \cap_{r>0,p>1} \mathcal{L}^1(\gamma_{r,p})$. Now, by Mehler's formula and the sublinearity of $q$, for any $x \in E$,

$$e^{-t/2} q(x) - \sqrt{1 - e^{-t}} \int q \, dm \le P_t q(x) \le e^{-t/2} q(x) + \sqrt{1 - e^{-t}} \int q \, dm$$

and therefore, $|e^{t/2} P_t q - q| \le \sqrt{e^t - 1} \int q \, dm$ on $Z$. Consequently, $e^{t/2} P_t q$ tends to $q$ uniformly on $Z$ and the result follows. $\square$

**Remark** Under the assumptions of the proposition, by [7], $q \in \cap_{p>1} \mathbb{D}_p^1$ also holds.

We now give a few corollaries (see [7]).

**Corollary 2.4.1** *For any $r > 0$ and $p > 1$, capacity $c_{r,p}$ satisfies tightness property $(H_4)$.*

*Proof.* Let $K$ be a convex symmetric compact subset of $E$ such that $m(K) > 0$ (such a set obviously exists) and let $q$ be the Minkowski functional associated with $K$:

$$q(x) = \inf\{\lambda > 0; \ x \in \lambda K\} \le +\infty.$$

Then $q$ is an l.s.c. sublinear symmetric function and $m(q < +\infty) > 0$. Therefore, by propositions 2.3 and 2.4, $q \in \mathcal{L}^1(\gamma_{r,p})$. Set $K_n = \{q \le n\}$. Then $K_n = nK$ is a compact set and $c_{r,p}(K_n^c) \le n^{-1} \gamma_{r,p}(q)$ tends to 0 as $n$ tends to $\infty$. $\square$

**Corollary 2.4.2** *For $1 \le j \le d$,*

$$\limsup_{t \to \infty} \frac{B_t^j}{\sqrt{2t \log \log t}} = 1$$

*out of a slim set.*

*Proof.* Define $q$ on $E$ by $q(\omega) = \limsup_{t \to \infty} \frac{B_t^j(\omega)}{\sqrt{2t \log \log t}}$. Then $q$ satisfies hypotheses of proposition 2.4. Hence $q \in \mathcal{L}^1(\gamma_{r,p})$ and consequently $q$ is $(r, p)$-quasi-continuous. As $q = 1$ a.s., $q = 1$ $(r, p)$-q.e. $\square$

**Corollary 2.4.3** *Let $\mathcal{H}$ be the Cameron-Martin space $(\mathcal{H} = \{\int_0^\cdot \varphi(s) \, ds; \ \varphi \in L^2(\mathbb{R}_+; \mathbb{R}^d)\})$. Then $\mathcal{H}$ is a slim set.*

*Proof.* As, for any $h \in \mathcal{H}$, $\limsup_{t \to \infty} \frac{|h(t)|}{\sqrt{2t \log \log t}} = 0$, it suffices to apply the previous corollary. $\square$

We now prove that the last assumption $(H_5)$ also holds.

**Proposition 2.5** *For any $r > 0$ and $p > 1$, $\mathbb{D}_p^r$ is dense in $L^1(\gamma_{r,p})$, which means that property $(H_5)$ is satisfied.*

*Proof.* There are many proofs of this fact. Here we use a general method based on the Hahn-Banach theorem. Let $L \in L^1(\gamma)'$ which vanishes on $\mathbb{D}_p^r$. By proposition 1.8, theorem 1.4 and proposition 2.2, there exist $l_1$ and $l_2$ $(r,p)$-finite energy measures such that, for any $t > 0$ and $\varphi \in C_b$, $\int P_t \varphi \, dl_1 = \int P_t \varphi \, dl_2$. Then, by dominated convergence, for any $\varphi \in C_b$, $\int \varphi \, dl_1 = \int \varphi \, dl_2$. Therefore $L$ vanishes on $C_b$ which is dense in $L^1(\gamma_{r,p})$. $\qquad\square$

As a consequence, an $(r,p)$-finite energy measure generating a given $(r,p)$-potential is uniquely determined by this potential.

Another example of slim set is the following.

**Proposition 2.6** *Any countable set is a slim set.*

*Proof.* It is enough to prove that, if $r > 0$, $p > 1$ and $x \in E$, then $\{x\}$ is $(r,p)$-polar. By corollary 1.5.2, we have to prove that $\nu(\{x\}) = 0$ for any $(r,p)$-finite energy measure $\nu$, or equivalently, that the Dirac measure $\delta_x$ is not an $(r,p)$-finite energy measure. Let $(l_n)_{n \geq 0}$ be an orthonormal sequence in $L^2(m)$ consisting of continuous linear functionals on $E$. We can assume $r \in \mathbb{N}$. By the differential characterization of the $(r,p)$-norm, we have, if $\delta_x$ is an $(r,p)$-finite energy measure,

$$\exists C \geq 0, \quad \forall N \in \mathbb{N}, \quad \forall \varphi \in \mathcal{D}(\mathbb{R}^N),$$

$$|\varphi(l_1(x), \cdots, l_N(x))|^p \leq C \left[ \int_{\mathbb{R}^N} |\varphi^{(r)}(y)|^p e^{-|y|^2/2} dy + \int_{\mathbb{R}^N} |\varphi(y)|^p e^{-|y|^2/2} dy \right].$$

Let $\psi \in \mathcal{D}(\mathbb{R}^N)$ with $\psi(0) = 1$. We apply the above inequality to $\varphi(y) = \psi(n(y_1 - l_1(x), \cdots, y_N - l_N(x)))$. Letting $n$ tend to infinity, we get, if $rp < N$, $1 \leq 0$. $\qquad\square$

We finish this section with some examples of finite energy measures and remarks. Let $\sigma$ be a measure on $\mathbb{R}^r$ such that there exists $0 \leq \alpha < 1/2$ with $\int e^{-\alpha|x|^2} d\sigma(x) < \infty$. Let $(l_1, \cdots, l_r)$ be a set of $r$ continuous linear functionals on $E$ which is an orthonormal system in $L^2(m)$. We can define by approximation $\sigma(l_1, \cdots, l_r)$ as an $(r,p)$-finite energy measure for any $p > \frac{1}{1-2\alpha}$ (cf. [2]). In particular, if $l \in E'$ and $\|l\|_2 = 1$, $\delta(l)$ is a $(1,2)$-finite energy measure ($\sqrt{2\pi}\delta(l)$ is the conditionning measure given $\{l = 0\}$). Clearly $\int l^2 \, d(\delta(l)) = 0$, therefore the support of $\delta(l)$ is contained in $\mathrm{Ker} l$ (which implies that $\delta(l)$ is singular with respect to $m$). Thus, $\mathrm{Ker} l$ is a closed subspace of $E$ which satisfies $m(\mathrm{Ker} l) = 0$ and $c_{1,2}(\mathrm{Ker} l) > 0$ (because $\delta(l)(\mathrm{Ker} l) = (2\pi)^{-1/2} > 0$). In another direction, it is proved in [8] that if a Borel linear subspace $G$ satisfies $m(G) = 0$ and $\mathcal{H} \subset G$, then $G$ is $(1, p)$-polar for any $p > 1$. Finally, we notice that according to [34], if $X$ is an $\mathbb{R}^d$-valued non degenerate (in Malliavin's sense) Wiener functional and if $\sigma$ is a temperated measure on $\mathbb{R}^d$, $\sigma(X)$ may be defined and it is a finite energy measure.

# 3 Multiparameter processes

We introduce,in this section (essentially based on [18]), a class of symmetric Markov multiparameter processes and we show that they allow us to interpret probabilistically some capacities. We shall see that, in particular, the capacities $c_{r,2}$ $(r > 0)$ defined in the previous section can be interpreted in such a way.

## $n$-parameter symmetric Markov processes

We fix a metric space $E$ and a Borel probability measure $m$ on $E$. We also fix a positive integer $n$. We consider $X$, an $n$-parameter $E$-valued measurable process defined on a probability space $(\Omega, \mathcal{A}, P)$. We begin with some notation.

**Notation 3.1** If $B \subset \{1, 2, \ldots, n\}$, we denote by $B^c$ the complement of $B$ in $\{1, 2, \ldots, n\}$. If $t \in \mathbb{R}^n_+$, we set $t_B = (t_i; \ i \in B)$ and we identify $\mathbb{R}^n_+$ with $\mathbb{R}^B_+ \times \mathbb{R}^{B^c}_+$ by identifying $t$ with $(t_B, t_{B^c})$. The order on $\mathbb{R}^B_+$ is the product order and $|t_B|$ denotes $\sum_{i \in B} t_i$.

**Definitions 3.1** Process $X$ is called an *$E$-valued $m$-symmetric $n$-parameter Markov process* if there exist $(\mathbb{P}^i; \ 1 \leq i \leq n)$, a family of $n$ strongly continuous semi-groups of sub-Markovian symmetric operators on $L^2(m)$, and $(\mathcal{F}^i; \ 1 \leq i \leq n)$, a family of $n$ filtrations on $(\Omega, \mathcal{A})$ such that

1. $\forall t \in \mathbb{R}^n_+$, $X_t \in \bigcap_{1 \leq i \leq n} \mathcal{F}^i_{t_i}$ and the law of $X_t$ is $m$.

2. $\forall 1 \leq i \leq n$, $\forall f \in L^2(m)$, $\forall u \in \mathbb{R}^{\{i\}^c}_+$, $\forall a, b \in \mathbb{R}^{\{i\}}_+$,

$$E[f(X_{a+b,u}) \mid \mathcal{F}^i_a] = \mathbb{P}^i_b f(X_{a,u}).$$

The semi-groups $(\mathbb{P}^i)$ are called the *transition semi-groups* of $X$.

**Remark** The problem to know for which family of semi-groups there exists a process admitting them as transition semi-groups is open. A necessary (but not sufficient, according to [30]) condition is that the semi-groups commute: $\forall i, j \in \{1, \cdots, n\}$, $\forall t_i, t_j \in \mathbb{R}_+$, $\mathbb{P}^i_{t_i} \mathbb{P}^j_{t_j} = \mathbb{P}^j_{t_j} \mathbb{P}^i_{t_i}$. Namely,

$$\mathbb{P}^i_{t_i} \mathbb{P}^j_{t_j} f(X_0) = E[\mathbb{P}^j_{t_j} f(X_{0,t_i}) \mid \mathcal{F}_0] = E[f(X_{0,t_i,t_j}) \mid \mathcal{F}_0]$$

where $\mathcal{F}_0$ denotes $\bigcap_i \mathcal{F}^i_0$.

**Notation 3.2** If $B \subset \{1, 2, \ldots, n\}$, we adopt the following notation:
$\mathrm{U} = \prod_{1 \le i \le n} \frac{1}{\sqrt{\pi}} \int_0^\infty a^{-1/2} e^{-a} \mathbb{P}_a^i \, da$. If $B \ne \emptyset$, $\mathrm{V}^B = \prod_{i \in B} \int_0^\infty e^{-a} \mathbb{P}_a^i \, da$ and $\mathrm{V}^\emptyset = I$ (identity in $L^2(m)$). When $B = \{1, \cdots, n\}$, we simply denote $\mathrm{V}^B$ by $\mathrm{V}$ ($\mathrm{V} = \mathrm{U}^2$). For $t \in \mathbb{R}_+^n$, $\mathcal{F}_t = \cap_{1 \le i \le n} \mathcal{F}_{t_i}^i$. Operator $\mathrm{U}$ is called the $1/2$- *potential operator* of $X$. Actually, $\mathrm{U} = \prod_{1 \le i \le n} (I - A_i)^{-1/2}$ where $A_i$ is the infinitesimal generator of $\mathbb{P}^i$.

If $g \in L^2(m)$, $s \in \mathbb{R}_+^n$ and $B \in \{1, \cdots, n\}$, we set

$$M_s^g = E[\int_{t \ge 0} e^{-|t|} g(X_t) \, dt \mid \mathcal{F}_s], \quad M_\infty^g = \int_{t \ge 0} e^{-|t|} g(X_t) \, dt,$$

if $B \ne \emptyset$, $\quad H_{B,s}^g = \int_{0 \le t_B \le s_B} e^{-|t_B|} \mathrm{V}^{B^c} g(X_{t_B, s_{B^c}}) \, dt_B$ and,

if $B = \emptyset$, $H_{B,s}^g = \mathrm{V} g(X_s)$.

In what follows, $N_2$ denotes the $L^2$-norm, with respect to $P$ as well as with respect to $m$.

We have the following easy consequences of the definitions.

**Theorem 3.1** *Assume that $X$ is an $E$-valued $m$-symmetric $n$-parameter Markov process. Then*

1. *(Doob's inequality)* $\forall g \in L^2(m)$, $\forall D$ *finite subset of* $\mathbb{R}_+^n$,

$$N_2[\sup_{s \in D} |M_s^g|] \le 2^n N_2[M_\infty^g]$$

2. *(Generalized Dynkin's formula)* $\forall g \in L^2(m)$, $\forall s \in \mathbb{R}_+^n$,

$$M_s^g = \sum_{B \subset \{1, 2, \cdots, n\}} e^{-|s_{B^c}|} H_{B,s}^g$$

3. $\forall g \in L^2(m)$, $\forall \varphi$ *non negative bounded Borel function on* $\mathbb{R}_+^n$,

$$N_2[M_\infty^g] = N_2[\mathrm{U}g] \quad \text{and} \quad N_2[\int_{t \ge 0} e^{-|t|} \varphi(t) g(X_t) \, dt] \le \|\varphi\|_\infty N_2[\mathrm{U}g].$$

## First inequality

In the remainder of this section, we assume that $X$ is an $E$-valued $m$-symmetric $n$-parameter Markov process and that the associated $1/2$-potential operator $\mathrm{U}$ is the natural extension to $L^2(m)$ of a Borel kernel $U(x, dy)$ satisfying $U(\mathcal{C}_b) \subset \mathcal{C}_b$. Then, fixing $p = 2$ and identifying $\mathrm{U}$ and $\mathrm{U}_2$ (cf. notation 1.1), hypotheses $(H_1)$, $(H_2)$, $(H_3)$ of section 1 are satisfied. We adopt henceforth the notation of section 1. In particular, we denote by $\gamma$ and $c$ the capacities associated with $U$ and $p = 2$, and $H$ denotes the space $\mathrm{U}(L^2(m))$. By the symmetry of $\mathrm{U}$ and corollary 1.7.1, we obtain directly:

**Proposition 3.1** *For any $u \in \mathcal{P}$, there exists a sequence $(p_k)$ in $L^2(m)^+$ such that $\lim_k \nabla p_k = u$ in $H$.*

In what follows, if $F$ is any non negative function on $\Omega$, we denote by $E(F)$ the upper integral with respect to $P$, which means:

$$E(F) = \inf\{E(G); \; G \text{ measurable and } G \geq F\}.$$

Finally, we also assume the following weak regularity property:
*Right continuity hypothesis*: $P$-a.s., $\forall t \; \lim_{t' \downarrow t} X_{t'} = X_t$.
The first inequality is then stated in the following theorem.

**Theorem 3.2** *For any function $f$ on $E$,*

$$E\left[\left[\sup_{t \geq 0} e^{-|t|}|f|(X_t)\right]^2\right] \leq 4^n[\gamma(f)]^2.$$

*Proof:* Suppose first $f = \nabla g$ with $g \in L^2(m)^+$. By theorem 3.1,
$e^{-|s|}\nabla g(X_s) \leq M_s^g$ and therefore $N_2\left[\sup_{s \in D} e^{-|s|}\nabla g(X_s)\right] \leq 2^n N_2(\mathbb{U}g) \leq 2^n\|\nabla g\|_H$,
or, $N_2\left[\sup_{s \in D} e^{-|s|}f(X_s)\right] \leq 2^n\|f\|_H$.
Suppose then $f \in \mathcal{P}$. By proposition 3.1, the same inequality holds.
Now, if $f \in \mathcal{F}^1(\gamma)$, then $|f| \leq \phi_f$ a.s. and $\|\phi_f\|_H = \gamma(f)$, therefore
$N_2\left[\sup_{s \in D} e^{-|s|}|f|(X_s)\right] \leq 2^n\gamma(f)$.
By right continuity hypothesis, if $f \in \mathcal{F}^1(\gamma) \cap \mathcal{C}$, $N_2\left[\sup_{t \geq 0} e^{-|t|}|f|(X_t)\right] \leq 2^n\gamma(f)$. This extends to any non negative l.s.c. function by increasing limit.
Finally the general result follows by the definition of $\gamma$. □

## Finite energy measures and additive functionals

Besides the previous hypotheses, we henceforth assume that hypothesis $(H_4)$ (tightness of $c$) and hypothesis $(H_5)$ (density of $H$ in $L^1(\gamma)$) are satisfied.

**Theorem 3.3** *Let $\nu \in \mathcal{E}$ be a finite energy measure and denote by $u_\nu$ the potential generated by $\nu$. Then there exists a unique random measure $A_\nu(dt)$ on $\mathbb{R}^n_+$ such that, for any sequence $(p_k)$ in $L^2(m)^+$ such that $\lim_k \nabla p_k = u_\nu$ in $H$, one has*

$$\forall \varphi \in \mathcal{C}_c(\mathbb{R}^n_+) \quad A_\nu(\varphi) = \lim_k \int \varphi(t)p_k(X_t)\, dt \text{ in } L^2(P).$$

*Then, for any non negative Borel function $\varphi$ (resp. $g$) on $\mathbb{R}^n_+$ (resp. $E$),*

$$E\left[\int \varphi(t)\, g(X_t)\, A_\nu(dt)\right] = \int \varphi(t)\, dt \int g\, d\nu.$$

*Sketch of the proof.* By theorem 3.1, for any bounded Borel function $\varphi$ on $\mathbb{R}_+^n$, for any $g \in L^2(m)$,

$$N_2[\int e^{-|t|}\varphi(t)g(X_t)\,dt] \leq 2\|\varphi\|_\infty \|\nabla g\|_H.$$

Then $A_\nu$ can be defined as a vague limit (weak limit) of some subsequence $p_{k'}(X_t)\,dt$ (in this sense, $T \longrightarrow A_\nu([0,T])$ is an "additive functional"). For the second part of the statement, we remark that if $p \in L^2(m)^+$,

$$E[\int \varphi(t)g(X_t)p(X_t)\,dt] = \int \varphi(t)\,dt \int g\,p\,dm.$$

We have to pass to the limit in this formula, but there are some technical difficulties, and, in particular, we need the right continuity hypothesis and additional assumption $(H_5)$. For details, we refer to [18]. □

## Second inequality

Under the same assumptions as in the previous theorem, we have:

**Theorem 3.4** *For any Borel function $f$ on $E$, for any $T \in \mathbb{R}_+^n$,*

$$\left(\int_{[0,T]} e^{-|t|}dt\right)^2 [\gamma(f)]^2 \leq E\left[\left[\sup_{t \in [0,T]} |f|(X_t)\right]^2\right].$$

*Proof.* By proposition 1.9, it suffices to consider the case where $f$ is a non negative u.s.c. function with compact support. Let then $\nu$ be the equilibrium measure of $f$ and let $A_\nu$ be the associated random measure (theorem 3.3). By theorem 1.5, we obtain

$$\left(\int_{[0,T]} e^{-|t|}dt\right)\gamma^2(f) = \int_{[0,T]} e^{-|t|}dt \in tf\,d\nu = E[\int_{[0,T]} e^{-|t|}f(X_t)\,A_\nu(dt)]$$

$$\leq N_2\left(\sup_{t \in [0,T]} |f(X_t)|\right) N_2\left(\int_{[0,T]} e^{-|t|}A_\nu(dt)\right).$$

Now, by the definition of $A_\nu$,

$$N_2\left(\int_{[0,T]} e^{-|t|}A_\nu(dt)\right) \leq \|u_\nu\|_H = \gamma(f).$$

The result follows. □

As a consequence of theorems 3.2 and 3.4, we have:

**Corollary 3.4.1** *For any Borel subset B of E, for any $T \in \mathbb{R}_+^n$,*

$$\left( \int_{[0,T]} e^{-|t|} dt \right)^2 [c(B)]^2 \leq P[\exists t \in [0,T]; \ X_t \in B] \leq 4^n e^{2|T|} [c(B)]^2.$$

*In particular, a Borel set B is polar if and only if, almost surely, $X_t \notin B$ for any t.*

These results can be used to give a probabilistic characterization of quasi-continuity (see [18]).

## Capacities $c_{n,2}$, $n \in \mathbb{N}^*$

We now consider the framework of section 2, and we fix $n \in \mathbb{N}^*$. In particular, $E$ denotes the Wiener space and $m$ is the Wiener measure. Let $W^{(n+1)}$ be an $\mathbb{R}^d$-valued $(n+1)$-parameter Brownian sheet defined on a probability space $(\Omega, \mathcal{A}, P)$. We set

$$X_t^{(n)} = e^{-|t|/2} W_{e^{t_1},\cdots,e^{t_n},}^{(n+1)}$$

Then $X^{(n)}$ is an $E$-valued $n$-parameter process, called the *E-valued n-parameter Ornstein-Uhlenbeck process* (see [25] for $n = 1$, [33] for $n = 2$, and [29], [14] for the general case). The following proposition follows easily from the definitions.

**Proposition 3.2** *Process $X^{(n)}$ is an E-valued m-symmetric n-parameter Markov process with continuous paths. Its transition semi-groups are given by:*

$$\forall 1 \leq i \leq n \ \forall t > 0 \quad \mathbb{P}_t^i = \mathbb{P}_{t,2} \ (\textit{Ornstein-Uhlenbeck semi-group on } L^2(m)).$$

The 1/2-potential operator of $X^{(n)}$ then is operator $\mathbb{U}_2^n$ and the corresponding capacity is $c_{n,2}$. Consequently, all assumptions of the previous paragraph are satisfied. In particular, corollary 3.4.1 is valid with $E =$ the Wiener space, $c = c_{n,2}$ and $X = X^{(n)}$. This situation was generalized in [1],[9].

## Capacities $c_{n+\alpha,2}$, $n \in \mathbb{N}$, $0 < \alpha < 1$

The framework and the notation are the same as in the previous paragraph. We set $r = n + \alpha$. Let $X^{(n+1)}$ be the $E$-valued $(n + 1)$-parameter Ornstein-Uhlenbeck process. We consider $(\tau_t)_{t \geq 0}$ a one-sided stable process of index $\alpha$, starting from 0, with càd-làg paths and independent of $W^{(n+2)}$. We set

$$Y_{t_1,\cdots,t_{n+1}}^{(r)} = X_{t_1,\cdots,t_n,\tau_{t_{n+1}}}^{(n+1)}$$

Process $Y^{(r)}$ is an $E$-valued $(n + 1)$-parameter right continuous process. We set, as before,

$$\forall 1 \leq i \leq n \quad \mathbb{P}_t^i = \mathbb{P}_{t,2}$$

and $\mathbb{P}_t^{n+1} = \int \mathbb{P}_{s,2} \, d\nu_t(s)$ where $\nu_t$ is the law of $\tau_t$.

**Proposition 3.3** *Process $Y^{(r)}$ is an $E$-valued $m$-symmetric $n + 1$-parameter Markov process of which the transition semi-groups are $(\mathbb{P}^i; 1 \leq i \leq n + 1)$.*

The proof is almost classical (cf. [17], [18]). The $1/2$-potential operator of $Y^{(r)}$ is $\mathbb{U}_2^n (I + (-A_2)^\alpha)^{-1/2}$, where $A_2$ is the Ornstein-Uhlenbeck operator in $L^2(m)$, and then the corresponding space $H$ is $\mathbb{D}_2^r$ with an equivalent norm. The associated capacity is therefore equivalent to $c_{r,2}$. The assumptions $(H_1)$ to $(H_5)$ are again satisfied. In particular, for any $T > 0$, there exist $a_T$ and $b_T$, positive constants, such that, for any Borel subset $B$ of $E$,

$$a_T [c_{r,2}(B)]^2 \leq P[\exists t \in [0, T]^{n+1}; \ Y_t^{(r)} \in B] \leq b_T [c_{r,2}(B)]^2.$$

**Remark** In the context of the previous examples, we can prove more precise properties of the random measure $A_\nu$ associated with a finite energy measure $\nu$, in particular $T \in \mathbb{R}_+^n \longrightarrow A_\nu([0, T])$ is almost surely continuous (see [19]).

# 4 Quasi-sure analysis

The quasi-sure analysis (terminology introduced by P. Malliavin) is the study of properties up to a slim set (in place of negligible set). The interest comes from the fact that the finite energy measures, which appear in particular in the context of conditionning, do not charge slim sets but may be singular with respect to the Wiener measure (see at the end of section 2). A useful tool is to consider spaces of Banach-valued functions (we refer to [7], [2], [4, 5], [24], [28]). In fact, there are different definitions which are not equivalent in general.

## Banach-valued functions

This paragraph is close to the work of L. Denis [4, 5], but the context is slightly different and we also adopt slightly different definitions.

The context here is that of the first section: We consider $(E, m, U)$ satisfying $(H_1)$ and $(H_2)$. We fix $1 < p < \infty$ and we assume $(H_3)$ too. The notation is that of section 1. We also fix a separable Banach space $\mathbb{B}$. It is clear that $U$ can be naturally extended to $L^p(m; \mathbb{B})$.

**Notation 4.1** We denote by $\overline{U}$ the natural extension of $U$ to $L^p(m; \mathbb{B})$ (operator $\overline{U}$ is a contraction in $L^p(m; \mathbb{B})$).

**Proposition 4.1** *Operator $\overline{U}$ is injective on $L^p(m; \mathbb{B})$.*

This is an easy consequence of the Hahn-Banach theorem and of the property:

$$\forall F \in L^p(m; \mathbb{B}) \; \forall \varphi \in \mathbb{B}' \quad \varphi(\overline{U}(F)) = \overline{U}(\varphi(F)).$$

**Notation 4.2** We denote by $H^{\mathbb{B}}$ the image $\overline{U}(L^p(m; \mathbb{B}))$ equipped with the norm

$$\|\overline{U}F\|_{H^{\mathbb{B}}} = \|F\|_{L^p(m;\mathbb{B})}.$$

Therefore, $H^{\mathbb{B}}$ is a separable Banach space isometric to $\mathcal{L}^p(m; \mathbb{B})$.

We then define $\mathcal{F}^1(\gamma; \mathbb{B})$ as the set of functions $f : E \longrightarrow \mathbb{B}$, defined q.e., such that $\|f\|_{\mathbb{B}} \in \mathcal{F}^1(\gamma)$. For such a function $f$, we set $\gamma(f) = \gamma(\|f\|_{\mathbb{B}})$.

We set

$$\mathcal{L}^1(\gamma; \mathbb{B}) = \{f \in \mathcal{F}^1(\gamma; \mathbb{B}); \; \exists (\varphi_k) \in \mathcal{C}_b(E; \mathbb{B}) \; \gamma(f - \varphi_k) \to 0\},$$

and we define $L^1(\gamma; \mathbb{B})$ as the quotient space with respect to the q.e. equality.

A function $f : E \longrightarrow \mathbb{B}$ is said to be quasi-continuous if there is a nest $(F_k)$ such that, for any $k$, $f|_{F_k} \in \mathcal{C}(F_k; \mathbb{B})$. We obtain easily, as in the scalar case, the following proposition.

**Proposition 4.2**     *1. Each element of $\mathcal{L}^1(\gamma; \mathbb{B})$ is quasi-continuous.*

*2. $L^1(\gamma; \mathbb{B})$ is a Banach space.*

One of the main results is the following extension of the scalar case (theorem 1.1).

**Theorem 4.1** *Any $h \in H^{\mathbb{B}}$ admits a quasi-continuous $m$-representative $\tilde{h}$, unique up to quasi-everywhere equality, and*

$$\tilde{h} \in \mathcal{L}^1(\gamma; \mathbb{B}) \text{ and } \gamma(\tilde{h}) \leq \|h\|_{H^{\mathbb{B}}}.$$

*Proof.* As $\mathcal{C}_b(E; \mathbb{B})$ is dense in $L^p(m; \mathbb{B})$, then $\mathcal{C}_b(E; \mathbb{B}) \cap H^{\mathbb{B}}$ is dense in $H^{\mathbb{B}}$. Now, for any $f = \overline{U}g$ in $\mathcal{C}_b(E; \mathbb{B}) \cap H^{\mathbb{B}}$, by theorem 1.1,

$$\gamma(f) \leq \gamma(\widetilde{\overline{U}\|g\|_{\mathbb{B}}}) \leq \|f\|_{H^{\mathbb{B}}}.$$

The result follows.                                                                                      $\square$

Hence, as in the scalar case, $H^{\mathbb{B}}$ may be considered as a subspace of $L^1(\gamma; \mathbb{B})$. We also have:

**Theorem 4.2** *Assume that* $(H_4)$ *is satisfied. Then* $f \in \mathcal{L}^1(\gamma; \mathbb{B})$ *if and only if* $f \in \mathcal{F}^1(\gamma; \mathbb{B})$ *and* $f$ *is quasi-continuous. If in addition* $(H_5)$ *is satisfied, then* $H^{\mathbb{B}}$ *is dense in* $L^1(\gamma; \mathbb{B})$.

*Proof.* Consider first a bounded $\mathbb{B}$-valued function $f$ which is quasi-continuous. By $(H_4)$, there exists a nest $(K_k)$ consisting of compact sets, such that, for any $k$, $f|_{K_k}$ is continuous. It is easy to see that the algebraic tensor product $\mathcal{C}(K_k) \otimes \mathbb{B}$ is dense in $\mathcal{C}(K_k; \mathbb{B})$. Therefore, there exists $\varphi_k \in \mathcal{C}(K_k) \otimes \mathbb{B}$ such that $\|f - \varphi_k\|_{\mathbb{B}} \leq \varepsilon \leq 1$ on $K_k$. By extension, we may assume that $\varphi_k \in \mathcal{C}_b(E) \otimes \mathbb{B}$ and $\|\varphi_k\|_\infty \leq \|f\|_\infty + 1$. If $(H_5)$ is satisfied, there exists $h_k \in H^{\mathbb{B}}$ such that $\gamma(\varphi_k - h_k) \leq \varepsilon$. We then have

$$\gamma(f - \varphi_k) \leq \varepsilon + (2\|f\|_\infty + 1)c(K_k^c) \text{ and } \gamma(f - h_k) \leq \gamma(f - \varphi_k) + \varepsilon.$$

This proves that $f \in \mathcal{L}^1(\gamma; \mathbb{B})$ and, if $(H_5)$ is satisfied, any bounded function in $\mathcal{L}^1(\gamma; \mathbb{B})$ can be approximated by elements of $H^{\mathbb{B}}$.

The general case may be obtained by the following remark which can be proved as in proposition 1.5: If $f \in \mathcal{F}^1(\gamma; \mathbb{B})$ and $f_n = n(n \vee \|f\|_{\mathbb{B}})^{-1}f$, then $\gamma(f - f_n)$ tends to 0 when $n$ tends to infinity. Now, $f_n$ is bounded, and quasi-continuous if so is $f$. □

Theorem 4.1 is a useful tool to transform almost-sure convergence results into quasi-sure convergence results. We follow L. Denis ([4, 5]).

**Theorem 4.3** *Let* $(f_n)_{n \geq 0}$ *be a sequence of* $H$. *Let, for any* $n$, $g = \mathbb{U}^{-1}f_n$. *We assume*

1. $\exists g_\infty$ *such that* $\lim_{n \to \infty} g_n = g_\infty$ *a.s.*

2. $\sup_n |g_n| \in L^p(m)$.

*Then, setting* $f_\infty = \mathbb{U}g_\infty$, *we have*

$$\lim_{n \to \infty} \widetilde{f_n} = \widetilde{f_\infty} \text{ q.e.}$$

*Proof.* Let $\overline{\mathbb{N}}$ be the compact set $\mathbb{N} \cup \{\infty\}$, and $\mathbb{B} = \mathcal{C}(\overline{\mathbb{N}})$. We may consider $g.$ as an element of $L^p(m; \mathbb{B})$ and $f. = \overline{\mathbb{U}}(g.)$. Therefore $f. \in H^{\mathbb{B}}$ and hence $f.$ admits a quasi-continuous representative $\widetilde{f.}$ . Clearly, $(\widetilde{f.})_n = \widetilde{f_n}$ q.e. for any $n \in \overline{\mathbb{N}}$. Therefore $\widetilde{f.}(\omega) \in \mathcal{C}(\overline{\mathbb{N}})$ for quasi-every $\omega$ and the result follows. □

In the same way, the continuous analogue holds too:

**Theorem 4.4** *Let* $T \in [0, \infty]$ *and let* $(f_t)_{t \in [0,T]}$ *be a family in* $H$. *We assume that there exists a family of functions* $(g_t)_{t \in [0,T]}$ *such that*

1. $\forall t \in [0, T]$, $g_t$ *is an* $m$-*representative of* $\mathbb{U}^{-1}f_t$,

2. *For almost all* $\omega \in E$, $t \longrightarrow g_t(\omega) \in \mathcal{C}([0, T]$,

3. $\sup_{t \in [0,T]} |g_t| \in \mathcal{L}^p(m)$.

*Then, there is a family* $(\tilde{f}_t)_{t \in [0,T]}$ *such that*

- *for any* $t \in [0, T]$, $\tilde{f}_t$ *is a quasi-continuous m-representative of* $f_t$,
- *for quasi-every* $\omega$, $t \longrightarrow \tilde{f}_t(\omega) \in \mathcal{C}([0, T])$.

*More precisely,* $\tilde{f}. \in \mathcal{L}^1(\gamma; \mathcal{C}([0, T]))$ *and*

$$\gamma(\sup_t |\tilde{f}_t|) \le \| \sup_t |g_t| \|_p.$$

Following the same ideas, we can prove a quasi-sure version of Kolmogorov's theorem.

**Theorem 4.5** *Let* $(X_t)_{t \in [0,1]^d}$ *be a family in H. We assume*

$$\exists C > 0, \; \exists \varepsilon > 0, \; \forall s, t \in [0, 1]^d \quad \|X_t - X_s\|_H^p \le C|t - s|^{d + \varepsilon}.$$

*Then there exists* $(\widetilde{X}_t)_{t \in [0,1]^d}$ *such that*

- $\forall t \in [0, 1]^d$ $\widetilde{X}_t = X_t$ *m-a.s.*
- $\forall t \in [0, 1]^d$ $\widetilde{X}_t$ *is quasi-continuous*
- *for quasi-every* $\omega \in E$, $t \longrightarrow \widetilde{X}_t(\omega)$ *is Hölder continuous of order* $\alpha$ *for any* $\alpha \in [0, \varepsilon/p[$.

*Proof.* Let for $\alpha \in [0, \varepsilon/p[$, $\mathcal{H}^\alpha$ be the space of continuous functions $f$ on $[0, 1]^d$, nul at 0, such that $\lim_{|s-t| \to 0, s \ne t} |t - s|^{-\alpha} |f(t) - f(s)| = 0$, equipped with the usual norm

$$\|f\|_{\mathcal{H}^\alpha} = \sup_{s \ne t} \frac{|f(t) - f(s)|}{|t - s|^\alpha}.$$

We can assume $X_0 = 0$. Define $Y_t = \mathbb{U}^{-1} X_t$. Then $Y_t \in L^p$ and $\|Y_t - Y_s\|_p^p \le C|t-s|^{d+\varepsilon}$. Therefore, by the classical Kolmogorov theorem, there exists a version of $Y$, still denoted by $Y$, which belongs to $L^p(m; \mathcal{H}^\alpha)$. Clearly $\hat{X} = \overline{\mathbb{U}}(Y)$ is a version of $X$ belonging to $H^{\mathcal{H}^\alpha} \subset \mathcal{L}^1(\gamma; \mathcal{H}^\alpha)$. $\qquad\square$

Still following L. Denis ([4, 5]), we now give applications to Wiener space.

## Applications to Wiener space

We use here the results of the previous paragraph in the framework of section 2. We denote by $(\mathcal{F}_t)_{t \ge 0}$ the natural filtration of $(B_t)_{t \ge 0}$.

## Brownian martingales

**Theorem 4.6** *Let $f \in \mathcal{L}^1(\gamma_{r,p})$. Then, there exists $F \in \mathcal{L}^1(\gamma_{r,p}; \mathcal{C}([0,\infty]))$ such that, for any $t \geq 0$, $F_t$ is an m-representative of $E(f \mid \mathcal{F}_t)$ and $F_\infty = f$ $(r,p)$-quasi-everywhere. Consequently, for any $t \in [0,\infty]$, $F_t$ is $(r,p)$-quasi-continuous, and, for $(r,p)$-quasi-every $\omega$, $t \in [0,\infty] \longrightarrow F_t(\omega)$ is continuous. Moreover, following capacitary Doob's inequality holds:*

$$\gamma_{r,p}(\sup_t |F_t|) \leq \frac{p}{p-1}\gamma_{r,p}(f).$$

(See [12] for a similar, slightly weaker, result by another method.)

*Proof:* First assume $f \in \mathbb{D}_p^r$. There exists $g \in L^p$ such that $f = U_p^r g$. It is easy to see that $U_p^r$ commutes with $E( \cdot \mid \mathcal{F}_t)$. Then, $E(f \mid \mathcal{F}_t) = U_p^r(E(g \mid \mathcal{F}_t))$. By Doob's inequality, there exists $G \in L^p(m; \mathcal{C}([0,+\infty]))$ such that, for any $t \geq 0$ $G_t = E(g \mid \mathcal{F}_t)$ a.s. and

$$\|G\|_{L^p(m;\mathcal{C}([0,+\infty]))} \leq \frac{p}{p-1}\|g\|_p.$$

Therefore theorem 4.4 applies and the first part of the statement of the theorem holds. Let $\varphi$ be the $(r,p)$-equilibrium potential of $f$, $\varphi = U_p^r \psi$. We can associate as before $\Phi$ with $\varphi$. Clearly, $(r,p)$-quasi-everywhere, for any $t$ $|F_t| \leq \Phi_t$. Therefore, by theorem 4.4

$$\gamma_{r,p}(\sup_t |F_t|) \leq \gamma_{r,p}(\sup_t \Phi_t) \leq \frac{p}{p-1}\|\psi\|_p = \frac{p}{p-1}\|\varphi\|_{r,p}$$

and $\|\varphi\|_{r,p} = \gamma_{r,p}(f)$.

The general result then follows from the density of $\mathbb{D}_p^r$ in $L^1(\gamma_{r,p})$. $\qquad\square$

## Quadratic variation

We fix $T > 0$. For any subdivision $\Delta = \{0 = t_0 \leq t_1 \cdots \leq t_n = T\}$ of the interval $[0,T]$, we denote by $S_\Delta$ the $d \times d$-matrix

$$S_\Delta = \sum_{i=0}^{n-1}(B_{t_{i+1}} - B_{t_i})(B_{t_{i+1}} - B_{t_i})^*$$

(where $(B_{t_{i+1}} - B_{t_i})^*$ denotes the transposed matrix of the column matrix $(B_{t_{i+1}} - B_{t_i})$. We denote by $I_d$ the $d \times d$-identity matrix.

**Theorem 4.7** *Let $(\Delta_n)_{n\geq 0}$ be a sequence of subdivisions of $[0,T]$ such that*

$$\lim_{n\to\infty} S_{\Delta_n} = T I_d \quad m\text{-a.s.}$$

*Then, $\lim_{n\to\infty} S_{\Delta_n} = T I_d$ outside a slim set.*

(The same result was proved by Feyel-de La Pradelle ([7]), using their result on sublinear functionals (proposition 2.4).)

*Proof.* It is easy to see that, for $r > 0$,

$$S_\Delta = U^r(2^{r/2}(S_\Delta - TI_d) + TI_d).$$

On the other hand, by Fernique's theorem, if $\lim_n S_{\Delta_n} = TI_d$ $m$-a.s., $\sup_n |S_{\Delta_n}| \in L^p$ (where $|\ \ |$ denotes a norm on $\mathbb{R}^{d\times d}$). We may then apply theorem 4.3. $\square$

## Stochastic differential equations

We fix $T > 0$, $n \in \mathbb{N}^*$, and we denote by $\mathbb{B}$ the Banach space $\mathcal{C}([0,T]; \mathbb{R}^n)$. We begin with a preliminary result (cf. [7]).

**Proposition 4.3** *Let* $\alpha. \in L^p([0,T]; (\mathbb{D}_p^r)^{n\times d})$ *be an* $\mathbb{R}^{n\times d}$-*valued adapted process. Then* $\int_0^. \alpha_s \, dB_s$ *belongs to* $(\mathbb{D}_p^r)^\mathbb{B}$.

*Sketch of the proof.* Set

$$\widehat{\mathbb{U}}^r = \frac{1}{\Gamma(r/2)} \int e^{-(3/2)t} t^{(r/2)-1} P_t \, dt.$$

There exists an adapted process $\beta \in L^p([0,T]); (L^p)^{n\times d})$ such that $\alpha_s = \widehat{\mathbb{U}}^r(\beta_s)$. Then $\int_0^. \beta_s \, dB_s \in L^p([0,T]; \mathbb{B})$ and

$$\int_0^. \alpha_s \, dB_s = \overline{\mathbb{U}}^r(\int_0^. \beta_s \, dB_s).$$

$\square$

We then obtain by successive approximations the following result which was obtained by other methods in [32], [27], ...

**Theorem 4.8** *Let* $\sigma$ *(resp.* $b$*) be a* $C_b^\infty$ *function from* $\mathbb{R}^n$ *into* $\mathbb{R}^{n\times d}$ *(resp.* $\mathbb{R}^n$*). For* $x \in \mathbb{R}^n$*, we consider the continuous strong solution* $X$ *of*

$$dX_t = \sigma(X_t)dB_t + b(X_t)dt, \quad X_0 = x.$$

*Then, for any* $r > 0$*,* $p > 1$*,* $X \in (\mathbb{D}_p^r)^\mathbb{B}$*. As a consequence, there exists a version of* $X$ *belonging to* $\cap_{r>0,p>1} \mathcal{L}^1(\gamma_{r,p}; \mathbb{B})$.

## Quasi-sure continuity of the Ornstein-Uhlenbeck semi-group

**Theorem 4.9** *Let* $f \in \mathcal{L}^1(\gamma_{r,p})$. *Then* $\lim_{t \to 0} P_t f = f$ *$(r,p)$-q.e. More precisely, for any* $T > 0$, $\omega \in E \longrightarrow (t \in [0,T] \to P_t f(\omega))$ *belongs to* $\mathcal{L}^1(\gamma_{r,p}; \mathcal{C}([0,T]))$.

*Proof:* Suppose first $f \in \mathbb{D}_p^r$. Then $f = \mathbb{U}_p^r g$ with $g \in L^p$ and $P_t f = U^r P_t g$ $(r,p)$-q.e. By [31], hypotheses of theorem 4.4 are satisfied with $f_t = P_t f$, $g_t = P_t g$. As, for any $t$, $f_t \in \mathcal{L}^1(\gamma_{r,p})$, the result is obtained in this case and $\gamma(\sup_t |P_t f|) \leq C\|f\|_{r,p}$. We can then proceed as in the proof of theorem 4.6. $\qquad\qquad\square$

Many other results of quasi-sure analysis can be found in [5]. Nevertheless, it is not always true that an $m$-a.s. convergence theorem admits a quasi-sure version. For example, if $d = 3$ or $4$, $\lim_{t \to \infty} |B_t(\omega)| = \infty$ $m$-a.s., but it is proved in [22] that, for any $\varepsilon > 0$,

$$c_{1,2}(\{\omega; \ \liminf_{t \to \infty} |B_t(\omega)| < \varepsilon\}) > 0.$$

# References

[1] J. BAUER Multiparameter processes associated with Ornstein-Uhlenbeck semi-groups, in *Classical and Modern Potential Theory and Applications*, p. 41-56, Kluwer Acad. Publ., Dortrecht-Boston-London, 1994

[2] N. BOULEAU, F. HIRSCH *Dirichlet forms and analysis on Wiener space*, Walter de Gruyter, Berlin-New York, 1991

[3] L. DENIS Analyse quasi-sûre de l'approximation d'Euler et du flot d'une E.D.S., C.R. Acad. Sc. Paris, 315 (1992), 599-602

[4] L. DENIS Convergence quasi-partout pour les capacités définies par un semigroupe sous-markovien, C.R. Acad. Sc. Paris, 315 (1992), 1033-1036

[5] L. DENIS Analyse quasi-sûre de certaines propriétés classiques sur l'espace de Wiener, Thèse Université Paris VI, 1994

[6] D. FEYEL Espaces de Banach adaptés. Quasi-topologie et balayage, in *Séminaire de théorie du potentiel Paris, No. 3*, p.81-102, Lecture N. in Math, Vol.681, Springer-Verlag, Berlin-Heidelberg-New York, 1978

[7] D. FEYEL, A. de LA PRADELLE Capacités gaussiennes, Ann. Inst. Fourier, 41-1 (1991), 49-76

[8] D. FEYEL, A. de LA PRADELLE Démonstration géométrique d'une loi du tout ou rien, C.R. Acad. Sci. Paris, 316 (1993), 229-232

[9] D. FEYEL, A. de LA PRADELLE On infinite dimensional sheets, Potential Analysis, 4-4 (1995), 345-359

[10] M. FUKUSHIMA Basic properties of Brownian motion and a capacity on the Wiener space, J. Math. Soc. Japan, 36-1 (1984), 161-175

[11] M. FUKUSHIMA A Dirichlet form on the Wiener space and properties of Brownian motion, in *Théorie du Potentiel Orsay 1983*, p. 290-300, Lecture N. in Math, Vol. 1096, Springer-Verlag, Berlin-Heidelberg-New York, 1984

[12] M. FUKUSHIMA Two topics related to Dirichlet forms: quasi everywhere convergences and additive functionals, in *CIME courses on Dirichlet forms 1992*, p. 21-53, Lecture N. in Math, Vol. 1563, Springer-Verlag, Berlin-Heidelberg-New York, 1994

[13] M. FUKUSHIMA, H. KANEKO On $(r, p)$-capacities for general Markovian semi-groups, in *Infinite dimensional analysis and stochastic processes*, p. 41-47, Pitman, Boston-London-Melbourne, 1985

[14] F. HIRSCH Représentation du processus d'Ornstein-Uhlenbeck à $n$ paramètres, in *Séminaire de Probabilités XXVII*, p. 302-303, Lecture N. in Math, Vol. 1557, Springer-Verlag, Berlin-Heidelberg-New York, 1993

[15] F. HIRSCH Potential theory related to some multiparameter processes, Potential Analysis, 4-3 (1995), 245-267

[16] F. HIRSCH, S. SONG Propriétés de Markov des processus à plusieurs paramètres et capacités, C.R. Acad. Sci. Paris, 319 (1994), 483-488

[17] F. HIRSCH, S. SONG Inequalities for Bochner's subordinates of two-parameter symmetric Markov processes, to appear in Ann. I.H.P.

[18] F. HIRSCH, S. SONG Markov properties of multiparameter processes and capacities, Probab. Th. Relat. Fields, 103-1 (1995), 45-71

[19] F. HIRSCH, S. SONG Symmetric Skorohod topology on $n$-variable functions and hierarchical Markov properties of $n$-parameter processes, Probab. Th. Relat. Fields, 103-1 (1995), 25-43

[20] Z. HUANG, J. REN Quasi sure stochastic flows, Stochastic and Stoch. Rep., 33 (1990), 149-157

[21] T. KASUMI, I. SHIGEKAWA Measures of finite $(r, p)$-energy and potentials on a separable metric space, in *Séminaire de Probabilités XXVI*, p. 415-444, Lecture N. in Math, Vol. 1526, Springer-Verlag, Berlin-Heidelberg-New York, 1992

[22] N. KÔNO 4-dimensional Brownian motion is recurrent with positive capacity, Proc. Japan Acad., 60-21 (1984), 57-59

[23] P. MALLIAVIN Implicit functions in finite corank on the Wiener space, in *Taniguchi Intern. Symp. on Stochast. Anal. Katata 1982*, p. 369-386, Kinokuniya, Tokyo, 1983

[24] P. MALLIAVIN, D. NUALART Quasi sure analysis of stochastic flows and Banach space valued smooth functionals on the Wiener space, J. Funct. Anal., 112 (1993), 287-317

[25] P.A. MEYER Note sur le processus d'Ornstein-Uhlenbeck, in *Séminaire de Probabilités XVI*, p. 142-163, Lecture N. in Math, Vol. 511, Springer-Verlag, Berlin-Heidelberg-New York, 1976

[26] J. REN Topologie P-fine sur l'espace de Wiener et théorème des fonctions implicites, Bull. Sc. math., 114 (1990), 99-114

[27] J. REN Analyse quasi-sûre des équations différentielles stochastiques, Bull. Sc. math., 114 (1990), 187-213

[28] I. SHIGEKAWA Sobolev spaces of Banach-valued functions associated with a Markov process, Probab. Th. Relat. Fields, 99 (1994), 425-441

[29] S. SONG Inégalités relatives aux processus d'Ornstein-Uhlenbeck à $n$-paramètres et capacités gaussiennes $c_{n,2}$, in *Séminaire de Probabilités XXVII*, p. 276-301, Lecture N. in Math, Vol. 1557, Springer-Verlag, Berlin-Heidelberg-New York, 1993

[30] S. SONG Construction d'un processus à deux paramètres à partir d'un semigroupe à un paramètre, in *Classical and Modern Potential Theory and Applications*, p. 419-452, Kluwer Acad. Publ., Dortrecht-Boston-London, 1994

[31] E.M. STEIN *Topics in Harmonic Analysis related to the Littlewood-Paley Theory*, Princeton Univ. Press, 1970

[32] H. SUGITA Positive generalized Wiener functionals and potential theory over abstract Wiener spaces, Osaka J. Math., 25 (1988), 665-696

[33] J.B. WALSH An introduction to stochastic partial differential equations, in *Ecole d'été de probabilités de Saint-Flour XIV-1984*, p. 266-437, Lecture N. in Math, Vol. 1180, Springer-Verlag, Berlin-Heidelberg-New York, 1986

[34] S. WATANABE *On stochastic differential equations and Malliavin calculus*, Tata Intitute of Fund. Research, Vol.73, Springer-Verlag, Berlin-Heidelberg-New York, 1984

# A model for loss of profits insurance

by

Knut K. Aase

Norwegian School of Economics and Business Administration
5035 Sandviken-Bergen, Norway

### Abstract

We consider a model for loss of profits insurance where we use the principle of equivalende in the actuarial sciences. Here we demonstrate a computationally efficient method for calculating expected discounted values when the underlying technological uncertainty is modeled by a time-continuous and homogeneous Markov process with a finite or possibly countable state space $E$. Such quantities may be interpreted as insurance premiums under certain assumptions, and an economic model is formulated where this is the case.

## 1 Introduction

Consider a firm which is either producing, i.e. in state 0, or it is down, i.e. in state 1, so that the state space $E = \{0, 1\}$. When in state 0 the firm pays a premium (rate) to an insurance company. When in state 1 the firm receives a compensation from the insurance company. This benefit we may think of as a rate for the moment. Let $Y_t$ = state of the firm at time $t$, and suppose that $Y$ is a Markov process which is time-homogeneous. Then the sojourn times in the various states are exponentially distributed, say with mean $1/\lambda$ in state 0 and mean $1/\mu$ in state 1.

We want to compute the premium rate for a given compensation rate, assuming that the parameters $\lambda$ and $\mu$ are known. Suppose that the continuous interest rate $\delta$ is a constant, and that each insurance company has a sufficiently large number of independent and identically distributed policies of this form, so that the exonomic risk premium is forced towards its limiting value of zero by competition in the insurance market. Alternatively, assume that the economic risk can be completely diversified in the reinsurance market. Abstracting from administrative expenses, under an infinite time horizon, by the principle of equivalence we then have that the expected discounted value of the premium rates must equal the expected discounted value of the compensations, or

$$E_i \left\{ \int_0^\infty e^{-\delta t} h(Y_t) dt \right\} = E_i \left\{ \int_0^\infty e^{-\delta t} g(Y_t) dt \right\}, \tag{1}$$

where $h(0) =$ the premium rate and $g(1) =$ the compensation rate, i.e., $h(0) > 0$, $h(1) = 0$ and $g(0) = 0$, $g(1) > 0$, and the symbol $E_i$ signifies conditional expected value given that $Y(0) = i$.

We may also think of $\delta$ as a risk adjusted rate of return in this line of business, in which case relation (1) also incorporates economic risk. More interestingly, in the last section of the paper it is discussed under what conditions the quantities appearing in (1) may actually be interpreted as market values of the insurance benefits in question. For the time being we shall assume that these quantities represent the relevant market values and thus are of interest.

Assuming that the firm produces at time zero when the contract is settled, $i = 0$, then by the Fubini theorem it follows that

$$E_0 \left\{ \int_0^\infty e^{-\delta t} h(Y_t) dt \right\} = \int_0^\infty e^{-\delta t} E(h(Y_t)|Y(0) = 0) dt$$
$$= \int_0^\infty e^{-\delta t} E(g(Y_t)|Y(0) = 0) dt = E_0 \left\{ \int_0^\infty e^{-\delta t} g(Y_t) dt \right\}. \tag{2}$$

Since

$$E(h(Y_t)|Y(0) = 0) = \sum_{j \in E} P_{0j}(t) h(j) \tag{3}$$

and

$$E(g(Y_t)|Y(0) = 0) = \sum_{j \in E} P_{0j}(t) g(j), \tag{4}$$

where $P_{ij}(t) = P\{Y(t+s) = j|Y(s) = i\}$ represent the transition (probability-)functions of the process $Y$, again by the Fubini theorem we have

$$\sum_{j \in E} \left( \int_0^\infty e^{-\delta t} P_{0j}(t) dt \right) h(j) = \sum_{j \in E} \left( \int_0^\infty e^{-\delta t} P_{0j}(t) dt \right) g(j), \tag{5}$$

or

$$h(0) \int_0^\infty e^{-\delta t} P_{00}(t) dt = g(1) \int_0^\infty e^{-\delta t} P_{01}(t) dt. \tag{6}$$

From this we obtain the relation we seek

$$h(0) = g(1) \cdot \frac{\int_0^\infty e^{-\delta t} P_{01}(t) dt}{\int_0^\infty e^{-\delta t} P_{00}(t) dt}, \tag{7}$$

a relation any actuary worth his salt would be able to derive. In order to solve this problem, we need to compute the quantities $\int_0^\infty e^{-\delta t} P_{ij}(t)dt$, i.e. find the Laplace transforms of $P_{ij}(t)$ expressed by the parameters $\delta, \lambda$ and $\mu$. A direct way to do this is by finding the transition functions $P_{ij}(t)$ and then perform the integrations. Since $Y$ is a birth and death process, these equations are

$$P'_{00}(t) = \lambda(P_{10}(t) - P_{00}(t)) \tag{8}$$
$$P'_{10}(t) = \mu(P_{00}(t) - P_{10}(t)) \tag{9}$$

Multiplying (8) by $\mu$ and (9) by $\lambda$ and adding, yields $\mu P'_{00}(t) + \lambda P'_{10}(t) = 0$ and $\mu P_{00}(t) + \lambda P_{10}(t) = c$. Since $P_{00}(0) = 1$ and $P_{10}(0) = 0$, (the firm produces at time $t = 0$), we find that $c = \mu$, so $\mu P_{00}(t) + \lambda P_{10}(t) = \mu$ or $\lambda P_{10}(t) = \mu(1 - P_{00}(t))$. Substituting this expression into (8), we obtain the following ordinary differential equation $P'_{00}(t) = \mu - (\mu + \lambda)P_{00}(t)$ with initial condition $P_{00}(0) = 1$, which can be solved to give

$$P_{00}(t) = \frac{\lambda}{\lambda + \mu} e^{-(\mu+\lambda)t} + \frac{\mu}{\lambda + \mu}. \tag{10}$$

From the relation $\mu P_{00}(t) + \lambda P_{10}(t) = \mu$ we also get

$$P_{10}(t) = \frac{\mu}{\mu + \lambda} - \frac{\mu}{\mu + \lambda} e^{-(\mu+\lambda)t}. \tag{11}$$

Finally we find

$$P_{01}(t) = 1 - P_{00}(t) \quad \text{and} \quad P_{11}(t) = 1 - P_{10}(t). \tag{12}$$

Thus we have managed to solve the coupled system of linear differential equations in the present model. It is now a simple task to compute the relevant Laplace transforms, and the solution to our problem is

$$h(0) = g(1)\frac{\lambda}{\delta + \mu}. \tag{13}$$

The expression in (13) gives us the premium rate $h(0)$ that must be required for a given compensation rate $g(1)$, interest rate $\delta$, parameters $\lambda$ and $\mu$ and initial condition $i = 0$. Our actuary worth his salt would still be able to write down this expression quite easily (see e.g., Aven and Myre (1988)). This is not the point with the present note.

Below we shall present a simple *principle* to derive relationships of this kind, where we avoid the problem of solving a coupled system of differential equations. This principle turns out to be powerful in more complicated situations, and, judging from the literature, it does not seem to be well-known among actuaries. Before we present this method, we briefly cosider the situation with a finite time horizon.

## 2 The finite time horizon case

Consider the time interval $[0, t]$ instead of $[0, \infty]$, where $t$ is fixed. In this case the principle of equivalence yields

$$E\left\{\int_0^t e^{-\delta u} h(Y_u) du | Y(0) = 0\right\} = E\left\{\int_0^t e^{-\delta u} g(Y_u) du | Y(0) = 0\right\}.$$

Since the insurance period is now $t$, we expect a relation of the type $h(0) = g(1) \cdot$ (function of the insurance period $t$). Indeed, we get

$$h(0) \int_0^t e^{-\delta u} P_{00}(u) du = g(1) \int_0^t e^{-\delta u} P_{01}(u) du,$$

yielding the relation

$$h(0) = g(1) \frac{\lambda(\delta + \lambda + \mu)(1 - e^{-\delta t}) - \lambda\delta(1 - e^{-(\delta+\lambda+\mu)t})}{\mu(\delta + \lambda + \mu)(1 - e^{-\delta t}) + \lambda\delta(1 - e^{-(\delta+\lambda+\mu)t})}. \qquad (14)$$

As $t \to \infty$ we see that $h(0) \to g(1)\frac{\lambda}{\delta+\mu}$ as it should.

## 3 Computation of expected, discounted payment streams

Many problems in the actuarial sciences consist in computing expected, discounted values of payment streams. Continuing our analysis of loss of profits insurance, consider a function $g : E \to R_+$, where $g(j)$ is a payment rate received when the process $Y$ is in state $j$. Again $Y$ is a time-homogeneous Markov process and we retain our economic assumptions. In this case the payment rate at time $t$ equals $g(Y_t)$, and the total discounted payment stream is $\int_0^\infty e^{-\delta t} g(Y_t) dt$. We seek the conditional expected value of this random variable given $i \in E$. We denote this function $V(i) = U^\delta g(i)$, given by

$$V(i) = E\left\{\int_0^\infty e^{-\delta t} g(Y_t) dt | Y_0 = i\right\}, \quad i \in E. \qquad (15)$$

This quantity is known as the $\delta$-potential of the function $g$ for the process $Y$. Alternatively, if the insurance period is $s$, we seek the quantity $E\left\{\int_0^s e^{-\delta t} g(Y_t) dt | Y_0 = i\right\}$. We concentrate on computing $V(i)$ given in (15). As above we derive

$$V(i) = \sum_{j \in E} \left[\int_0^\infty e^{-\delta t} P_{ij}(t) dt\right] g(j) = U^\delta g(i). \qquad (16)$$

We use the notation

$$U_{ij}^{\delta} = \int_0^{\infty} e^{-\delta t} P_{ij}(t)dt, \tag{17}$$

where $U_{ij}^{\delta}$ is the Laplace transform of $P_{ij}(t)$. Consider now the stopping time $T(\omega)$. Then we know that

$$
\begin{aligned}
V(i) &= E\left\{\int_0^{\infty} e^{-\delta t} g(Y_t)dt | Y_0 = i\right\} \\
&= E\left\{\int_0^{T} e^{-\delta t} g(Y_t)dt | Y_0 = i\right\} + E\{e^{-\delta T} V(Y_T) | Y_0 = i\} \tag{18}
\end{aligned}
$$

for all $i \in E$, $\delta \geq 0$ and $g \geq 0$ (see e.g. Cinlar (1975)). An important use of this result enables us to transform the problem of finding $V(i)$ from the domain of solving systems of coupled differential equations to the algebraic problem of solving a system of linear equations. First recall that the *generator* $A$ of a Markov process is defined as

$$A_{ij} = \begin{cases} -\mu(i) & \text{if } i = j \\ \mu(i)p_{ij} & \text{if } i \neq j \end{cases}$$

where $\mu(i)$ is the exponential parameter associated with state $i$, and where $p_{ij}$ are the transition probabilities of the embedded Markov chain $X = \{X_n; n = 0, 1, 2, \ldots\}$ signifying the states visited by the Markov process $Y$. These parameters we consider known and part of the input data. Assuming that the process $Y$ is regular, the knowledge of $A$ gives us the parameters of the process and vice versa. Furthermore, from the knowledge of the parameters of the process we are able to obtain all the joint probability distributions of $Y$ if we in addition know the initial distribution of $Y(0)$. This is of course the reason why there parameters constitute the input data, and the parameters happen to be estimable from observations. The following result is now crucial:

*Let $g$ be bounded and let $\delta > 0$. Then $V$ is the unique solution of the following system of linear equations*

$$(\delta I - A)V = g \tag{19}$$

*where $I$ is the identity matrix.*

The proof of this result follows from (18) and the strong Markov property by conditioning on the first time $T(\omega)$ of transition, where this $T$ is now a Markov time. Again Cinlar (1975) is a good reference. See also Kemeny et al. (1966). It may be noticed that equation (20) may also be interpreted as a special version of the Thiele differential equation in the case of time-homogeneity.

Notice that if $g = 1_j$, then $V$ equals the $j$-th column of the matrix $U^\delta$, which means that we can find this matrix by solving $(\delta I - A)U^\delta = I$. In the case where $E$ is finite, this further means that

$$U^\delta = (\delta I - A)^{-1}. \tag{20}$$

# 4 Loss of profits insurance by potential theory

Returning to our model of section 1, the parameters are given as follows: $\mu(0) = \lambda$, $\mu(1) = \mu$, $p_{01} = 1$, $p_{00} = 0$, $p_{10} = 1$, $p_{11} = 0$ and the generator is given as

$$A = \begin{pmatrix} -\lambda & \lambda \\ \mu & -\mu \end{pmatrix}$$

The compensation rates are $g(0) = 0$ and $g(1) = g > 0$, and the expected present value $f(i)$ of the compensation stream, given the various initial states, satisfy $(\delta I - A)V = g$, where $V(i) = E_i \left\{ \int_0^\infty e^{-\delta t} g(Y_t) dt \right\}$, $i = 0, 1$. The corresponding expected present value $\widetilde{V}(i) = E_i \left\{ \int_0^\infty e^{-\delta t} h(Y_t) dt \right\}$ of the premium stream, where $h(0) = h > 0$, $h(1) = 0$, is given by the solution of $(\delta I - A)\widetilde{V} = h$. The first relation is found by solving

$$\begin{pmatrix} \delta + \lambda & -\lambda \\ -\mu & \delta + \mu \end{pmatrix} \begin{pmatrix} V(0) \\ V(1) \end{pmatrix} = \begin{pmatrix} 0 \\ g \end{pmatrix},$$

from which we find $V(0) = \frac{\lambda g}{\delta(\delta + \mu + \lambda)}$. Solving the other system of linear equations we get $\widetilde{V}(0) = \frac{(\delta + \mu)h}{\delta(\delta + \mu + \lambda)}$. The principle of equivalence now gives $\widetilde{V}(0) = V(0)$, which means that $h = g\frac{\lambda}{\delta + \mu}$, as found earlier.

From this expression we observe that as $\lambda \to \infty$ (state 0 becomes instantaneous), then the firm spends less and less time producing, or the firm is down most of the time. Thus the sojourn times available to pay the premium rate shrink towards zero, so $h \to +\infty$. On the other hand, if $\mu \to +\infty$, state 1 tends towards an instantaneous stae, production stops happen with an ever lower frequency and the time lengths of the down periods become smaller and smaller. Thus the compensation tends towards zero and the premium rate $h \to 0$. Note also that even if $\lambda = \mu$, we do not get $h = g$ unless $\delta = 0$.

From the above we observe that it is considerably simpler to use the potential theory to solve this kind of problems. In the case where $E$ has

more than two states, the Kolmogorov differential equations may be difficult to solve, but the system of linear equations is routine sailing. Imagine
a situation where the production firm has $n$ different machines giving rise
to $k_n$ different levels of production depending on what machines are working. State $0 =$ full production, state $1 =$ one level down,..., $k_n =$ production is down. The compensation function must satisfy $0 \leq g(0) \leq g(1) \leq$
$\ldots \leq g(k_n)$. Then we first compute $V(i) = E_i \left\{ \int_0^\infty e^{-\delta t} g(Y_t) dt \right\}$ for $i =$
$0, 1, 2, \ldots, k_n$. The premium rates are paid in the states $\{0, 1, 2, \ldots, m\}$,
where $m < k_n$, and we may require various schemes for the function $h$, e.g.
that $h(0) > h(1) > h(2) > \ldots > h(m)$ or that this function is a constant.
Then we compute $\tilde{V}(i) = E_i \left\{ \int_0^\infty e^{-\delta t} h(Y_t) dt \right\}$ for $i = 0, 1, 2, \ldots, k_n$. Finally the various premium rates are determined according to the principle
of equivalence. In this case the computations involved are simple by the
potential method, given that the parameters of the process $Y$ are known,
the function $g$ is specified and the design for $h$ is set, whereas it may be
a rather tricky task to solve the system of coupled differential equations
and then compute the relevant Laplace transforms by the direct method.

The situation with a finite time horizon may also be analyzed by
the use of potential theory, but the savings in complexity is no longer
so substantial. Suppose $t$ is a fixed insurance period. Then it is trivially a stopping time, so by equation (18) the present value $V(i) =$
$E_i \left( \int_0^t e^{-\delta u} g(Y_u) du \right) + E_i(e^{-\delta t} V(Y(t)))$. The last term cam be written

$$E_i\{e^{-\delta t} V(Y_t)\} = e^{-\delta t} E\{V(Y_t) | Y_0 = i\} = e^{-\delta t} \sum_{j \in E} P_{ij}(t) V(j),$$

so that

$$E_i \left( \int_0^t e^{-\delta u} g(Y_u) du \right) = V(i) - e^{-\delta t} \sum_{j \in E} P_{ij}(t) V(j). \qquad (21)$$

We are seeking the quantity on the left-hand side of (21). The quantity
$V(i)$ can be computed as earlier, but in order to compute the last quantity
on the right-hand side we need the transition functions $P_{ij}(t)$. In other
words, it appears that we need to solve the Kolmogorov equations. The
only savings in effort is that there is no need to compute the integrals
$\int_0^t e^{-\delta s} P_{ij}(s) ds$.

**Example 1.** Consider again our model for loss of profits insurance.
By the potential method we must first compute the quantities $V(0) =$
$\frac{\lambda g}{\delta(\delta + \mu + \lambda)}$, $V(1) = \frac{(\delta + \lambda) g}{\delta(\delta + \mu + \lambda)}$, $\tilde{V}(0) = \frac{(\delta + \mu) h}{\delta(\delta + \mu + \lambda)}$ and $\tilde{V}(1) = \frac{\mu h}{\delta(\delta + \lambda + \mu)}$. From

this we see that

$$E_0 \left( \int_0^t e^{-\delta u} g(Y_u) du \right) = \frac{\lambda g}{\delta(\delta + \mu + \lambda)} - e^{-\delta t}(P_{00}(t)V(0) + P_{01}(t)V(1)).$$

Plugging in our expressions for $P_{00}(t)$ and $P_{01}(t)$ from our earlier work, we get

$$E_0 \left( \int_0^t e^{-\delta u} g(Y_u) du \right)$$

$$= \frac{\lambda g}{\delta(\delta + \mu + \lambda)} - \frac{e^{-\delta t} \lambda g}{(\lambda + \mu)\delta(\delta + \lambda + \mu)} \{ \mu + (\delta + \lambda) - \delta e^{-(\mu + \lambda)t} \}.$$

Similarly

$$E_0 \left( \int_0^t e^{-\delta u} h(Y_u) du \right) = \frac{h(\delta + \mu)}{\delta(\delta + \lambda + \mu)} - e^{-\delta t} \{ P_{00}(t)\tilde{V}(0) + P_{01}(t)\tilde{V}(1) \}$$

$$= \frac{h(\delta + \mu)}{\delta(\delta + \lambda + \mu)} - \frac{e^{-\delta t} h}{(\lambda + \mu)(\delta + \lambda + \mu)\delta} \{ \mu(\delta + \mu) + \lambda \mu + \delta \lambda e^{-(\mu + \lambda)t} \}.$$

The principle of equivalence finally gives

$$E_0 \left\{ \int_0^t e^{-\delta u} h(Y_u) du \right\} = E_0 \left\{ \int_0^t e^{-\delta u} g(Y_u) du \right\},$$

implying that $h = g \frac{\lambda(\mu + \delta + \lambda)(1 - e^{-\delta t}) - \lambda \delta(1 - e^{-(\delta + \lambda + \mu)t})}{\mu(\delta + \lambda + \mu)(1 - e^{-\delta t}) + \delta \lambda(1 - e^{-(\delta + \mu + \lambda)t})}$, which is the same expression as we found earlier by the direct method.

# 5    The actuarial value of an assurance

So far we have only considered discounted values of payment streams. In the present section we shall study assurance against transition from a certain state $j$ to another state $k$. If such a transition occurs at time $t$, a benefit of the amount $B_{jk}(t)$ is paid immediately afterwards. If more than one such transition may take place in a time interval, payment is made upon each transition. Given that $Y(s) = i$, the expected discounted value at time $s$ of the benefits paid in the time interval $(s, u]$, is given by the formula

$$A_s^{i,jk} = \int_s^u P_{ij}(t - s) \mu_j p_{jk} B_{jk}(t) e^{-\delta(t-s)} dt. \tag{22}$$

In order for a payment to be possible at time $t > s$, the process must have left state $i$ and arrived at state $j$ by time $t$; the probability of this event being $P_{ij}(s,t) = P_{ij}(t-s)$ from time-homogeneity. Furthermore a transition has to happen at time $t$, the probability of this event being $\mu_j dt$, and the conditional probability that this transition is to state $k$ is $p_{jk}$. The benefit is then $B_{jk}(t)$ and the discounted value at time $s$ is $B_{jk}(t)e^{-\delta(t-s)}$. Integrating all this from time $t = s$ to $t = u$ we obtain (22).

If $B_{jk}(t) = B_{jk} \, \forall t$, i.e., a family of constants, $s = 0$ and $u = +\infty$, then

$$A_0^{i,jk} = \int_0^\infty P_{ij}(t)\mu_j p_{jk} B_{jk} e^{-\delta t} dt = \mu_j p_{jk} B_{jk} \int_0^\infty P_{ij}(t)e^{-\delta t} dt. \qquad (23)$$

From equation (20) we already have a simple method to compute integrals of the type $\int_0^\infty P_{ij}(t)e^{-\delta t} dt = U_{ij}^\delta$, and this leaves us with a correspondingly simple method to compute $A_0^{i,jk}$ given in (23).

If the insurance period $u$ is finite, we use (18) with $T = u$ and $g = 1_j$ in which case we get

$$\int_0^u e^{-\delta t} P_{ij}(t)dt = \int_0^\infty e^{-\delta t} P_{ij}(t)dt - e^{-\delta u} \sum_{k \in E} P_{ik}(u) \int_0^\infty e^{-\delta t} P_{kj}(t)dt. \qquad (24)$$

It is no longer sufficient only to compute $U_{ik}^\delta$ for all $i$ and $k$; in addition we need the transition functions $P_{ik}(u)$ for all $i$ and $k$.

**Example 2.** Returning to our model of loss of profits insurance, suppose now that the payments from the insurance company to the firm take place just after the times of breakdown of production. The payments can be characterized by the constants $B_{01} = B, B_{10} = 0$. We want to compute the actuarial value $A_0^{0,01}$ of this insurance contract. It is

$$A_0^{0,01} = \mu_0 p_{01} B \int_0^\infty P_{01}(t)e^{-\delta t} dt = \lambda B \, U_{01}^\delta.$$

In order to compute the last term, we only need to solve

$$\begin{pmatrix} \delta + \lambda & -\lambda \\ -\mu & \delta + \mu \end{pmatrix} \begin{pmatrix} U_{01}^\delta \\ U_{11}^\delta \end{pmatrix} = \begin{pmatrix} 0 \\ 1 \end{pmatrix},$$

giving $U_{01}^\delta = \frac{\lambda}{\delta(\delta+\lambda+\mu)}$, or $A_0^{0,01} = \frac{\lambda^2 B}{\delta(\delta+\lambda+\mu)}$. If this premium is to be paid as a stream, then $\tilde{V}(0) = A_0^{0,01}$, or $h(0) = B\lambda \frac{\lambda}{\delta+\lambda}$. It follows from this that if the insurance benefit $B$ is to be actuarially equivalent to a compensation rate $g(1)$, then $g(1) = B\lambda$.

# 6 An economic model supporting the actuarial analysis

We now return to the equestion under what conditions the actuarial values computed in this paper can be considered as market values of the relevant insurance contracts. This is by no means a simple question, and we shall only present one economic model and give a set of sufficient conditions under which this is possible. First we reformulate the principle of equivalence as follows:

> The market value of the premium payments = the market value of the compensations, or the benefits.

We now use an economic equilibrium model of consumption involving jumps in the aggregate consumption in the market (see Aase (1993–94). According to the general theory of valuation under uncertainty, the above principle can be written

$$E_i\left\{\int_0^\infty u_k'(c(s),s)h(Y(s))ds|\mathcal{F}_0\right\} = E_i\left\{\int_0^\infty u_k'(c(s),s)g(Y(s))ds|\mathcal{F}_0\right\},$$

where $g$ and $h$ have the same interpretations as earlier, and where $u_k'(c,t)$ equals the marginal utility of the market (or of the representative agent) at the point of aggregate consumption $c$ at time $t$. The symbol $\mathcal{F}_0$ means a $\sigma$-field giving the possible remaining information available at time zero. We now make the following assumptions:

- The aggregate consumption process is independent of the technology process $Y$.

- The marginal utility function is given as $u_k'(x,t) = \alpha x^{\alpha-1}e^{-rt}$, $r > 0$, $\alpha \in (0,1)$, $x \geq 0$. Here $(1-\alpha) =$ coefficient of relative risk aversion in the market, and $r =$ subjective discount rate of the market.

- $c(s) = c(0)\prod_{n=1}^{N(t)}(1 + U^{(n)})$, where $N(t) =$ number of jumps in the aggregate consumption process $c$ by time $t$ and $U^{(1)}, U^{(2)},\ldots$ is a sequence of independent and identically distributed jumps in the aggregate consumption process $c$. The random variables $U^{(n)}$ take values in $(-1,\infty)$ and we assume that $(1+U^{(n)}) > 0$ almost surely.

- $N$ is a homogeneous Poisson process with rate $\gamma > 0$.

- $r > \gamma(v(\alpha-1)-1)$ where the function $v(\alpha-1) = E(1+U^{(n)})^{(\alpha-1)}$.

Under these assumptions and taking for granted the economic equilibrium analysis referred to above, our principle of equivalence can be expressed as follows:

$$\int_0^\infty E\{u_k'(c(s),s)|\mathcal{F}_0\}E_i\{h(Y(s))\}ds = \int_0^\infty E\{u_k'(c(s),s)|\mathcal{F}_0\}E_i\{g(Y(s))\}ds.$$

Under the above assumptions we have

$$E\{u_k'(c(s),s)|\mathcal{F}_0\} = \alpha e^{-rs}E\{c(s)^{\alpha-1}|\mathcal{F}_0\},$$

where

$$E\{c(s)^{\alpha-1}|\mathcal{F}_0\} = c(0)^{\alpha-1}e^{(v(\alpha-1)-1)\gamma s}.$$

Denoting $\beta = r - (v(\alpha-1)-1)\gamma$, our principle reads

$$\int_0^\infty e^{-\beta s}E_i\{h(Y(s))\}ds = \int_0^\infty e^{-\beta s}E_i\{g(Y(s))\}ds.$$

We are now back in the framework of this paper, where $\beta$ corresponds to $\rho$. Hence, for the loss of profits insurance discussed in this paper, the premium can be expressed as $h_0 = g_1\frac{\lambda}{\beta+\mu}$ where $\beta = r + (1 - v(\alpha-1))\gamma$. From this correspondence we notice in particular that the equilibrium premium rate depends upon:

- preferences in the market represented by $\alpha$ and $r$.

- both the jump sizes in the aggregate consumption process through the function $v(\alpha-1)$, and the corresponding jump frequency $\gamma$.

In the actuarial sciences it is rarely discussed how premiums must depend upon economic factors, since "premium principles" are usually taken to be exogenous to the model. Usually risk neutrality is implicitly assumed in actuarial models, but this appears to be incosistent with the very existence of insurance, since if all the agents in the economy are risk neutral, nobody will demand insurance of anything. On the contrary, in the above model it is assumed that the agents in the economy are risk averse, resulting in a risk aversion in the market as a whole. The quantity $(1-\alpha)$ here represents the coefficient of relative risk aversion in this market, stemming from the various individual attitudes towards risk (not necessarily all equal) of the agents in this market.

# References

[1] Aase, Knut K. (1993), A jump/diffusion consumption-based capital asset princing model and the equity premium puzzle, *Mathematical Finance*, vol. 3, no 2, 65–84.

[2] Aase, Knut K. (1993), Continuous trading in an exchange economy under discontinuous dynamics – A resolution of the equity premium puzzle, *Scand. J. Management*, V. 9S, 1–28.

[3] Aven, T. and Myre (1988), A model for loss of profits insurance, *Scandinavian Actuarial Journal*.

[4] Blumenthal, R. M. and R. K. Getoor (1968), *Markov Processes and Potential Theory*, Academic Press, New York.

[5] Cinlar, Erhan (1975), *Introduction to stochastic processes*, Prentice-Hall, Inc. Englewood Cliffs, New Jersey.

[6] Kemeny, J. G., J. L. Snell, and A. W. Knapp (1966), *Denumerable Markov chains*, Van Nostrand, New York.

# A Note On Population Growth In A Crowded Stochastic Environment

Fred Espen Benth
Department of Mathematics
University of Oslo
Box 1053 Blindern, N-0316 Oslo
Norway
&
Lehrstuhl für Mathematik V
Universität Mannheim A5
D-68131 Mannheim
Germany

## Abstract

We find an explicit unique solution in the space of Kondratiev distributions, $(\mathcal{S})^{-1}$, to a stochastic differential equation modelling population growth in a crowded stochastic environment.

## 1   Introduction

In this paper we are going to study a stochastic version of the Verhulst model for population growth,

$$X_t = X_0 + r \int_0^t X_s \diamond (N - X_s) ds + \alpha \cdot \int_0^t X_s \diamond (N - X_s) \delta B_s \qquad (1)$$

where $r, \alpha, N$ are constants, $N, r$ positive. $\delta B_s$ denotes the (generalized) Skorohod integral. A precise meaning of this integral will be given in the next section. We denote by $\diamond$ the Wick product.

(1) was first proposed by Lindstrøm et. al. [LØU] as a modell for population growth in a crowded stochastic environment. For deterministic initial conditions $X_0$, where $0 \leq X_0 \leq 1$ and $X_0 \neq \frac{1}{2}$, they found an explicit solution to (1) using white noise methods. Their solution is a "true" stochastic process. The case $X_0 = \frac{1}{2}$ represents some kind of "stochastic bifurcation point", since no stochastic variable exists as a solution for this initial condition *(see Lindstrøm et. al. [LØU] for their remark.)* The main motivation for this paper is to give an explicit solution also for the case $X_0 = \frac{1}{2}$. In section 4 we show that for this initial condition, we do not even have a solution in the space of Hida distributions, $(\mathcal{S})^*$. This suggests that the space of Kondratiev distributions, $(\mathcal{S})^{-1}$, is the natural space for this problem. Moreover, using Wick Calculus on the space of Kondratiev distributions, $(\mathcal{S})^{-1}$, we are able to find an explicit solution of (1 ) for general initial conditions with positive expectation. Now,

112

however, the solution is no longer a stochastic variable, but a *generalized* stochastic variable living in the abstract space $(S)^{-1}$.

## 2 Some Preliminaries

We start by recalling some of the basic definitions and features of *the white noise analysis*. For a more complete account, see Hida et. al. [HKPS] and Gjessing et. al. [GHLØUZ].

As usual, let $S'(\mathbb{R}^d)$ denote the space of tempered distributions on $\mathbb{R}^d$, which is the dual of the well-known Schwartz space $S(\mathbb{R}^d)$. By the Bochner-Minlos theorem there exists a measure $\mu$ on $S'$ such that

$$\int_{S'} e^{i<\omega,\phi>} d\mu(\omega) = e^{-\frac{1}{2}\|\phi\|^2}, \phi \in S$$

where $\|.\|$ is the $L^2(\mathbb{R}^d)$-norm. $\langle\cdot,\cdot\rangle$ is the dual pairing between $S'$ and $S$. Let $\mathcal{B}$ denote the Borel sets on $S'$ (equipped with the weak star topology). Then the triple $(S',\mathcal{B},\mu)$ is called *the white noise probability space*.

If we define

$$\tilde{B}_x(\omega) := \tilde{B}_{x_1,\ldots,x_d}(\omega) := \langle\omega, \mathcal{X}_{[0,x_1]\times\ldots\times[0,x_d]}(\cdot)\rangle$$

then $\tilde{B}_x$ has an x-continuous version $B_x$ which becomes a *d-parameter Brownian motion*. The *d-parameter Wiener-Ito integral* of $\phi \in L^2$ is defined by

$$\int_{\mathbb{R}^d} \phi(y) dB_y(\omega) = \langle\omega,\phi\rangle$$

Of special interest will be the space $L^2(S'(\mathbb{R}^d),\mu)$, or $L^2(\mu)$ for short. *The Wiener-Ito chaos expansion theorem* says that every $F \in L^2(\mu)$ has the form

$$F(\omega) = \sum_{n=0}^{\infty} \int_{(\mathbb{R}^d)^n} f_n(u) dB_u^{\otimes n}(\omega) \tag{2}$$

where $f_n \in L^2(\mathbb{R}^{nd})$ and $f_n$ is symmetric in its $n$ variables (in the sense that $f_n(u_{\sigma_1},\ldots,u_{\sigma_n}) = f_n(u_1,\ldots,u_n)$ for all permutations $\sigma$, where $u_i \in \mathbb{R}^d$). The right hand side of (2) are the *multiple Ito integrals*.

There is an equivalent expansion of $F \in L^2(\mu)$ in terms of the Hermite polynomials:

$$h_n(x) = (-1)^n e^{\frac{x^2}{2}} \frac{d^n}{dx^n}(e^{-\frac{x^2}{2}}); n = 0, 1, 2, \ldots$$

We explain this more closely: Define the *Hermite function* $\xi_n(x)$ of order $n$ as

$$\xi_n(x) = \pi^{-1/4}((n-1)!)^{-1/2} e^{-\frac{x^2}{2}} h_{n-1}(\sqrt{2}x)$$

where $x \in \mathbb{R}, n = 1, 2, \ldots$. $\{\xi_n\}_{n=1}^{\infty}$ forms an orthonormal basis for $L^2(\mathbb{R})$. Therefore the family $\{e_\alpha\}$ of tensor products

$$e_\alpha := e_{\alpha_1,\ldots,\alpha_m} := \xi_{\alpha_1} \otimes \ldots \otimes \xi_{\alpha_d}$$

(where $\alpha$ denotes the multi-index $(\alpha_1, \ldots, \alpha_d)$) forms an orthonormal basis for $L^2(\mathbb{R}^d)$. Assume that the family of all multi-indices $\beta = (\beta_1, \ldots, \beta_d)$ is given a fixed ordering

$$(\beta^{(1)}, \beta^{(2)}, \ldots, \beta^{(n)}, \ldots)$$

where $\beta^{(k)} = (\beta_1^{(k)}, \ldots, \beta_d^{(k)})$. Put

$$e_n = e_{\beta^{(n)}}; n = 1, 2, \ldots$$

Let $\alpha = (\alpha_1, \ldots, \alpha_m)$ be a multi-index. It was shown by Ito that

$$\int_{(\mathbb{R}^d)^n} e_1^{\widehat{\otimes}\alpha_1} \widehat{\otimes} \ldots \widehat{\otimes} e_m^{\widehat{\otimes}\alpha_m} dB^{\otimes n} = \prod_{j=1}^m h_{\alpha_j}(\theta_j) \tag{3}$$

where $\theta_j(\omega) = \int_{\mathbb{R}^d} e_j(x) dB_x(\omega), n = |\alpha|$ and $\widehat{\otimes}$ denotes the *symmetrized tensor product* (e.g., $f\widehat{\otimes}g(x,y) = \frac{1}{2}[f(x)g(y) + f(y)g(x)]$ if $x, y \in \mathbb{R}$ and similarly for more than two variables). If we define, for each multiindex $\alpha = (\alpha_1, \ldots, \alpha_m)$,

$$H_\alpha(\omega) = \prod_{j=1}^m h_{\alpha_j}(\theta_j)$$

then we see that (3) can be written

$$\int_{(\mathbb{R}^d)^n} e^{\widehat{\otimes}\alpha} dB^{\otimes|\alpha|} = H_\alpha(\omega) \tag{4}$$

using multi-index notation: $e^{\widehat{\otimes}\alpha} = e_1^{\widehat{\otimes}\alpha_1} \widehat{\otimes} \ldots \widehat{\otimes} e_m^{\widehat{\otimes}\alpha_m}$ if $e = (e_1, e_2, \ldots)$. Since the family $\{e^{\widehat{\otimes}\alpha}; |\alpha| = n\}$ forms an orthonormal basis for the symmetric functions in $L^2((\mathbb{R}^d)^n)$, we see by combining (2) and (4) that we have the representation

$$F(\omega) = \sum_\alpha c_\alpha H_\alpha(\omega) \tag{5}$$

(the sum being taken over all multi-indices $\alpha$ of nonnegative integers). Moreover, it can be proved that

$$\|F\|^2_{L^2(\mu)} = \sum_\alpha \alpha! c_\alpha^2$$

where $\alpha! = \alpha_1! \ldots \alpha_m!$.

The Hida test function space $(S)$ and the Hida distribution space $(S)^*$ can be given the following characterization, due to Zhang [Z].

**Theorem 2.1:** Let $\psi \in L^2(\mu)$ have the chaos expansion

$$\psi(\omega) = \sum_\alpha c_\alpha H_\alpha(\omega)$$

Then $\psi$ is a *Hida test function*, i.e. $\psi \in (S)$, if

$$\sup_\alpha c_\alpha^2 \alpha! (2\mathbb{N})^{\alpha k} < \infty, \ \forall \ \text{natural numbers } k < \infty$$

where

$$(2\mathbb{N})^\alpha := \prod_{j=1}^m (2^d \beta_1^{(j)} \ldots \beta_d^{(j)})^{\alpha_j} \text{ for } \alpha = (\alpha_1, \ldots, \alpha_m)$$

A *Hida distibution* $\Psi$, $\Psi \in (S)^*$, is a formal series

$$\Psi = \sum_\alpha b_\alpha H_\alpha \tag{6}$$

where

$$\sup_\alpha b_\alpha^2 \alpha! ((2I\!N)^{-\alpha})^q < \infty \text{ for some } q > 0$$

If $\Psi \in (S)^*$ and $\psi \in (S)$ is given as in the theorem, the action of $\Psi$ on $\psi$ is given by

$$\langle\langle \Psi, \psi \rangle\rangle = \sum_\alpha \alpha! b_\alpha c_\alpha \tag{7}$$

Note that no assumptions are made regarding the convergence of the formal series in (6). We can in a natural way regard $L^2(\mu)$ as a subspace of $(S)^*$. In particular, if $F \in L^2(\mu)$ then by (7) the action of $F$ on $\psi \in (S)$ is given by

$$\langle\langle F, \psi \rangle\rangle = E[F \cdot \psi]$$

Since 1 is an element of $(S)$, the expectation function can be extended to $(S)^*$:

$$E[\Psi] = \langle\langle \Psi, 1 \rangle\rangle$$

We will now introduce the spaces $(S)^1$ and $(S)^{-1}$ which were first constructed by Kondratiev [K]. For a complete account on the following results, see Albeverio et. al. [ADKS] and Kondratiev et. al. [KLS]:

**Definiton 2.2:** Define $(S)^\rho$ and $(S)^{-\rho}$ as follows:

**Part a):** For $0 \leq \rho \leq 1$ let $(S)^\rho$ consist of all

$$\psi = \sum_\alpha c_\alpha H_\alpha \in L^2(\mu)$$

such that

$$\|\psi\|_{\rho,k}^2 := \sum_\alpha c_\alpha^2 (\alpha!)^{1+\rho} (2I\!N)^{\alpha k} < \infty \text{ for all } k < \infty$$

**Part b):** The space $(S)^{-\rho}$ consists of all formal expansions

$$\Psi = \sum_\alpha b_\alpha H_\alpha$$

such that

$$\sum_\alpha b_\alpha^2 (\alpha!)^{1-\rho} (2I\!N)^{-\alpha q} < \infty \text{ for some } q < \infty$$

The family of seminorms $\|f\|_{\rho,k}^2$; $k = 1, 2, \dots$ defines a topology on $(S)^\rho$.

We remark that $(S) = (S)^0$ and $(S)^* = (S)^{-0}$ in the above construction. $(S)^{-1}$ will be called the space of Kondratiev distributions. **Definition 2.3:** Let $\Phi = \sum_\alpha a_\alpha H_\alpha$, $\Psi = \sum_\alpha b_\beta H_\beta$ be two elements of $(S)^{-\rho}$. Then the *Wick product* of $\Phi$ and $\Psi$ is the element $\Phi \diamond \Psi$ in $(S)^{-\rho}$ given by

$$\Phi \diamond \Psi = \sum_{\alpha,\beta} a_\alpha b_\beta H_{\alpha+\beta}$$

It can be shown that $(\mathcal{S})^1$ is closed under the Wick product.

The *Hermite Transform*, see Lindstrøm et. al. [LØU], has a natural extension to $(\mathcal{S})^{-1}$, the space of Kondratiev distributions:

**Definition 2.4:** If $\Psi = \sum_\alpha b_\alpha H_\alpha \in (\mathcal{S})^{-1}$ then the Hermite Transform of $F$, $\mathcal{H}\Psi = \tilde{\Psi}$, is defined by

$$\tilde{\Psi}(z) = \mathcal{H}\Psi(z) = \sum_\alpha b_\alpha z^\alpha$$

where $z = (z_1, z_2, ....) \in \mathbb{C}_0^N$, and

$$z^\alpha = z_1^\alpha z_2^\alpha \ldots z_m^\alpha$$

for $\alpha = (\alpha_1, \ldots, \alpha_m)$.

The Hermite Transform characterizes $(\mathcal{S})^{-1}$ in the following way:

**Lemma 2.5:** $\Psi \in (\mathcal{S})^{-1}$ if and only if there exist some $\epsilon > 0, q < \infty$ such that the Hermite transform of $\Psi$, $\mathcal{H}\Psi$, is a bounded analytic function on $B_q(0, \epsilon)$.

Convergence of sequences in $(\mathcal{S})^{-1}$ can be characterized in terms of the Hermite Transform as follows:

**Lemma 2.6:** The following are equivalent **I:** $\Psi_n \to \Psi$ in $(\mathcal{S})^{-1}$ **II:** There exist $\epsilon > 0, q < \infty, M < \infty$ such that

$$\mathcal{H}\Psi_n(z) \to \mathcal{H}\Psi(z) \text{ as } n \to \infty \text{ for } z \in \mathbf{B}_q(0, \epsilon)$$

and

$$|\mathcal{H}\Psi_n(z)| \le M \text{ for all } n = 1, 2, ...; z \in \mathbf{B}_q(0, \epsilon)$$

where

$$\mathbf{B}_q(0, \epsilon) = \{z = (z_1, z_2, \ldots) \in \mathbb{C}_0^N; \sum_\alpha |z^\alpha|^2 (2N)^{\alpha q} < \epsilon^2\}$$

Note that the Hermite Transform transforms the Wick product into an ordinary product.

The Wick product gives a nice relation between functional integration in $(\mathcal{S})^{-1}$ and Skorohod/Ito integration. We define integration in $(\mathcal{S})^{-1}$ as follows:

**Definition 2.7:** Assume $\Psi_s \in (\mathcal{S})^{-1}$ for each $s \in [0, T]$, where $0 < T \le \infty$. If

$$\langle\langle \Psi_s, \psi \rangle\rangle \in L^1([0, T], ds)$$

for all $\psi \in (\mathcal{S})^1$, we define the unique $(\mathcal{S})^{-1}$-element $\int_0^T \Psi_s ds$ by

$$\langle\langle \int_0^T \Psi_s ds, \psi \rangle\rangle = \int_0^T \langle\langle \Psi_s, \psi \rangle\rangle ds$$

Consider the case $d = 1$, i.e the probability space $\mathcal{S}'(\mathbb{R})$: Define the element

$$W_t = \sum_{k=1}^\infty \xi_k(t) H_{\epsilon_k}$$

where $\epsilon_k$ is the multi-index with zeros except at position $k$, where it has value 1. It can be shown, see for instance Gjessing et. al. [GHLØUZ], that $W_t \in (S)^*$. Moreover, for a Skorohod integrable element $\Psi_s \in L^2(\mu)$ it can be shown that

$$\int_0^t \Psi_s \diamond W_s ds = \int_0^t \Psi_s \delta B_s$$

where the integral on the right hand side is the Skorohod integral. See Lindstrøm et. al. [LØU], Hida et. al. [HKPS] and Benth [B] for a discussion of this relation. We can say that functional integration in $(S)^{-1}$ involving Wick product with $W_t$ generalizes the Skorohod/Ito integration. This connection motivates the following interpretation of (1): We look for an element $X_t$ in $(S)^{-1}$ which satisfies

$$X_t = X_0 + r \int_0^t X_s \diamond (N - X_s) ds + \alpha \int_0^t X_s \diamond (N - X_s) \diamond W_s ds \qquad (8)$$

We end this section with a nice property of the $(S)^{-1}$ space, the so-called Wick Calculus theorem, see theorem 12 in Kondratiev et. al. [KLS]:

**Theorem 2.8:** Let $\Psi \in (S)^{-1}$. Assume $f : \mathbb{C} \to \mathbb{C}$ is analytic in a neighborhood of $E[\Psi]$. Then

$$f(\tilde{\Psi}(z)) = \tilde{\Phi}(z)$$

is the Hermite Transform of an element $\Phi$ of $(S)^{-1}$. ∎

We remark that the definitions and results presented in the above language can be found in Holden et. al. [HLØUZ].

# 3    The Solution Of The Population Model

For simplicity we will assume that $N = 1$ in model (8). We also assume that $X_0 \in (S)^{-1}$ and that

$$E[X_0] > 0$$

Hermite transforming the stochastic equation (8) into an ordinary complex differential equation, and solving, we obtain the candidate

$$X_t = (1 + \Theta_0 \diamond \exp^\diamond(-rt - \alpha B_t))^{\diamond(-1)} \qquad (9)$$

where

$$\Theta_0 = (1 - X_0) \diamond X_0^{\diamond(-1)} = X_0^{\diamond(-1)} - 1 \qquad (10)$$

We have written $\exp^\diamond$ for the *Wick exponential*, i.e the element defined by

$$\exp^\diamond \Phi = \sum_{n=1}^\infty \frac{1}{n!} \Phi^{\diamond n}$$

(see theorem 2.8 above.) We show that $X_t$ is an element in $(S)^{-1}$:

Define

$$g(z) = z^{-1} - 1$$

Obviously, $g(z)$ is analytic in a neighborhood around $E[X_0] > 0$. Hence, $\Phi_0$ is an element of $(S)^{-1}$. Furthermore, define

$$f(z) = (1 + z)^{-1}$$

We have

$$E[\Phi_0 \diamond \exp^\diamond(-rt - \alpha B_t)] = (E[X_0]^{-1} - 1)e^{-rt}$$

When $0 < E[X_0] \le 1$ we have

$$0 \le (E[X_0]^{-1} - 1)e^{-rt} \le (E[X_0]^{-1} - 1)$$

and, when $E[X_0] > 1$,

$$(E[X_0]^{-1} - 1) \le (E[X_0]^{-1} - 1)e^{-rt} < 0$$

for all $t \ge 0$. In both cases is the expectation bounded away from $-1$ for all $t \ge 0$. Hence, there exist constants $q, \epsilon$ such that

$$f(\tilde{\Phi}_0(z) \exp(-rt - \alpha \tilde{B}_t(z)))$$

is analytic and bounded for $z \in B_q(0, \epsilon)$, for all $t \ge 0$. This implies by theorem 2.8 that $X_t$ is an element of $(S)^{-1}$ for all $t \ge 0$.

To show that $X_t$ is a solution of equation (8), we must prove that $X_t$ satisfies

$$\frac{dX_t}{dt} = (r + \alpha W_t) \diamond X_t \diamond (1 - X_t)$$

in $(S)^{-1}$. But by lemma 2.6, II this is equivalent with showing that

$$\frac{\tilde{X}_{t+h}(z) - \tilde{X}_t(z)}{h} \to (r + \alpha \tilde{W}_t(z))\tilde{X}_t(z)(1 - \tilde{X}_t(z))$$

pointwise boundedly for $z \in B_q(0, \epsilon)$ when $h \to 0$. This can be seen to hold by direct calculation: The Hermite transform of (9) is:

$$\tilde{X}_t(z) = (1 + \tilde{\Theta}_0(z)e^{-rt - \alpha \tilde{B}_t(z)})^{-1}$$

Hence, after some manipulation,

$$\frac{\tilde{X}_{t+h}(z) - \tilde{X}_t(z)}{h} = \left( \frac{1 - e^{-rh}}{h} e^{-\alpha \tilde{B}_t(z)} + e^{-rh} \frac{(e^{-\alpha \tilde{B}_t(z)} - e^{-\alpha \tilde{B}_{t+h}(z)})}{h} \right)$$

$$\times e^{-rt} \tilde{\Theta}_0(z) \tilde{X}_t(z) \tilde{X}_{t+h}(z)$$

We see that $\tilde{X}_{t+h}(z) \to \tilde{X}_t(z)$ pointwise boundedly for $z \in \mathbf{B}_q(0, \epsilon)$. By definition we have

$$\frac{d}{dt} e^{-\alpha \tilde{B}_t(z)} = \lim_{h \to 0} \frac{e^{-\alpha \tilde{B}_t(z)} - e^{-\alpha \tilde{B}_t(z)}}{h} = \alpha \tilde{W}_t(z)e^{-\alpha \tilde{B}_t(z)}$$

for every $z \in \mathbf{B}_q(0, \epsilon)$. Moreover, we can show that this convergence is bounded on $\mathbf{B}_q(0, \epsilon)$. Hence

$$\frac{\tilde{X}_{t+h}(z) - \tilde{X}_t(z)}{h} \to (r + \alpha \tilde{W}_t(z))\tilde{X}_t(z)(1 - \tilde{X}_t(z))$$

pointwise boundedly on $\mathbf{B}_q(0, \epsilon)$.

Since $\tilde{X}_t(z)$ is the unique solution of the Hermite transformed version of equation (8), it follows by injectivity of the Hermite transform that $X_t$ is unique. We have the conclusion:

**Theorem 3.1:** Assume $X_0 \in (S)^{-1}$ with $E[X_0] > 0$. Then

$$X_t = (1 + \Theta_0 \diamond \exp^\diamond(-rt - \alpha B_t))^{\diamond(-1)}$$

where

$$\Theta_0 = X_0^{\diamond(-1)} - 1$$

is the unique $(S)^{-1}$ solution of (8) with $N = 1$. ∎

118

# 4    Some Concluding Remarks

By the uniqueness of the Hermite Transform, the $(S)^{-1}$ element $X_t$ given in (9) and (10) has to coincide with the solution found by Lindstrøm et. al. [LØU] for constant initial conditions $X_0 = x \neq \frac{1}{2}$. As we have seen, the results above are worked out for general initial conditions

$$X_0 \in (S)^{-1}$$

where $E[X_0] > 0$. This means that for stochastic variables as initial conditions we have a solution as well. Note that the case of anticipating initial conditions is also included.

Lindstrøm et. al. [LØU] prove that their solution for constant initial conditions $0 < X_0 < 1, X_0 \neq \frac{1}{2}$, is a stochastic process. We will here give an argument which shows that $X_t$ is *not* an element of the Hida distribution space $(S)^*$ for such initial conditions. We let $X_0 = x > 0$, and allow $x = \frac{1}{2}$:

In Hida et. al. [HKPS] the S-transform of an element of $(S)^*$ is defined as

$$SF(\xi) = \langle\langle F, \exp^\circ W_\xi\rangle\rangle$$

for $\xi \in S(\mathbb{R})$. The S-transform of $X_t$ given in (9) and (10) is

$$SX_t(\xi) = (1 + (\frac{1}{x} - 1)\exp(-rt - \alpha \int_0^t \xi(s)ds))^{-1}$$

This object is well defined for all $\xi \in S(\mathbb{R})$, and $v_t = SX_t(\xi)$ is the unique solution of the problem

$$v_t = x + r\int_0^t v_s(1 - v_s)ds + \alpha \int_0^t v_s(1 - v_s)\xi(s)ds$$

for each $\xi$. However, $SX_t(\xi)$ can *not* be extended to an analytic function

$$z \to SX_t(z\xi)$$

on the complex plane $\mathbb{C}$. Hence, $X_t$ is not an element of $(S)^*$. (See Hida et. al. [HKPS] for a characterization of $(S)^*$-elements in terms of the S-transform.) We see that for $\xi, \eta \in S(\mathbb{R})$ the mapping

$$\lambda \to SX_t(\xi + \lambda\eta)$$

can only be analytic in a *neighborhood* of zero in $\mathbb{C}$. This tells us that $X_t$ is not contained in any of the spaces $(S)^{-\rho}$, $\rho \in [0,1)$. (See Albeverio et. al. [ADKS] for the characterization of these spaces by the S-transform.) Hence, in order to give a rigorous treatment of equation (1) which covers the intial condition $X_0 = \frac{1}{2}$, we have to work in the space of Kondratiev distributions $(S)^{-1}$.

**Acknowledgements:** This work has been supported by the Norwegian Research Council(NFR) under grant (NAVF)100549/410. The author would like to thank professor Bernt Øksendal for fruitful suggestions and inspiring discussions. Professor Yuri G. Kondratiev is acknowledged for encouraging talks.

# References

[ADKS] S.Albeverio, Yu. L. Daletsky, Yu. G. Kondratiev and L. Streit: *Non-Gaussian Infinite Dimensional Analysis*; Preprint, University of Bielefeld (1994)

[B] F. E. Benth: *Integrals in the Hida Distribution Space $(S)^*$*; In T.Lindstrøm, B.Øksendal and A.S.Ustunel (eds): Stochastic Analysis and Related Topics. Vol. 8, Gordon and Breach (1993)

[GHLØUZ] H. Gjessing, H. Holden, T. Lindstrøm, B. Øksendal, J. Ubøe and T.-S. Zhang: *The Wick Product*; In H.Niemi, G.Högnäs, A.N.Shiryaev and A.Melnikov (eds.): Frontiers in pure and applied mathematics Vol 1, TVP Publishers, Moscow (1993)

[HKPS] T. Hida, H.-H. Kuo, J. Potthoff and L. Streit: *White Noise -An Infinite Dimensional Calculus*; Kluwer Academic Publishers (1993)

[HLØUZ] H. Holden, T. Lindstrøm, B. Øksendal, J. Ubøe and T. -S. Zhang: *The Pressure Equation for Fluid Flow in a Stochastic Medium*; Preprint, University of Oslo (1994)

[K] Yu. G. Kondratiev: *Generalized Functions in Problems of Infinite Dimensional Analysis*; Ph. D. Thesis, Kiev University, Ukraine, (1978)

[KLS] Yu. G. Kondratiev, P. Leukert and L. Streit: *Wick Calculus in Gaussian Analysis*; Preprint, University of Bielefeld, (1994)

[LØU] T.Lindstrøm, B.Øksendal, J. Ubøe: *Wick Multiplication And Ito-Skorohod Stochastic Differential Equations*; In S.Albeverio et.al. (eds.): Ideas and Methods in Mathematic Analysis. Cambridge University Press (1992)

[Z] T.-S.Zhang: *Characterizations of White Noise Test Functions and Hida Distributions*; Stochastics and Stochastics Reports **41**, (1992)

# A Generalized Feynman-Kac Formula for the Stochastic Heat Problem with Anticipating Initial Conditions

Fred Espen Benth
University of Oslo, N-0316 Oslo, Norway
&
University of Mannheim A5, D-68131 Mannheim, Germany

### Abstract

Using White Noise Analysis, we construct a Feynman-Kac formula for the stochastic heat equation with anticipating initial conditions. The obtained solution has applications to nonlinear filtering and heat transport with noise.

## 1   Introduction.

We will in this paper study a partial differential equation of parabolic type disturbed by time-dependent white noise. Suggestively, we write

$$\frac{\partial u}{\partial t}(t,x,\omega) = L_x u(t,x,\omega) + h(t,x)u(t,x,\omega)W_t(\omega)$$

$W_t(\omega)$ is white noise in time, and $L_x$ is a second order uniformly elliptic partial differential operator. Time will be supposed to run over a finite intervall $[0,T]$, and the space variable $x$ is in $\mathbb{R}^d$. An Ito interpetation of the above equation is

$$du(t,x,\omega) = L_x u(t,x,\omega)dt + h(t,x)u(t,x,\omega)dB_t(\omega) \qquad (1)$$

where $dB_t$ denotes Ito integration.

(1) has many interesting applications. In nonlinear filtering, the equation is known as the Zakai-equation, see Zakai [Z], and models the so called unnormalized conditional probability density. As a physical model, (1) can be considered as heat transport in a medium, where we have a stochastic source or potential. Stochastic heat problems in the above two contexts have been considered by many different authors. We will only mention a few works: From a filtering point of view: Pardoux [Pa], Krylov and Rozovskii [KR], Gyöngy and Krylov [GK], Léandre and Russo [LR] and Benth [B2] have studied (1). For problem (1) considered as a physical model, we quote the papers of Bertini and Cancrini [BC], Holden et. al. [HLØUZ] and Benth [B3].

The purpose of this paper is to present a (generalized) Feynman-Kac formula for (1) using White Noise Calculus. This means that the problem of solving (1) will be translated by the $S$-transform into a determinstic problem. In section 3 we will see that the deterministic problem becomes a heat equation with a (deterministic) potential.

Moreover, we can represent its solution with a Feynman-Kac formula. After showing that this solution has the correct properties, we can apply the invers $S$-transform to obtain a solution for our original stochastic problem. Such a procedure for studying stochastic partial differential equations has been followed by many authors. See for instance the works by Potthoff [P], Lindstrøm et. al. [LØU2] and Benth [B2].

In White Noise Analysis one can show a nice property of the singular white noise $W_t$ in connection with Ito/Skorohod integration. If $\delta B_t$ denotes Skorohod integration, and $\diamond$ the Wick product, one can show that

$$\int_0^t X_s \diamond W_s ds = \int_0^t X_s \delta B_s$$

for Skorohod integrable processes $X_s$. For a discussion of this relation, see Hida et al. [HKPS], Lindstrøm et. al. [LØU2] and Benth [B1]. A definition of the Skorohod integral is given in Nualart and Zakai [NZ]. We here only remark that in the case of an Ito integrable process $X_s$, the Skorohod and Ito integral coincide. Using the above representation of the Ito/Skorohod integral, we can rewrite (1) as

$$\frac{\partial u}{\partial t}(t,x,\omega) = L_x u(t,x,\omega) + h(t,x)u(t,x,\omega) \diamond W_t(\omega) \tag{2}$$

In Benth [B2], (2) was considered with a "smooth" noise. I. e. , instead of $W_t$ in (2), a smooth version was considered. Such a smoothing procedure has physical reasons, see Lindstrøm et. al. [LØU1].

We remark that our method allows for anticipating initial conditions.

The paper is organized as follows: In section 2 we present some theory from the White Noise Analysis. The solution of (2) is constructed and studied in section 3. Finally, in section 4, we consider some applications of our results.

# 2 Mathematical Preliminaries.

In this section some of the basic elements of White Noise Analysis are introduced. For a more complete account, see the book by Hida et.al. [HKPS].

Let $S(\mathbb{R})$ be the Schwartz space of rapidly decreasing functions, and denote by $S'(\mathbb{R})$ its dual, which is the space of tempered distributions. Equip $S'(\mathbb{R})$ with the $\sigma$-algebra $\mathcal{B}$ generated by the cylinder sets in $S'(\mathbb{R})$. By the Bochner-Minlos theorem, we define a unique measure $\mu$ on $(S'(\mathbb{R}),\mathcal{B})$ via its characteristical functional

$$\int_{S'(\mathbb{R})} \exp(i\langle\omega,\xi\rangle)d\mu(x) = \exp(-\frac{1}{2}|\xi|_2^2) \tag{3}$$

$<\cdot,\cdot>$ is the dual pairing between $S'(\mathbb{R})$ and $S(\mathbb{R})$, and $|\cdot|_2$ is the norm in $L^2(\mathbb{R})$. $(S'(\mathbb{R}),\mathcal{B},\mu)$ becomes a probability space.

We will denote the space $L^2(S'(\mathbb{R}),\mathcal{B},\mu)$ by $(L^2)$ and its norm by $\|\cdot\|_2$. For each $\xi \in S(\mathbb{R})$ we define the coordinate process

$$W_\xi(\omega) = \langle\omega,\xi\rangle$$

which belongs to $(L^2)$. These coordinate processes can be extended to $\xi \in L^2(\mathbb{R})$ by choosing a sequence $\{\xi_n\}_n \subset S(\mathbb{R})$ which converges to $\xi$ in $L^2(\mathbb{R})$, and using a standard limiting argument. In particular, $\xi = \mathbf{1}_{[0,t)}$ defines a Brownian motion denoted

$$B_t(\omega) = \langle \omega, \mathbf{1}_{[0,t)} \rangle$$

We proceed with defining the Hida test function space and the space of Hida distributions. Consider the self-adjoint extension of the harmonic Oscillator $H$ on $L^2(\mathbb{R})$, and let $\Gamma(H)$ be its second quantization. Let $\mathcal{P}$ denote the algebra of polynomials in $(L^2)$, i. e. the algebra generated by the coordinate processes. We define for $p \in \mathbb{R}$ the norm $\| \cdot \|_{2,p}$ on $\mathcal{P}$ by

$$\|\phi\|_{2,p} := \|\Gamma(H)^p \phi\|_2 .$$

Denote by $(S)_p$ the completion of $\mathcal{P}$ in this norm. The space of Hida test functions, $(S)$, is now defined as the projective limit of the $(S)_p$ spaces. The dual space, the space of Hida distributions, is denoted $(S)^*$ and is equal to the union of the duals of $(S)_p$, $p \in \mathbb{N}_0$. The dual pairing between $(S)^*$ and $(S)$ will be denoted by $\langle\langle \Phi, \phi \rangle\rangle$, $\Phi \in (S)^*$, $\phi \in (S)$.

With the spaces $(S)$ and $(S)^*$ at hand, we are in the position to define the $S$-transform. The $S$-transform maps an element in $(S)^*$ into a complex valued function on $S(\mathbb{R})$ in the following way: Let $\Phi \in (S)^*$, $\xi \in S(\mathbb{R})$,

$$S\Phi(\xi) = \langle\langle \Phi, \mathrm{Exp}W_\xi \rangle\rangle \tag{4}$$

where

$$\mathrm{Exp}W_\xi(\omega) = e^{\langle \omega, \xi \rangle - \frac{1}{2}|\xi|_2^2}$$

In order to say something more about the $S$-transform, we define the notion of *U-functional*.

**Definition 1** Assume that $F$ is an everywhere defined, complex valued function on $S(\mathbb{R})$. If $F$ satisfies the following two properties it is called a *U-functional*:

**(A)** $F$ has an entire analytic extension in the following sense: for every $\eta, \xi \in S(\mathbb{R})$, the function $\lambda \to F(\eta + \lambda\xi)$ defined on the real line has an entire analytic extension to $\mathbb{C}$. This extension is denoted by $z \to F(\eta + z\xi)$, $z \in \mathbb{C}$.

**(B)** For every $\xi \in S(\mathbb{R})$, the entire function $z \to F(z\xi)$ is at most of order 2, i. e., there exist constants $K_1, K_2 > 0$ and $p \in \mathbb{R}$, so that for all $z \in \mathbb{C}, \xi \in S(\mathbb{R})$, the following bound holds:

$$|F(z\xi)| \leq K_1 e^{K_2 |z|^2 |\xi|_{2,p}^2} \tag{5}$$

where $|\xi|_{2,p} = |H^p \xi|_2$.

The $S$-transform characterizes the space $(S)^*$. See Potthoff and Streit [PS], Hida et. al. [HKPS] and Kondratiev et. al. [KLPSW] for a proof of the following characterization theorem:

**Theorem 2** Assume that $\Phi \in (S)^*$, then its $S$-transform is a U-functional. Conversely, if $F$ is a U-functional, then there exists a unique element $\Phi \in (S)^*$ such that $S\Phi = F$.

In Timpel and Benth, [TB], it is shown that there is a natural topology on the space of U-Functionals such that the $S$-transform is a homeomorphism.

Consider $\Phi, \Psi \in (S)^*$ with $S$-transforms $F, G$ respectively. Then the pointwise product of $F$ and $G$ is again a U-functional. Hence, there exists an element in $(S)^*$ denoted by $\Phi \diamond \Psi$ whose $S$-transform is $F \cdot G$, i. e.,

$$S(\Phi \diamond \Psi)(\xi) = F(\xi) \cdot G(\xi) \tag{6}$$

$\Phi \diamond \Psi$ is called the *Wick product* of $\Phi$ and $\Psi$. If we define the *Wick exponential* of a coordinate process $W_\phi$ to be

$$\mathrm{Exp}W_\phi = \sum_{n=0}^{\infty} \frac{1}{n!} W_\phi^{\diamond n} \tag{7}$$

one can show, using the $S$-transform, that

$$e^{W_\phi - \frac{1}{2}|\phi|^2} = \mathrm{Exp}W_\phi$$

as elements in $(S)^*$.

One has the following important convergence result for sequences in $(S)^*$.

**Theorem 3** Assume $\{F_n, n \in \mathbb{N}\}$ is a sequence of U-functionals, such that for every $\xi \in S(\mathbb{R})$, the sequence $\{F_n(\xi), n \in \mathbb{N}\}$ is Cauchy. If the bound (5) holds for all $F_n$ uniformly in $n$, then there exists a unique element $\Phi \in (S)^*$, such that $\Phi_n = S^{-1}F_n$ converges strongly to $\Phi$.

Integration in $(S)^*$ is defined in the following way:

**Theorem 4** Let $(X, \nu)$ be a measure space, and $x \to \Phi(x)$ be a mapping from $X$ into $(S)^*$. If $\langle\langle \Phi(x), \phi \rangle\rangle \in L^1(X, \nu)$ for every $\phi \in (S)$, then the *Pettis integral*

$$\int_X \Phi(x)dx$$

defined as $\phi \to \int_X \langle\langle \Phi(x), \phi \rangle\rangle dx$ belongs to $(S)^*$. In particular, for every $\phi \in (S)$

$$\langle\langle \int_X \Phi(x)dx, \phi \rangle\rangle = \int_X \langle\langle \Phi(x), \phi \rangle\rangle dx$$

Note that the $S$-transform and the Pettis integral commute:

$$S(\int_X \Phi(x)dx)(\xi) = \int_X S\Phi(x)(\xi)dx$$

We can also characterize integrability in $(S)^*$ through the $S$-transform:

**Theorem 5** Let $(X, \nu)$ be a measure space, and $x \to \Phi(x)$ be a mapping from $X$ into $(S)^*$. We assume that the $S$-transform $F_x = S\Phi(x)$ satisfies the following conditions:

**1** For every $\xi \in S(\mathbb{R})$ the mapping $x \to F_x(\xi)$ is measurable,

**2** There exists a $p \in \mathbb{N}_0$ such that for all $x \in X, F_x$ satisfies the bound

$$|F_x(z\xi)| \leq C_x e^{K_x|z|^2|\xi|_2^2}$$

where $x \to K_x$ is bounded $\nu$-a.e., and $x \to C_x$ is integrable with respect to $\nu$.

Then there exists a $q \in I\!N_0$ such that $\Phi(\cdot)$ is Bochner integrable on $(S)_{-q}$. In particular,

$$\int_X \Phi(x)d\nu(x) \in (S)^*$$

and

$$S\left(\int_X \Phi(x)d\nu(x)\right)(\xi) = \int_X S\Phi(x)(\xi)d\nu(x)$$

For the proof of this theorem, see Hida et. al. [HKPS] and Kondratiev et. al. [KLPSW].

We close this section with the definition of the *singular white noise*: The singular white noise is the element $W_t$ defined through the $S$-transform as

$$SW_t(\xi) = \xi(t) \tag{8}$$

By the characterization theorem we see that $W_t \in (S)^*$.

# 3 Construction of the Solution.

Consider (2), i. e.,

$$\phi_t(x,\omega) = \phi_0(x,\omega) + \int_0^t L_x \phi_s(x,\omega)ds + \int_0^t h(s,x)\phi_s(x,\omega) \diamond W_s(\omega)ds \tag{9}$$

where

$$L_x = \sum_{i=1}^d b_i(x)\frac{\partial}{\partial x_i} + \sum_{i,j=1}^d \frac{1}{2}\sigma_i\sigma_j(x)\frac{\partial^2}{\partial x_i \partial x_j} \tag{10}$$

$L_x$ is assumed to be uniformly elliptic. Hence, $L_x$ is the infinitesimal generator to the Ito diffusion

$$dX_t = b(X_t)dt + \sigma(X_t)dU_t$$

$U_t$ is a d-dimensional Brownian motion. We remark that $X_t$ is written on a compact vector form.

We state the conditions on $\phi_0$ and $h$:

(I) $\phi_0(x,\cdot) \in (S)^*$ for all $x \in I\!R^d$. Moreover, we assume that the mapping

$$x \to S\phi_0(x;\xi)$$

is in $C_b^2(I\!R^d)$ for each $\xi \in S(I\!R)$.

(II) For each multi-index $\alpha = (\alpha_1, \ldots, \alpha_d)$, where $|\alpha| \le 2$, we assume that

$$|\frac{\partial^{|\alpha|}}{\partial^{\alpha_1}x_1 \cdots \partial^{\alpha_d}x_d}S\phi_0(x;z\xi)| \le K_1^{(\alpha)}(x)e^{K_2^{(\alpha)}(x)|z|^2|\xi|^2_{2,q(\alpha)}}$$

where $K_2^{(\alpha)}$ is bounded $x$-a.e., and $K_1^{(\alpha)}(\cdot) \in L^p(I\!R^d)$ for a $p \in [1,\infty]$ and $q^{(\alpha)} \in I\!N_0$.

(III) $h$ is bounded continuous in both variables. In addition $\frac{\partial h}{\partial t}$ exists and is integrable on $[0,T]$.

If we informally apply the $S$-transform to the above stochastic partial differential equation, we obtain

$$S\phi_t(x)(\xi) = S\phi_0(x;\xi) + \int_0^t L_x S\phi_s(x)(\xi)ds + \int_0^t h(s,x)\cdot\xi(s)\cdot S\phi_s(x)(\xi)ds \quad (11)$$

where, as usual, $\xi \in S(\mathbb{R})$. For a fixed $\xi$, we recognize (11) as a heat transport problem on the form

$$\frac{\partial u_t}{\partial t} = L_x u_t + h(t,x)\xi(t)u_t$$

$$u_0(x) = S\phi_0(x;\xi)$$

There exists a Feynman-Kac formula to this problem. Moreover, we have the proposition:

**Proposition 6** Assume $k(t,x) : [0,T] \times \mathbb{R}^n \to \mathbb{R}$ is bounded continuous. Furthermore, assume that $\frac{\partial k}{\partial t}$ exists and is integrable in $t$ on $[0,T]$. Then

$$u(t,x) = E^x\left[f(X_t)\exp(-\int_0^t k(t-s,X_s)ds)\right] \quad (12)$$

solves the problem

$$\frac{\partial u_t}{\partial t} = L_x u_t - k(t,x)u_t \quad (13)$$

$$u_0(x) = f(x) \quad (14)$$

where $f \in C_0^2(\mathbb{R}^n)$. $E^x$ denotes the expectation with respect to the diffusion $X_t$ where $X_0 = x$.

**Proof:** By condition I, it can be shown by Ito's formula that the integrand in $E[\cdot]$ can be written as a stochastic process. Hence, $u(t,x;\xi)$ will be differentiable with respect to $t$.

By definition we have

$$L_x u(t,x) = \lim_{v\downarrow 0}\frac{1}{v}\{E^x\left[u(t,X_v)\right] - u(t,x)\} \quad (15)$$

Using the Markov Property, a direct calculation yields

$$E^x\left[u(t,X_v)\right] = E^x\left[E^{X_v}\left[f(X_t)\exp(-\int_0^t k(t-s,X_s)ds)\right]\right]$$

$$= E^x\left[E^x\left[f(X_{t+v})\exp(-\int_0^t k(t-s,X_{s+v})ds)\,|\mathcal{F}_v\right]\right]$$

$$= E^x\left[f(X_{t+v})\exp(-\int_0^t k(t-s,X_{s+v})ds\right]$$

By a change of variables in $\int_0^t k(t-s,X_{s+v})ds$ we get

$$\int_0^t k(t-s,X_{s+v})ds = \int_v^{t+v} k(t-(\theta-v),X_\theta)d\theta$$

127

$$= \int_0^{t+v} k(t+v-\theta, X_\theta)d\theta - \int_0^v k(t+v-\theta, X_\theta)d\theta$$

Inserted in (15) we obtain

$$L_x u(t,x) = \lim_{v\downarrow 0} \frac{1}{v} \left\{ E^x \left[ f(X_{t+v}) \exp(-\int_0^{t+v} k(t+v-s, X_s)ds) \times \right. \right.$$

$$\left. \left. \exp(\int_0^v k(t+v-s, X_s)ds) \right] - u(t,x) \right\}$$

Hence,

$$= \lim_{v\downarrow 0} \frac{1}{v} \{u(t+v,x) - u(t,x)\} +$$

$$\lim_{v\downarrow 0} \frac{1}{v} \left\{ E^x \left[ f(X_{t+v}) \exp(-\int_0^{t+v} k(t+v-s, X_s)ds) \exp(\int_0^v k(t+v-s, X_s)ds) \right] \right.$$

$$\left. - E^x \left[ f(X_{t+v}) \exp(-\int_0^{t+v} k(t+v-s, X_s)ds) \right] \right\}$$

$$= \frac{\partial u}{\partial t}(t,x) + E^x \left[ f(X_t) \exp(-\int_0^t k(t-s, X_s)ds) \times \right.$$

$$\left. \lim_{v\downarrow 0} \frac{1}{v} \left\{ \exp(\int_0^v k(t+v-s, X_s)ds) - 1 \right\} \right]$$

To treat the last limit, define

$$g(\theta) = \exp(\int_0^\theta k(t+\theta-s, X_s)ds)$$

The derivative in zero for this function is

$$g'(0) = \lim_{v\to 0} \frac{1}{v} \{\exp(\int_0^v k(t+v-s, X_s)ds) - 1\}$$

Differentiation gives

$$g'(\theta) = g(\theta)\{k(t, X_\theta) - \int_0^\theta \frac{\partial}{\partial\theta} k(t+\theta-s, X_s)ds\}$$

Hence,

$$g'(0) = k(t,x)$$

which completes the proof. ∎
Let $k(t,x)$ be given by

$$k(t,x) = -\xi(t)h(t,x)$$

which satisfies the conditions on $k$ in proposition 6. Consider the following function

$$\phi_t(x,\omega) = E^x \left[ \phi_0(X_t,\omega) \diamond \text{Exp}(\int_0^t h(s, X_{t-s})W_s(\omega)ds) \right] \tag{16}$$

We will show that $\phi_t(x,\omega) \in (S)^*$ and is a solution of (9):

**Proposition 7** Let $\phi_0$ and $h$ satisfy the conditions I-III. Then $\phi_t$ given in (16) is an element of $(S)^*$. Moreover, the $S$-transform of $\phi_t$ is given by

$$S\phi_t(x)(\xi) = E_{\tilde{\omega}}^x\left[S\phi_0(X_t(\tilde{\omega});\xi)\exp(\int_0^t h(s,X_{t-s}(\tilde{\omega}))\xi(s)ds)\right]$$

**Proof:** By theorem 5, it suffices to prove that

$$|S\phi_0(X_t(\tilde{\omega});z\xi)\exp(\int_0^t h(s,X_{t-s}(\tilde{\omega}))z\xi(s)ds)| \leq C_1(\tilde{\omega})e^{C_2(\tilde{\omega})|z|^2|\xi|_{2,q}^2}$$

for a $q \in \mathbb{N}_0$, where $C_2$ is bounded $\mu$-a.e., and $C_1$ is $\mu$-integrable. By condition II,

$$|S\phi_0(X_t(\tilde{\omega});z\xi)\exp(\int_0^t h(s,X_{t-s}(\tilde{\omega}))z\xi(s)ds)|$$

$$\leq K_1(X_t(\tilde{\omega}))e^{K_2(X_t(\tilde{\omega}))|z|^2|\xi|_{2,q}^2} \cdot e^{|h|_\infty|z||\xi|_\infty t}$$

$$\leq K_1(X_t(\tilde{\omega}))e^{\frac{1}{2}|h|_\infty Ct} \cdot e^{(K_2(X_t(\tilde{\omega}))+\frac{1}{2}Ct|h|_\infty)|z|^2|\xi|_{2,r}^2}$$

Since $K_2$ is bounded for a.e. x, we have that

$$C_2(\tilde{\omega}) = K_2(X_t(\tilde{\omega})) + \frac{1}{2}Ct|h|_\infty$$

is bounded $\mu$-a.s. Let

$$C_1(\tilde{\omega}) = K_1(X_t(\tilde{\omega}))e^{\frac{1}{2}Ct|h|_\infty}$$

With $p_t(x,y)$ as the density for the diffusion $X_t$, we get by the assumption $K_1 \in L^p(\mathbb{R}^d)$:

$$\int_{S'(\mathbb{R})} K_1(X_t(\tilde{\omega}))d\mu(\tilde{\omega}) = \int_{\mathbb{R}^d} K_1(y)p_t(x,y)dy < \infty$$

for a.e. $(t,x) \in [0,T] \times \mathbb{R}^d$. Hence $C_1$ belongs to $(L^1)$ for a.e. $(t,x)$. From theorem 5 it then follows that $\phi_t \in (S)^*$ and that

$$S\phi_t(x)(\xi) = E^x\left[S\phi_0(X_t;\xi)\exp(\int_0^t h(s,X_{t-s})\xi(s)ds)\right]$$

∎

We proceed with showing that $\phi_t(x,\omega)$ given in (16) solves (9): Using theorem 5, we see that

$$\int_0^t h(s,x)S\phi_s(x)(\xi) \cdot \xi(s)ds = \int_0^t h(s,x)S\phi_s(x)(\xi) \cdot SW_s(\xi)ds$$

$$= \int_0^t h(s,x)S(\phi_s(x) \diamond W_s)(\xi)ds = S(\int_0^t h(s,x)\phi_s(x) \diamond W_s ds)(\xi)$$

We must show that the infinitesimal generator $L_x$ commutes with the $S$-transform. This is proved in the following proposition:

**Proposition 8**

$$S(L_x\phi_t(x))(\xi) = L_xS(\phi_t(x))(\xi) \tag{17}$$

**Proof:** What we must prove, is that

$$S\left(\lim_{v\downarrow 0}\frac{1}{v}\left\{E^x\left[\phi_t(X_v)\right]-\phi_t(x)\right\}\right)(\xi)=\lim_{v\downarrow 0}\frac{1}{v}\left\{E^x\left[S\phi_t(X_v)(\xi)\right]-S\phi_t(x)(\xi)\right\}$$

By the proof of prop.(6) the latter limit exists for each $\xi$. We must show that we have an exponential bound on the right hand side: From the proof of prop. (6), we can write the right hand side as

$$\frac{u(t+v,x;\xi)-u(t,x;\xi)}{v}$$

$$+E^x\left[u_0(X_{t+v};\xi)e^{\int_0^{t+v}k(t+v-s,X_s)ds}\cdot\frac{1}{v}(e^{-\int_0^v k(t+v-s,X_s)ds}-1)\right]$$

By Ito's formula, we have

$$u(t,x;\xi)=u_0(x;\xi)+\int_0^t E^x[Y(s,X_s;\xi)]ds$$

where $Y(s,x;\xi)$ depends on $b$, $\sigma$ and $\partial_\alpha u_0$ for $|\alpha|\leq 2$. Hence, by condition II and III we obtain a uniform exponential bound in $\xi$ using the mean value theorem. Th. (3) then gives the proposition. ∎

We conclude what we have proved so far:

**Theorem 9** Let $\phi_0$ and $h$ satisfy conditions I-III. Then

$$\phi_t(x,\omega)=E^x\left[\phi_0(X_t,\omega)\diamond\mathrm{Exp}(\int_0^t h(s,X_{t-s})W_s(\omega)ds)\right]$$

is a $(S)^*$ solution of (9).

We remark that

$$\int_0^t h(s,X_{t-s})W_s(\omega)ds=\int_0^t h(s,X_{t-s})dB_s(\omega)$$

Note that in the Ito integral, $h$ is not dependent on $\omega$. If we define the function

$$\gamma_{t,\tilde{\omega}}(s)=1_{[0,t)}(s)h(s,X_{t-s}(\tilde{\omega}))$$

we can see that the stochastic integral actually is a coordinate process

$$\int_0^t h(s,X_{t-s}(\tilde{\omega}))W_s(\omega)ds=\langle\omega,\gamma_{t,\tilde{\omega}}\rangle$$

Whence, we have the following representation of the solution (16):

$$\phi_t(x,\omega)=E^x\left[\phi_0(X_t,\omega)\diamond\exp(\int_0^t h(s,X_{t-s})dB_s(\omega)-\frac{1}{2}\int_0^t h^2(s,X_{t-s})ds)\right]\quad(18)$$

# 4   Applications.

We end this paper by looking at some applications of the results developed in the preceeding section. The stochastic heat transport problem and nonlinear filtering will be discussed. However, we start with considering initial data:

Recall the boundedness condition of the $S$-transform of $\phi_0(x,\omega)$, condition II in section 3. At first sight, such a condition might look difficult to verify, since one actually has to perform the $S$-transform. However, for a large class of initial data, this condition holds: Assume $\tilde{\phi}_0(\cdot) \in C_b^2(\mathbb{R}^d) \cap L^p(\mathbb{R}^d)$. If $\Phi(\omega) \in (S)^*$, we have that the initial data

$$\phi_0(x,\omega) = \tilde{\phi}_0(x) \cdot \Phi(\omega)$$

satisfies condition II. This is easily seen by application of the $S$-transform:

$$S\phi_0(x;\xi) = \tilde{\phi}_0(x) \cdot S\Phi(\xi)$$

We turn our attention to deterministic initial conditions. This special choice of initial data has interest in nonlinear filtering. In fact, in nonlinear filtering theory, the initial data is a $C_0^\infty$ function. The following lemma holds true:

**Lemma 10** Let $h$ satisfy condition II. If $\phi_0$ is deterministic, where $\phi_0 \in C_0^\infty(\mathbb{R}^d)$, then $\phi_t$ is adapted and

$$\int_0^T E[\phi_t^2(x,\cdot)]dt < \infty$$

**Proof:** Using the Hölder inequality, we obtain

$$E_\omega\left[\phi_t(x,\omega)^2\right] \le E_\omega\left[E_{\tilde{\omega}}^x\left[\phi_0^2(X_t(\tilde{\omega}))(\text{Exp}(\int_0^t h(s,X_{t-s}(\tilde{\omega}))dB_s(\omega)))^2\right]\right]$$

$$= E_{\tilde{\omega}}^x\left[\phi_0^2(X_t(\tilde{\omega}))E_\omega\left[(\text{Exp}(\int_0^t h(s,X_{t-s}(\tilde{\omega}))dB_s(\omega)))^2\right]\right] \le K \cdot t < \infty$$

by the boundedness of $h$ and $\phi_0$.

$\int_0^t h(s,X_{t-s})dB_s(\omega)$ is adapted. This implies that $\phi_t$ is adapted, since it is the limit of sums and products of adapted processes. Hence, the lemma follows. ∎

**The Heat Equation:**   The heat problem with potential or source is to find a function $u_t(x)$ which satisfies the partial differential equation

$$\frac{\partial u_t}{\partial t} = \frac{1}{2}\Delta_x u_t + h(t,x)u_t \tag{19}$$

$$u_0(x) = f(x) \tag{20}$$

where $f$ is some "nice" initial function. $\Delta$ is the Laplace operator. If the system has a stochastic potential, a possible model could be

$$\frac{\partial u_t}{\partial t}(x,\omega) = \frac{1}{2}\Delta_x u_t + h(t,x)u_t(x,\omega) \diamond W_t(\omega) \tag{21}$$

$$u_0(x) = f(x,\omega) \tag{22}$$

We have allowed the initial condition $f$ to be stochastic (and possibly anticipating). The integral formulation of this problem is

$$u_t(x,\omega) = f(x,\omega) + \frac{1}{2}\int_0^t \Delta_x u_s(x,\omega)ds + \int_0^t h(s,x)u_s(x,\omega)\delta B_s(\omega)$$

¿From the theory developed in section 3, the solution is

$$u_t(x,\omega) = E^x\left[f(\mathbf{x}_t,\omega) \diamond \text{Exp}\{\int_0^t h(s,\mathbf{x}_{t-s})W_s(\omega)ds\}\right] \qquad (23)$$

where $\mathbf{x}_t$ is a d-dimensional Brownian motion. An equivalent representation is

$$u_t(x,\omega) = E^x\left[f(\mathbf{x}_t,\omega) \diamond \exp(\int_0^t h(s,\mathbf{x}_{t-s})dB_s(\omega) - \frac{1}{2}\int_0^t h^2(s,\mathbf{x}_{t-s})ds)\right] \qquad (24)$$

It is well known that the deterministic heat problem with potential is positivity preserving. That means, if the initial data is positive, the solution is positive. An interesting question is if this holds for the heat problem with stochastic potential and initial condition. We see that if the initial condition is deterministic, positivity is obviously preserved. For the case of stochastic initial condition, this is not so easy to see, since the solution roughly is a Wick product between two random variables. The Wick product is unfortunately not in general positivity preserving. See Kuo [K] for a counterexample. However, in Benth [B3], it is proved that the above solution is positive in the case of positive random initial data. Moreover, for sufficiently smooth $f$, that means $f(x,\cdot) \in (L^p), p > 2$, the solution can be written

$$u_t(x,\omega) = E^x\left[f(\mathbf{x}_t,\omega - 1_{[0,t)}(\cdot)h(\cdot,\mathbf{x}_{t-.}))\times\right.$$

$$\left.\exp(\int_0^t h(s,\mathbf{x}_{t-s})dB_s(\omega) - \frac{1}{2}\int_0^t h^2(s,\mathbf{x}_{t-s})ds)\right]$$

**Nonlinear Filtering:** In nonlinear filtering, the famous Zakai equation models *the unnormalized conditional probabilty density*, $\phi_t(x,\omega)$, see Zakai [Z]:

$$d\phi_t(x,\omega) = A_x^*\phi_t(x,\omega) + h(t,x)\phi_t(x,\omega)dY_t(\omega) \qquad (25)$$

$$\phi_0(x,\omega) = p_0(x) \qquad (26)$$

$Y_t$ is the observation process of the filtering problem, which in the context of filtering theory is a standard Brownian motion. $A_x^*$ is the adjoint of the infinitesimal generator associated to the system process

$$dZ_t = b(Z_t)ds + \sigma(Z_t)dU_t$$

In addition, $p_0(x)$ is the probability density of $Z_0$. Of course, (25) is interpreted in the Ito-sense.

Using the theory developed above, we can state an explicit solution to this problem. However, since the Brownian motion $B_t$ used in our context is the "white noise Brownian motion", see section 2, the solution has to be interpreted as a *weak* one.

Restricting ourselves to one space dimension, i. e. $d = 1$, $A_x$ considered as a differential operator equals

$$A_x = b(x)\frac{d}{dx} + \frac{1}{2}\sigma^2(x)\frac{d^2}{dx^2}$$

132

Under appropriate differential hypotheses on $b, \sigma$ this yields

$$A_x^* = (\rho(x)\frac{d}{dx} + \frac{1}{2}\sigma^2(x)\frac{d^2}{dx^2}) - c(x) = L_x - c(x)$$

where

$$c(x) = b^{'}(x) - (\sigma^{'}(x))^2 - \sigma(x)\sigma^{''}(x)$$

and

$$\rho(x) = 2\sigma(x)\sigma^{'}(x) - b(x)$$

Hence, (25) and (26) can be formulated

$$\phi_t(x,\omega) = p_0(x) + \int_0^t \{L_x\phi_s(x,\omega) - c(x)\phi_s(x,\omega)\}ds + \int_0^t h(s,x)\phi_s(x,\omega)dY_s(\omega) \quad (27)$$

We substitute $Y_t$ with $B_t$, and use the relation

$$\int_0^t \phi_s(x,\omega)dB_s(\omega) = \int_0^t \phi_s(x,\omega) \diamond W_s ds$$

which holds for Ito-integrable processes $\phi_s$, see section 1. By slightly extending the results from section 3, we obtain the weak solution

$$\phi_t(x,\omega) = E^x\left[p_0(X_t)\exp\{-\int_0^t c(X_s)ds\}\text{Exp}\{\int_0^t h(s,X_{t-s})dB_s\}\right] \quad (28)$$

Since we have smooth initial data, the solution $\phi_t$ is an Ito integrable ($L^2$) process.

**Acknowledgements:** The author would like to thank professor Bernt Øksendal and professor Jürgen Potthoff for inspiring and fruitful talks. This work has been supported by The Norwegian Research Council (NAVF; 100549/410).

# References

[B1]  F. E. Benth: *Integrals In The Hida Distribution Space* $(S)^*$. In T. Lindstrøm, B. Øksendal and A. S. Ustunel (eds.): Stochastic Analysis and Related Topics, Stochastic Monographs Vol 8. Gordon and Breach Science Publishers. 1993.

[B2]  F. E. Benth: *A Functional Process Solution to a Stochastic Partial Differential Equation with Applications to Nonlinear Filtering.* To appear in Stochastics.

[B3]  F. E. Benth: *On the Positivity of the Stochastic Heat Equation.* Manuscript, University of Mannheim, 1994.

[BC]  L. Bertini and N. Cancrini: *The Stochastic heat equation and Intermittence.* Preprint no. 1032, University of Rome (La Sapienza), 1994.

[GK]  I. Gyöngy & N. V. Krylov: *SPDEs With Unbounded Coefficients I & II;* Stochastics 32-33, 1990.

[HLØUZ]  H. Holden, T. Lindstrøm, B. Øksendal, J. Ubøe and T. -S. Zhang: *The Burgers Equation with a Noisy Force.* Comm. PDE. 19, 1994.

[HKPS] T. Hida, H.-H. Kuo, J. Potthoff and L. Streit: *White Noise -An Infinite Dimensional Calculus.* Kluwer Academic Press, 1993.

[K] H.-H. Kuo: *Convolution and Fourier Transform of Hida Distributions.* Manuscript, Lousiana State University, USA, 1990.

[KLPSW] Yu. G. Kondratiev, P. Leukert. J. Potthoff, L. Streit, W. Westerkamp: *Generalized Functionals in Gaussian Spaces - The Characterization Theorem Revisited.* Preprint, University of Mannheim, 1994.

[KS] I. Karatzas & S. Shreve: *Brownian Motion and Stochastic Calculus.* Springer 1988.

[KR] N. V. Krylov and B. L. Rozovskii: *Stochastic Partial Differential Equations and Diffusions Processes.* Russian Math. Surveys. **37**, 1982.

[LØU1] T. Lindstrøm, B. Øksendal and J. Ubøe: *Stochastic Modelling of Fluid Flow in a Porous Medium.* In S. Chen and J. Yong (eds.): Control Theory, Stochastic Analysis and Applications. World Scientific. 1991.

[LØU2] T. Lindstrøm, B. Øksendal and J. Ubøe: *Wick Multiplication and Ito-Skorohod Stochastic Differential Equations.* In S. Albeverio et. al. (eds.): Ideas and Methods in Mathematical Analysis, Stochastics and Applications. Cambridge University Press. 1992.

[LR] R. Léandre and F. Russo: *Estimation de la Densité de la Solution de l'Équation de Zakai Robuste.* To appear in Journal of Potential Analysis.

[NZ] D. Nualart and M. Zakai: *Generalized Stochastic Integrals and the Malliavin Calculus.* Prob. Th. Rel. Fields **73**, 1986.

[Pa] E. Pardoux: *Stochastic Partial Differential Equations and Filtering of Diffusion Processes.* Stochastics **3**, 1979.

[P] J. Potthoff: *White Noise Approach to Parabolic Stochastic Partial Differential Equations.* Preprint, University of Mannheim, 1994.

[PS] J. Potthoff and L. Streit: *A Characterization of Hida Distributions.* J. Funct. Anal. **101**, 1991.

[TB] M. Timpel and F. E. Benth:*Topological Aspects of the Characterization of Hida Distributions - A Remark.* To appear in Stochastics.

[Z] M. Zakai: *On the Optimal Filtering of Diffusion Processes.* Z. Wahrschein. verw. Geb.. **11**, 1969.

# Wick products of complex valued random variables

**Fred Espen Benth[1], Bernt Øksendal[2], Jan Ubøe[3] and Tusheng Zhang[3]**

1. Department of Mathematics V, University of Mannheim A5,

D-68131 Mannheim, Germany

2. Department of Mathematics, University of Oslo, Box 1053, Blindern

N-0316 Oslo 3, Norway

3. Department of Engineering, Stord/Haugesund College, Skåregaten 103

N-5 500 Haugesund, Norway

**ABSTRACT.** In this paper we consider Wick products of complex valued random variables. We prove that Wick products of such variables coincide with the ordinary product in a variety of cases. Ordinary SDEs are considered in relation to their Wick versions. We present examples where these notions are equivalent in the complex case.

## 1. Introduction

The relationship between stochastic integration and complex analysis has been a topic of several authors. Analytic functions are conformal mappings and they will always map Brownian paths into new Brownian paths. The area is thus characterized by a number of phenomena which do not appear in the real case. The basis for many of these issues can be found from the complex version of the Ito formula, see e.g. [14]:

(1.1)
$$df(Z_t) = \frac{\partial f}{\partial z} dZ_t + \frac{\partial f}{\partial \overline{z}} d\overline{Z}_t + \frac{1}{2} \frac{\partial^2 f}{\partial z^2} dZ_t dZ_t + \frac{\partial^2 f}{\partial z \partial \overline{z}} dZ_t d\overline{Z}_t + \frac{1}{2} \frac{\partial^2 f}{\partial \overline{z}^2} d\overline{Z}_t d\overline{Z}_t$$

If $f$ happens to be an analytic function, this simplifies to:

(1.2)
$$df(Z_t) = \frac{\partial f}{\partial z} dZ_t + \frac{1}{2} \frac{\partial^2 f}{\partial z^2} dZ_t dZ_t$$

Usually one only wants to consider processes $Z_t$ with some kind of holomorphic structure, e.g. conformal martingales see [5] or [18]. In these cases the quadratic variation term $dZ_t dZ_t$ vanish, and we end up with the ordinary chain rule:

$$(1.3) \qquad df(Z_t) = \frac{\partial f}{\partial z} dZ_t$$

Once we have a chain rule of this form, we are able to solve various problems in stochastic calculus using simple techniques from classical calculus. The awkward correction terms from the usual Ito calculus are no longer present, and the basic intuition from ordinary differential equation applies without change. By contrast, one can achieve more or less the same effect using Wick products and Wick calculus. For references to the theory of Wick calculus see [6], [8], [10] or [12]. This way of approach applies already in the real variable case.

**Some notation**

Let $S(\mathbb{R})$ be denote the usual Schwartz space of rapidly decreasing smooth $(C^\infty)$ functions on $\mathbb{R}$ with its dual space $S'(\mathbb{R})$ equipped with the weak star topology, and let $S'_\mathbb{C}(\mathbb{R})$ denote the complexification of $S'(\mathbb{R})$. On $S'_\mathbb{C}(\mathbb{R})$ we define a probability measure $\mu$ as the product of two white noise measures, see [7]. To be more precise, the complexification of the real white noise probability space is carried out as follows. Put:

$$(1.4) \qquad S_\mathbb{C}(\mathbb{R}) = S(\mathbb{R}) + i S(\mathbb{R}) \qquad \text{and:} \qquad S'_\mathbb{C}(\mathbb{R}) = S'(\mathbb{R}) + i S'(\mathbb{R})$$

By the Bochner-Minlos theorem, define two measures $\mu_1$ and $\mu_2$ on $S'(\mathbb{R})$ with:

$$(1.5) \qquad \int_{S'(\mathbb{R})} \exp(i\langle \omega, \phi \rangle) d\mu_j(\omega) = \exp(-\frac{1}{4}\|\phi\|^2_{L^2(\mathbb{R})}), \quad j = 1, 2.$$

With $\mathcal{B}$ the Borel $\sigma$-algebra on $S'_\mathbb{C}(\mathbb{R})$, introduce the product measure $\nu = \mu_1 \times \mu_2$. Then the triplet:

$$(1.6) \qquad (S'_\mathbb{C}(\mathbb{R}), \mathcal{B}, \nu)$$

is called the complex white noise probability space. From the expression (1.5) we get the familiar isometry $E[|<\cdot,\phi>|^2] = ||\phi||^2_{L^2(\mathbb{R})}$ for all $\phi \in S(\mathbb{R})$ where $<\omega,\phi> = \omega(\phi)$ is the dual action. Using this isometry, we can define $<\omega,\phi> := \lim_{k\to\infty} <\omega,\phi_k>$ for all $\phi \in L^2(\mathbb{R})$ ($\phi_k$ is any sequence in $S(\mathbb{R})$ s.t. $\phi_k \to \phi$ in $L^2(\mathbb{R})$). This allows us to define:

$$(1.7) \qquad \tilde{\mathbb{B}}_t(\omega) := \langle \omega, 1_{[0,t)} \rangle = \langle \omega_{real}, 1_{[0,t)} \rangle + i \langle \omega_{imaginary}, 1_{[0,t)} \rangle$$

$\tilde{\mathbb{B}}_t$ is then essentially a Brownian motion in the complex plane in the sense that there exist a $t$-continuous version $\mathbb{B}_t$ of $\tilde{\mathbb{B}}_t$ such that $\mathbb{B}_t$ is a Brownian motion in the complex plane. We let $B_{1t}$ and $B_{2t}$ denote the real and the imaginary components of $\mathbb{B}_t$. We also need the corresponding white noise processes and indicate these as $W_t, W_{1t}$ etc. The familiar constructions of white noise analysis now carry over to the complex case with some minor modifications.

Following Hida [7], we introduce the complex Hermite polynomials $H_{n,m}(z,\bar{z})$ as:

$$(1.8) \qquad H_{n,m}(z,\bar{z}) = (-1)^{n+m} \exp(z\bar{z}) \frac{\partial^{n+m}}{\partial \bar{z}^n \partial z^m} \exp(-z\bar{z})$$

where $n, m$ are non-negative integers. With this definition, we see that our Brownian motion can be written:

$$(1.9) \qquad \mathbb{B}_t(\omega) = H_{1,0}(<\omega, 1_{[0,t)}>, \overline{<\omega, 1_{[0,t)}>})$$

Denote by $(L^2_{\mathbb{C}}) := L^2(\nu)$, and let $\mathcal{H}_{(n,m)}$ be the subspace spanned by the functions $\{H_{n,m}(<\omega, e_i>, \overline{<\omega, e_i>})\}_i$, where $\{e_i\}_i$ is a CONS in $L^2_{\mathbb{C}}(\mathbb{R})$. We will make the assumption that $e_i \in S_{\mathbb{C}}(\mathbb{R})$ for all $i = 1, 2, \dots$. From [7], proof of prop. 6.11, we have the orthogonality relation:

$$(1.10) \quad \begin{aligned} &\int_{S'_{\mathbb{C}}(\mathbb{R})} H_{p,k}(<\omega, \psi>, \overline{<\omega, \psi>}) \overline{H_{n,m}(<\omega, y>, \overline{<\omega, y>})}\, d\nu(\omega) \\ &= \delta_{p,n}\delta_{q,m} p!q! (y, \psi)^p \overline{(y, \psi)}^q \end{aligned}$$

In [7] it is shown that we have a Wiener-Ito-Segal decomposition for every $\phi \in (L^2_{\mathbb{C}})$:

138

$$\phi(\omega) = \sum_{n,m} \phi_{n,m}(\omega) \tag{1.11}$$

where $\phi_{n,m} \in \mathcal{H}_{(n,m)}$. We obtain a Fock space structure, i.e. $\phi$ is in a one-to-one correspondence with a sequence of functions $\{f^{(n,m)}\}_{n,m}$, with $f^{(n,m)} \in L^2_{\mathbb{C}}(\mathbb{R}^{n+m})$. Moreover:

$$\|\phi\|^2_{(L^2_{\mathbb{C}})} = \sum_{n=1}^{\infty} \sum_{m=1}^{n} n!m! |f^{(n,m)}|^2_{L^2_{\mathbb{C}}(\mathbb{R}^{n+m})} \tag{1.12}$$

We introduce the complex Kondratiev spaces of random test functions and distributions. Our construction follows closely the one found in [1]. Let $\mathbf{P} := \mathbf{P}(\{e_i\})$ be the space of polynomials as defined in [7], Ch. 6.3. Every element $\phi \in \mathbf{P}$ is expressible in the form:

$$\phi(\omega) = \sum_{n=0}^{N} \sum_{m=0}^{n} \phi_{n,m}(\omega) \tag{1.13}$$

where $\phi_{n,m} \in \mathcal{H}_{(n,m)}$. Let the space $(S_{\mathbb{C}})^1_p$ be the completion of $\mathbf{P}$ in the norm:

$$\|\phi\|^2_{2,p,\mathbb{C}} = \sum_{n=0}^{\infty} \sum_{m=0}^{n} (n!m!)^2 |f^{(n,m)}|^2_{2,p,\mathbb{C}} \tag{1.14}$$

where $|\cdot|_{2,p,\mathbb{C}}$ is the complexification of the norm $|\cdot|_{2,p} := |A^p \cdot|_2$. $A$ is the harmonic oscillator. The complex Kondratiev space of random test functions is the projective limit of the spaces $(S_{\mathbb{C}})^1_p$, and is denoted $(S_{\mathbb{C}})^1$. Its dual, the space of complex Kondratiev distributions, is denoted $(S_{\mathbb{C}})^{-1}$. All elements $\Phi \in (S_{\mathbb{C}})^{-1}$ is in a one-to-one correspondence with a sequence of functions $\{F^{(n,m)}\}_{n,m}$, with $F^{(n,m)} \in S'_{\mathbb{C}}(\mathbb{R}^{n+m})$, such that for a $p > 0$:

$$\|\Phi\|^2_{-2,-p,\mathbb{C}} := \sum_{n=0}^{\infty} \sum_{m=0}^{n} |F^{(n,m)}|^2_{2,-p,\mathbb{C}} < \infty \tag{1.15}$$

On the space of complex Kondratiev distributions, we want to introduce the $S$-transform: For $\Phi \in (S_{\mathbb{C}})^{-1}$ and $\xi \in S_{\mathbb{C}}(\mathbb{R})$, let:

$$(1.16) \qquad S\Phi(\xi) := \langle \overline{\Phi}, \exp(<\cdot, \xi> + \overline{<\cdot, \xi>} - |\xi|_2^2) \rangle$$

where $\langle \cdot, \cdot \rangle$ is the dual pairing between $(S_{\mathbb{C}})^{-1}$ and $(S_{\mathbb{C}})^1$. It is easy to see that for $\phi \in (L_{\mathbb{C}}^2)$:

$$(1.17) \qquad S\phi(\xi) = \int_{S'(\mathbb{R})} \overline{\phi(\omega)} \exp(<\omega, \xi> + \overline{<\omega, \xi>} - |\xi|_2^2) \, dv(\omega)$$

From formula A.40 in [7], we have:

$$(1.18) \qquad \begin{aligned} &\exp(<\omega, \xi> + \overline{<\omega, \xi>} - |\xi|_2^2) \\ &= \sum_{n,m=0}^{\infty} \frac{|\xi|_2^{n+m}}{n!m!} H_{n,m}(<\omega, \xi \cdot |\xi|_2^{-1}>, \overline{<\omega, \xi \cdot |\xi|_2^{-1}>}) \end{aligned}$$

Using the orthogonality relation for the complex Hermite polynomials, we obtain:

$$(1.19) \qquad SB_t(\xi) = |\xi|_2(1_{[0,t)}, \xi \cdot |\xi|_2^{-1}) = \int_0^t \overline{\xi(s)} \, ds$$

From [1] it is known that the $S$-transform characterizes the Kondratiev distributions. Consider a function $G : \mathcal{U} \to \mathbb{C}$, where $\mathcal{U}$ is a neighborhood around zero in $S_{\mathbb{C}}(\mathbb{R})$. If $G$ is locally bounded on $\mathcal{U}$, and the mapping $z \to G(\xi + z\eta)$ is analytic in a neighborhood around zero in $\mathbb{C}$ for each pair $\xi, \eta \in \mathcal{U}$, then there exists a $\Phi \in (S_{\mathbb{C}})^{-1}$ such that $S\Phi = G$. Opposite, every element in $(S_{\mathbb{C}})^{-1}$ has a $S$-transform which is of this type. We refer the reader to the papers [12] and [1], and the contribution of F. E. Benth in this volume, for more about the Kondratiev distribution space. The Wick product of two complex Kondratiev distributions is defined as follows: Let $\Phi, \Psi \in (S_{\mathbb{C}})^{-1}$, then:

$$(1.20) \qquad \Phi \diamond \Psi = S^{-1}(S\Phi \cdot S\Psi)$$

With this definition, we can easily calculate the $S$-transform of $\mathbb{B}_t^{\diamond k}$, for a integer $k$:

$$(1.21) \qquad S\mathbb{B}_t^{\diamond k}(\xi) = \left( \int_0^t \overline{\xi(s)} \, ds \right)^k$$

A straightforward calculation shows that:

$$(1.22) \qquad S\left( H_{k,0}(<\omega, \xi>, \overline{<\omega, \xi>}) \right)(\xi) = \left( \int_0^t \overline{\xi(s)} \, ds \right)^k$$

Hence, we find:

$$(1.23) \quad \mathbb{B}_t^{\diamond k}(\omega) = H_{k,0}(<\omega, 1_{[0,t)}>, \overline{<\omega, 1_{[0,t)}>}) = <\omega, 1_{[0,t)}>^k = \mathbb{B}_t(\omega)^k$$

## 2. Complex Wick multiplication

We now let $f(z) = \sum_{n=0}^{\infty} a_n z^n$ be an entire function. If $X$ is a random variable, we define the Wick version of $f(X)$ by the expression:

$$f^{\diamond}(X) = \sum_{n=0}^{\infty} a_n X^{\diamond n} = \lim_{N \to \infty} \sum_{n=0}^{N} a_n X^{\diamond n}$$

the limit being taken in $(S_{\mathbb{C}})^{-1}$, see [10] or [12]. If $X \in (S_{\mathbb{C}})^{-1}$, this limit always exists. With these conventions the following theorem follows trivially from (1.23).

THEOREM 2.1

Let $f : \mathbb{C} \to \mathbb{C}$ be an entire function, and let $f^{\diamond}$ denote the Wick version. Then:

$$(2.1) \qquad f(\mathbb{B}_t) = f^{\diamond}(\mathbb{B}_t)$$

We have proved that Wick-powers of complex Brownian motion coincide with usual powers. In the following we want to extend this property to other random variables as well. We first observe the following lemma:

## LEMMA 2.2

$$(2.2) \qquad \mathbb{B}_s^{\diamond m} \diamond \mathbb{B}_t^{\diamond n} = \mathbb{B}_s^m \cdot \mathbb{B}_t^n$$

## PROOF

Assume that $t \geq s$, then we have:

$$
\begin{aligned}
\mathbb{B}_s^{\diamond m} \diamond \mathbb{B}_t^{\diamond n} &= \mathbb{B}_s^{\diamond m} \diamond (\mathbb{B}_t - \mathbb{B}_s + \mathbb{B}_s)^{\diamond n} \\
&= \mathbb{B}_s^{\diamond m} \diamond \sum_{k=0}^{n} \binom{n}{k} (\mathbb{B}_t - \mathbb{B}_s)^{\diamond k} \diamond \mathbb{B}_s^{\diamond(n-k)} \\
&= \sum_{k=0}^{n} \binom{n}{k} (\mathbb{B}_t - \mathbb{B}_s)^{\diamond k} \diamond \mathbb{B}_s^{\diamond(n-k+m)} \\
&= \sum_{k=0}^{n} \binom{n}{k} (\mathbb{B}_t - \mathbb{B}_s)^k \diamond \mathbb{B}_s^{n-k+m} \\
&= \sum_{k=0}^{n} \binom{n}{k} (\mathbb{B}_t - \mathbb{B}_s)^k \cdot \mathbb{B}_s^{n-k+m} \\
&= \mathbb{B}_s^m \cdot \mathbb{B}_t^n
\end{aligned}
$$

(2.3)

In the fifth equality we have used that $\mathbb{B}_t - \mathbb{B}_s$ and $\mathbb{B}_s$ are strongly independent. In this case the Wick product always coincide with the ordinary product, see [6].

□

## PROPOSITION 2.3

Let $p$ be a polynomial in $k$ complex variables, then:

$$(2.4) \qquad p^{\diamond}(\mathbb{B}_{t_1}, \mathbb{B}_{t_2}, \ldots, \mathbb{B}_{t_k}) = p(\mathbb{B}_{t_1}, \mathbb{B}_{t_2}, \ldots, \mathbb{B}_{t_k})$$

where $p^{\diamond}$ is interpreted in the sense that all powers are Wick powers.

## PROOF

Use lemma 1.4 repeatedly to see that:

$$(2.5) \qquad \mathbb{B}_{t_1}^{\diamond n_1} \diamond \mathbb{B}_{t_2}^{\diamond n_2} \diamond \cdots \diamond \mathbb{B}_{t_k}^{\diamond n_k} = \mathbb{B}_{t_1}^{n_1} \cdot \mathbb{B}_{t_2}^{n_2} \cdots \mathbb{B}_{t_k}^{n_k}$$

The general result then follows by linearity.

□

For easy reference we will call any expression which can be written on the form $\Pi_k = p(\mathbb{B}_{t_1}, \mathbb{B}_{t_2}, \ldots, \mathbb{B}_{t_k})$ a $\mathbb{B}$-analytic polynomial. The multiplicative property (2.4) can now be extended to limits of $\mathbb{B}$-analytic polynomials. A convenient space to work in is then the space $(S_\mathbb{C})^{-1}$ of Kondratiev distributions, see [10] or [12]. We start out with some definitions:

## DEFINITION 2.4

$X \in (S_\mathbb{C})^{-1}$ is called $\mathbb{B}$-analytic if there exists a sequence $X_n$ of $\mathbb{B}$-analytic polynomials such that $X_n \to X$ strongly in $(S_\mathbb{C})^{-1}$.

## DEFINITION 2.5

$X \in (L^p)$, $p > 1$, is $\mathbb{B}_p$-analytic if there exists a sequence $X_n$ of $\mathbb{B}$-analytic polynomials such that $X_n \to X$ in $(L^p)$.

From these definitions, we have:

## COROLLARY 2.6

Let $X \in (L^p)$, $p > 1$, be $\mathbb{B}_p$-analytic. Then $X$ is $\mathbb{B}$-analytic.

## PROOF

By assumption we have a sequence $X_n$ of $\mathbb{B}$-analytic polynomials converging to $X$ in $(L^p)$. But convergence in $(L^p)$ implies strong convergence in $(S_\mathbb{C})^{-1}$, see [12]. Hence, $X$ is $\mathbb{B}$-analytic.

$\square$

## COROLLARY 2.7

If $\{X_n\}$ is a sequence of $\mathbb{B}$-analytic elements which converges strongly to $X$ in $(S_\mathbb{C})^{-1}$, then $X$ is $\mathbb{B}$-analytic.

## PROOF

The proof is straightforward: Since $X_n$ converges strongly to $X$ in $(S_\mathbb{C})^{-1}$, there exists a $p > 0$ such that:

$$(2.6) \qquad \|X_n - X\|_{2,-p,\mathbb{C}} \to 0$$

For each $n$, let $\{Y_n^m\}$ be a sequence of $\mathbb{B}$-analytic polynomials converging strongly to $X_n$ in $(S_\mathbb{C})^{-1}$. (Such a sequence exists by definition of $\mathbb{B}$-analyticity). Since $X_n$ is an element in $(S_\mathbb{C})^{-1}_{-p}$, we have:

$$(2.7) \qquad \|Y_n^m - X_n\|_{2,-p,\mathbb{C}} \to 0, \quad m \to \infty$$

This yields that for each $n$, there exists a natural number $N_n$ such that:

$$(2.8) \qquad \|Y_n^m - X_n\|_{2,-p,\mathbb{C}} < \frac{1}{n}, \quad \text{for } m \geq N_n$$

It is then easy to see that the sequence $\{Y_n^{N_n}\}$ of $\mathbb{B}$-analytic polynomials converges strongly in $(S_\mathbb{C})^{-1}$ to $X$: For a given $\epsilon > 0$, we find a $M_\epsilon$ such that $1/n < \epsilon/2$ and:

$$(2.9) \qquad \|X - X_n\|_{2,-p,\mathbb{C}} < \epsilon/2$$

for $n \geq M_\epsilon$. Hence, by the triangle inequality:

$$(2.10) \qquad \|X - Y_n^{N_n}\|_{2,-p,\mathbb{C}} \leq \|X - X_n\|_{2,-p,\mathbb{C}} + \|X_n - Y_n^{N_n}\|_{2,-p,\mathbb{C}} < \epsilon$$

$$\square$$

For $\mathbb{B}_p$-analyticity, we have the same result:

COROLLARY 2.8

If $\{X_n\}$ is a sequence of $\mathbb{B}_p$-analytic elements which converges to $X$ in $(L^p)$, then $X$ is $\mathbb{B}_p$-analytic.

$$\square$$

PROPOSITION 2.9

If $X, Y$ are $\mathbb{B}$-analytic, then $X \diamond Y$ is $\mathbb{B}$-analytic.

PROOF

Let $\{X_n\}$ and $\{Y_n\}$ be two sequences of $\mathbb{B}$-analytic polynomials converging strongly in $(S_\mathbb{C})^{-1}$ to $X$ and $Y$ respectively. That means, for $p, q > 0$:

$$(2.11) \qquad \|X_n - X\|_{2,-p,\mathbb{C}} \to 0, \quad n \to \infty$$

144

and:

(2.12)
$$\|Y_n - Y\|_{2,-q,\mathbb{C}} \to 0, \quad n \to \infty$$

Define, for an $\alpha > \frac{1}{2}$:

(2.13)
$$r := \alpha + \max(p,q)$$

From the triangle inequality and Corollary 4.22 in [9], we have:

(2.14) $\|X \diamond Y - X_n \diamond Y_n\|_{2,-r,\mathbb{C}} \le \|X \diamond (Y - Y_n)\|_{2,-r,\mathbb{C}} + \|Y_n \diamond (X - X_n)\|_{2,-r,\mathbb{C}}$

(2.15) $\qquad \le K_1 \|X\|_{2,-p\mathbb{C}} \|Y - Y_n\|_{2,-q,\mathbb{C}} + K_2 \|Y_n\|_{2,-q,\mathbb{C}} \|X - X_n\|_{2,-p,\mathbb{C}}$

We see that $X_n \diamond Y_n$ converges strongly to $X \diamond Y$, and since $X_n \diamond Y_n$ is a W–analytic polynomial, the proposition follows.

$\square$

THEOREM 2.10

Let $p > 1$ and $q > p$. Assume that $X \in (L^q)$ is $\mathbb{B}_q$-analytic and that $Y \in (L^{\frac{qp}{q-p}})$ is $\mathbb{B}_{\frac{qp}{q-p}}$-analytic. Then $X \cdot Y \in (L^p)$ is $\mathbb{B}_p$-analytic. Moreover,

(2.16)
$$X \diamond Y = X \cdot Y$$

PROOF

Let $X_n$ and $Y_n$ be the $\mathbb{B}$-analytic sequences converging to $X$ and $Y$ respectively. $X_n \cdot Y_n$ is of course again a $\mathbb{B}$-analytic polynomial. Observe that by the Cauchy-Schwartz inequality we have:

(2.17)
$$\|f \cdot g\|_p \le \|f\|_q \|g\|_{\frac{qp}{q-p}}.$$

Hence, by the triangle inequality it follows that:

(2.18)
$$X_n \cdot Y_n \to X \cdot Y$$

in $(L^p)$. Thus, $X \cdot Y$ is $\mathbb{B}_p$-analytic. Since $X_n \cdot Y_n$ is a $\mathbb{B}$-analytic polynomial, we have:

$$(2.19) \qquad\qquad X_n \cdot Y_n = X_n \diamond Y_n.$$

Since $X_n$ and $Y_n$ converge in $(L^q)$ and $(L^{\frac{qp}{q-p}})$ respectively, we can show that:

$$(2.20) \qquad\qquad S(X_n \diamond Y_n)(\xi) \to SX(\xi) \cdot SY(\xi)$$

pointwise, and:

$$(2.21) \qquad\qquad |S(X_n \diamond Y_n)(\xi)| \le K,$$

uniformly in $n$. Here, $\xi$ is in a neighbourhood around zero in $S_{\mathbb{C}}(\mathbb{R})$. Hence, by Theorem 5 in [12], it follows that:

$$(2.22) \qquad\qquad X_n \diamond Y_n \to X \diamond Y$$

weakly in $(S_{\mathbb{C}})^{-1}$. By the corollary above, $X_n \cdot Y_n \to X \cdot Y$ strongly in $(S_{\mathbb{C}})^{-1}$. Hence:

$$(2.23) \qquad X \diamond Y = \lim_n X_n \diamond Y_n = \lim_n X_n \cdot Y_n = X \cdot Y.$$

where $\overset{\cdot}{\lim}_n$ denotes the strong limit in $(S_{\mathbb{C}})^{-1}$.

$\square$

Remark

Note if $X$ is $\mathbb{B}_q$ analytic for some $q > 1$ and $Y$ is $\mathbb{B}_r$-analytic for all $r < \infty$, then it is always possible to find $p > 1$ s.t. the conditions in theorem 2.9 are satisfied.

We now go on to consider stochastic processes. Here we call a stochastic process $X_t$ $\mathbb{B}$-analytic if $X_t$ is $\mathbb{B}$-analytic for every fixed $t$ and similarly for $\mathbb{B}_p$-analyticity. We want to consider certain elementary operations on $\mathbb{B}$-analytic processes and start out with some observations.

## PROPOSITION 2.11

Assume $X_t$ is an Itô integrable and that for each $t$ it can be approximated in $L^2$ by adapted $\mathbb{B}$-analytic polynomials. Then then Itô integral:

$$\text{(2.24)} \qquad \int_0^T X_t \, d\mathbb{B}_t$$

is $\mathbb{B}_2$-analytic.

## PROOF

By definition of the Itô integral:

$$\text{(2.25)} \qquad \int_0^T X_t \, d\mathbb{B}_t = \lim_j \sum_j X_{t_j} \cdot (\mathbb{B}_{t_{j+1}} - \mathbb{B}_{t_j})$$

where the limit is taken in $(L^2)$. By assumption, $X_{t_j}$ is $\mathbb{B}_2$-analytic. Since $X_t$ is adapted, we have:

$$\text{(2.26)} \qquad E[|X_{t_j}|^2 |\mathbb{B}_{t_{j+1}} - \mathbb{B}_{t_j}|^2] = E[|X_{t_j}|^2] \cdot E[|\mathbb{B}_{t_{j+1}} - \mathbb{B}_{t_j}|^2].$$

Hence, we see that $X_{t_j}(\mathbb{B}_{t_{j+1}} - \mathbb{B}_{t_j})$ is $\mathbb{B}_2$-analytic. The Itô integral is then the $(L^2)$-limit of $\mathbb{B}_2$-analytic elements, which imply the proposition.

□

## PROPOSITION 2.12

Let $\{x_i(t)\}_{i=1}^\infty$ be any sequence of $L^2(\mathbb{R})$-functions. Then all the integrals $\int x_i(t) \, d\mathbb{B}_t$, $i = 1, 2, \ldots$ are $\mathbb{B}_q$-analytic for any $q$ and any combination of Wick powers and Wick products of these random variables coincide with the corresponding expressions defined in terms of the ordinary product.

## PROOF

It is an easy application of the Burkholder-Gundy inequalities to see that the integrals $\int x_i(t) \, d\mathbb{B}_t$, $i = 1, 2, \ldots$ are $\mathbb{B}_q$-analytic for any $q$. The second part follows from theorem 2.10.

□

PROPOSITION 2.13

Let $X_t$ be a $\mathbb{B}$-analytic process where $\langle X_t, \phi \rangle$ is measurable on $[0, T]$ for all $\phi \in (S)^1$. Assume there exists a $p \geq 0$ such that:

$$(2.27) \qquad \int_0^T \|X_t\|_{2,-p,c}\, dt < \infty$$

Then the Bochner integral $\int_0^T X_t\, dt \in (S_C)^{-1}$ is $\mathbb{B}$-analytic.

PROOF

By Pettis' Theorem (see e.g. [22]), the measurability of $\langle X_t, \phi \rangle$ implies strong measurability, i.e. the existence of a sequence $\{X_t^n\}_n$ in $(S_C)^{-1}$ converging strongly to $X_t$. By inspection of the proof of Pettis' theorem in [22], this sequence can be chosen in the following manner:

$$(2.28) \qquad X_s^n = Y_i, \quad \text{when } s \in B_i^n,$$

where $\{B_i^n\}_{i=1}^{N_n}$ are disjoint measurable sets in $[0, T]$, and $Y_i = X_{s_i}$ for some $s_i \in [0, T]$. Hence, by assumption, $X_t^n$ is $\mathbb{B}$-analytic.

By the condition $\int_0^T \|X_t\|_{2,-p}\, dt < \infty$ Bochner integrability of $X_t$ follows. We have:

$$(2.29) \qquad \int_0^T X_t\, dt = \lim_{n \to \infty} \int_0^T X_t^n\, dt = \lim_{n \to \infty} \sum_{i=1}^{\infty} Y_i m(B_i^n).$$

The limit is strongly in $(S_C)^{-1}$. Hence, the Bochner integral is the strong limit of $\mathbb{B}$-analytic elements, and the proposition follows.

$\square$

In the Kondratiev space we have a generalization of Itô/Skorohod integration. If $X_t$ is an Itô integrable process, then:

$$(2.30) \qquad \int_0^T X_t \diamond W_t\, dt = \int_0^T X_t\, d\mathbb{B}_t$$

We have the following result about $\mathbb{B}$-analyticity of this integral:

PROPOSITION 2.14

Let $X_t$ be a $\mathbb{B}$-analytic process such that:

i) $SX_t(\xi)$ is measurable for $\xi \in \mathcal{U}$

ii) $|SX_t(\xi)| \leq C(t)$, where $C(t) \in L^1([0, T], dt)$ for $\xi \in \mathcal{U}$

Then $X_t \diamond \mathbb{W}_t$ is Bochner integrable, and:

$$(2.31) \qquad \int_0^T X_t \diamond \mathbb{W}_t dt$$

is $\mathbb{B}$-analytic.

PROOF

We have:

$$(2.32) \qquad \mathbb{W}_t = \lim_{h \to 0} \frac{1}{h}(\mathbb{B}_{t+h} - \mathbb{B}_t),$$

where the limit is strong in $(S_\mathbb{C})^{-1}$. Hence, $\mathbb{W}_t$ is a $\mathbb{B}$- analytic process. This implies by proposition (above) that $X_t \diamond \mathbb{W}_t$ is $\mathbb{B}$-analytic. By Theorem 6 in [12], we have the Bochner integrability of $X_t \diamond \mathbb{W}_t$. Hence, the proposition follows.

$\square$

## 3. Applications to SDEs

We now want to compare SDEs of the from:

$$(3.1) \qquad dZ_t = (X_t \cdot Z_t + Y_t)d\mathbb{B}_t + (U_t \cdot Z_t + V_t)dt$$

$$(3.2) \qquad dZ_t = (X_t \diamond Z_t + Y_t)d\mathbb{B}_t + (U_t \diamond Z_t + V_t)dt$$

The equation (3.2) can be solved under very mild conditions on the coefficients. We will first consider the properties of this equation. If in addition, the coefficients are $\mathbb{B}$-analytic and sufficiently nice for (3.1) to make sense, we can expect the two solutions to coincide.

PROPOSITION 3.1

Assume that for all $\xi \in S_\mathbb{C}(\mathbb{R})$, the $S$-transforms $S(X_t)(\xi), S(Y_t)(\xi), S(U_t)(\xi)$ and $S(V_t)(\xi)$ are locally Lipschitz functions (as functions of $t$). If (3.2) has a $(S_\mathbb{C})^{-1}$-valued solution defined for all $t \geq 0$, this solution is unique.

Remark: We call a function $f = f(z)$ locally Lipschitz if there for every $z_0$ exists a constant $C < \infty$ s.t. $|f(z) - f(z_0)| \leq C|z - z_0|$ for every $z$ in a neighbourhood of $z_0$.

PROOF

Apply the $S$-transform to both sides of (3.2) to see that that the $S$-transform of $Z_t$ is uniquely defined. Since any element in $(S_C)^{-1}$ is uniquely defined in terms of its $S$-transform, the proposition follows.

$\square$

PROPOSITION 3.2

Assume that:

i) $S(X_t)(\xi), S(Y_t)(\xi), S(U_t)(\xi)$ and $S(V_t)(\xi)$ are measurable for $\xi \in \mathcal{U}$

ii) $|S(X_t)(\xi)|, |S(Y_t)(\xi)|, |S(U_t)(\xi)|, |S(V_t)(\xi)| \leq C_T(t)$, where

$e^{C_T(t)} \in L^p([0,T], dt)$ for every $\xi \in \mathcal{U}$, all $p > 0$ and all $T < \infty$

iii) $Z_0 \in (S_C)^{-1}$ then (3.2) has a $(S_C)^{-1}$-valued solution $Z_t$ given by the expression:

$$(3.3) \quad \begin{aligned} Z_t &= Z_0 \diamond \mathrm{Exp}[\int_0^t X_r d\mathbb{B}_r + \int_0^t U_r dr] \\ &+ \int_0^t \mathrm{Exp}[\int_s^t X_r d\mathbb{B}_r + \int_s^t U_r dr] \diamond Y_s d\mathbb{B}_s \\ &+ \int_0^t \mathrm{Exp}[\int_s^t X_r d\mathbb{B}_r + \int_s^t U_r dr] \diamond V_s ds \end{aligned}$$

PROOF

The idea is to use the analogy with the differential equation $y' = f(t)y + g(t)$. This equation has the solution $y = y_0 e^{\int_0^t f(r)dr} + \int_0^t e^{\int_s^t f(r)dr} g(s)ds$. Formally we write that $f(t) = X_t \frac{d\mathbb{B}_t}{dt} + U_t$ and $g(t) = Y_t \frac{d\mathbb{B}_t}{dt} + V_t$. If we insert this in the solution formula and replace all the ordinary products with Wick products, we get (3.3). From the arguments of proposition 2.14, we see that all the necessary expressions are Bochner integrable and that (3.3) makes sense as an element of $(S_C)^{-1}$. If we insert this expression in (3.2) all the operations in (3.2) are well defined. Hence the ordinary chain rule applies and $Z_t$ is a solution.

□

As we remarked earlier the interesting question is to compare the equation (3.1) and (3.2). To make sense out of (3.1) we must put quite strong growth conditions on the coefficients. We have to work with elements in $(L^p)$. The idea is then to apply theorem 2.10 to see that the Wick products coincide with the ordinary product.

THEOREM 3.3

Assume that:

    i) $X_t = x(t)$ where $x(t) \in L^2(\mathbb{R})$

    ii) $U_t = \int_0^t u_1(s)d\mathbb{B}_s + u_2(t)$ where $u_1(t), u_2(t) \in L^2(\mathbb{R})$

    iii) $Y_t, V_t$ are adapted and $\mathbb{B}_q$-analytic for some $q > 2$ and satisfies the conditions in 3.2.

    iv) $Z_0$ is $\mathbb{B}_r$-analytic for some $r > 1$ (If $Z_0$ is non-constant, the meaning of (3.1) is interpreted in the Hitsuda-Skorohod sense, se [8]).

Then the solutions of (3.1) and (3.2) coincide and are both given by the expression:

(3.4)
$$Z_t = Z_0 \cdot e^{\int_0^t X_r d\mathbb{B}_r + \int_0^t U_r dr} + \int_0^t e^{\int_s^t X_r d\mathbb{B}_r + \int_s^t U_r dr} \cdot Y_s d\mathbb{B}_s + \int_0^t e^{\int_s^t X_r d\mathbb{B}_r + \int_s^t U_r dr} \cdot V_s ds$$

Before we turn to the proof of this theorem, we need to prove two technical lemmas.

LEMMA 3.4

If $e^{|X|} \in (L^p)$ for all $p$ and $X$ is $\mathbb{B}_p$-analytic for all $p$, then $e^X$ is also $\mathbb{B}_p$-analytic for all $p$.

PROOF

Put $f_N = e^{|X|} - \sum_{n=0}^{N} \frac{1}{n!}|X|^n$. By the monotone convergence theorem $f_N \to 0$ in every $(L^p)$. Observe that:

$$(3.5) \quad E[|e^X - \sum_{n=0}^{N} \frac{1}{n!} X^n|^p] = E[|\sum_{n=N+1}^{\infty} \frac{1}{n!} X^n|^p]$$

$$\leq E[|\sum_{n=N+1}^{\infty} \frac{1}{n!} |X|^n|^p] = E[|f_N|^p] \to 0$$

Hence $e^X$ can be approximated as well as we please by $\sum_{n=0}^{N} \frac{1}{n!} X^n$ in any $(L^p)$. Since $X$ is $\mathbb{B}_q$-analytic for every $q$, clearly each $\sum_{n=0}^{N} \frac{1}{n!} X^n$ is $\mathbb{B}_p$-analytic and this proves the lemma.

□

## LEMMA 3.5

$e^{\int_0^t X_r d\mathbb{B}_r}$ and $e^{\int_0^t U_r dr}$ are adapted and $\mathbb{B}_p$-analytic for every $p$.

## PROOF

The adaptedness is trivial. Fix $t$ and put $X = \int_0^t x(s) d\mathbb{B}_r$. By proposition 2.12 $X$ is $\mathbb{B}_q$-analytic for every $q$. Choose any $p$. We must prove that $e^{|X|} \in (L^p)$. For simplicity we will replace $x$ and $\mathbb{B}$ by the corresponding real expressions in the rest of the argument. Since $X$ is gaussian, we then have:

$$(3.6) \quad E[|X|^{2n}] = \frac{(2n)! E[|X|^2]}{2^n n!} = \frac{(2n)! \int_0^t x(s)^2 ds}{2^n n!}$$

From this we get using Stirlings formula:

$$E[e^{p|X|}] = \sum_{n=0}^{\infty} \frac{p^n}{n!} E[|X|^n] \leq \sum_{n=0}^{\infty} \frac{p^n}{n!} \left( E[|X|^{2n}] \right)^{\frac{1}{2}}$$

$$(3.7)$$

$$= \sum_{n=0}^{\infty} \frac{p^n}{n!} \left( \frac{(2n)! \int_0^t x(s)^2 ds}{2^n n!} \right)^{\frac{1}{2}} < \infty$$

Hence $X$ satisfies the hypothesis of the previous lemma and the conclusion follows. The proof for the expression $e^{\int_0^t U_r dr}$ is similar.

□

We now turn to the proof of theorem 3.4.

PROOF

We first want to prove that the expression in (3.3) equals the expression in (3.4). Since $\int_0^t X_r d\mathbb{B}_r$ and $\int_0^t U_r dr$ are $\mathbb{B}_q$-analytic for every $q$, we have:

$$(3.8) \qquad \text{Exp}[\int_s^t X_r d\mathbb{B}_r + \int_s^t U_r dr] = e^{\int_s^t X_r d\mathbb{B}_r + \int_s^t U_r dr}$$

It then follows by lemma 3.6, theorem 2.10 and proposition 2.11 that:

$$(3.9) \qquad \int_0^t \text{Exp}[\int_s^t X_r d\mathbb{B}_r + \int_s^t U_r dr] \diamond Y_s ds = \int_0^t e^{\int_s^t X_r d\mathbb{B}_r + \int_s^t U_r dr} \cdot Y_s ds$$

The same arguments works to prove that all the terms in (3.3) and (3.4) are equal. If we insert the expression (3.4) on the right side of (3.1), we may replace all the the ordinary products by Wick products. Hence the right side of (3.1) equals $dZ_t$ and this completes the proof of theorem 3.4.

$\square$

LEMMA 3.6

Let $f$ be analytic in a neighbourhood of $y_0$. Then the differential equation:

$$(3.10) \qquad y' = f(y) \qquad y(z_0) = y_0$$

has a unique solution $y = y(z)$ analytic in a neighbourhood of $z_0$.

PROOF

Local uniqueness follows from the Lipschitz continuity of $f$ at $y_0$. If $f(y_0) = 0$, then $y(z) = y_0$ is the solution. If on the other hand $f(y_0) \neq 0$, then $f(z) \neq 0$ in a neighbourhood of $y_0$ and the function $\frac{1}{f(z)}$ is analytic on this neighbourhood. Hence there exists a neighbourhood of $y_0$ and an analytic function $g(z)$ on this neighbourhood s.t. $g'(z) = \frac{1}{f(z)}$. Since $g'(z) \neq 0$ it follows from the inverse function theorem, see [16] theorem 1.3.7, that $g$ has an inverse function $h$ which is analytic in some neighbourhood of $g(y_0)$. Put $y(z) = h(z - z_0 + g(y_0))$. Then $y(z_0) = y_0$ and:

$$(3.11) \qquad \begin{aligned} y' &= h'(z - z_0 + g(y_0)) = h'(g(h(z - z_0 + g(y_0)))) \\ &= \frac{1}{g'(h(z - z_0 + g(y_0)))} = f(h(z - z_0 + g(y_0))) = f(y) \end{aligned}$$

Hence $y = y(z)$ is analytic in a neighbourhood of $z_0$ and is a solution of (3.10).

$\square$

THEOREM 3.7

Let $f$ be analytic in a neighbourhood $D$ and let $z_0 \in D$. Then:

$$(3.12) \qquad dZ_t = f^\diamond(Z_t)d\mathbb{B}_t \qquad Z_0 = z_0$$

has a unique $(S_C)^{-1}$-valued solution $Z_t$ defined for all $t \geq 0$. Moreover there exists a stopping time $\tau > 0$ s.t. $Z_{t \wedge \tau}$ is a local solution to:

$$(3.13) \qquad dZ_t = f(Z_t)d\mathbb{B}_t \qquad Z_0 = z_0$$

PROOF

Uniqueness. Choose and fix $\xi \in S_C(\mathbb{R})$. Then apply the $S$-transform to both sides of (3.12) to see that $y = S(Z_t)(\xi)$ is a solution to the ODE:

$$(3.14) \qquad y' = f(y)\overline{\xi}(t) \qquad y(0) = z_0$$

Since (3.14) has a unique solution $y$ and the $S$-transform uniquely character-izes every element in $(S_C)^{-1}$, the solution $Z_t$ of (3.12) is unique.

Existence. First use the previous lemma to find an analytic function $y(z) = \sum_{k=0}^{\infty} a_k z^k$ s.t. $y$ is a solution to the problem:

$$(3.15) \qquad y' = f(y) \qquad y(0) = z_0$$

Since $y$ is analytic in a neighbourhood of the origin, there exists two positive constants $M < \infty, r < \infty$ s.t. $|a_k| \leq Mr^k$. The expression:

$$(3.16) \qquad Z_t = \sum_{k=0}^{\infty} a_k \mathbb{B}_t^{\diamond k} = \sum_{k=0}^{\infty} a_k \mathbb{B}_t^k$$

then makes sense as an element of $(S_C)^{-1}$. The ordinary chain rule applies, and hence $Z_t$ is a solution of (3.12). Now let $\tau_d$ be the first exit time of $\mathbb{B}_t$ from a small neighbourhood of the origin. If we put $Y_t = Z_{t \wedge \tau_d} = \sum_{k=0}^{\infty} a_k \mathbb{B}_{t \wedge \tau_d}^k$, it follows from the complex Ito formula that the ordinary chain rule applies. $Y_t$ is then a local solution to (3.13) in the sense that it solves the problem $Z_{t \wedge \tau_d} = z_0 + \int_0^{t \wedge \tau_d} f(Z_s) d\mathbb{B}_s$.

$\square$

## ACKNOWLEDGEMENT

This work is supported by VISTA, a research cooperation between The Norwegian Academy of Science and Letters and Den Norske Stats Oljeselskap (Statoil).

The first author also wishes to express his gratitude to Norges Forskningsråd for the support by NAVF grant 100549/410.

## REFERENCES

1. S.Albeverio, Y.L.Daletsky, Y.G.Kondratiev, and L.Streit (1994). Non-Gaussian Infinite Dimensional Analysis. Preprint nr. 217, University of Bochum, Germany.
2. H.Arai (1986). On the algebra of bounded holomorphic martingales. *Proc.Amer. Math.Soc. 97*, 616-620.
3. F.E.Benth (1993). Integrals in the Hida distribution space $(S)^*$. T. Lindstrøm, B.Øksendal and A.S.Ustunel (editors): *Stochastic Analysis and Related Topics*, Gordon & Breach, 89-99.
4. M.Fukushima and M.Okada (1984). On conformal martingale diffusions and pluripolar sets. *J.Func.Anal. 55*, 377-388.
5. R.K.Getoor and M.J.Sharpe (1972). Conformal martingales. *Invent.Math. 16*, 271-308.
6. H.Gjessing, H.Holden, T.Lindstrøm, B.Øksendal, J.Ubøe and T.-S.Zhang (1993). Wick multiplication, H.Niemi et al. (editors): *Frontiers in Pure and Applied Probability*, TVP Publishers, Moskow, 29-67.
7. T.Hida (1980), *Brownian Motion*, Springer.
8. T.Hida, H.-H.Kuo, J.Potthoff and L.Streit (1993). *White Noise: An infinite dimensional calculus*, Mathematics and its applications 253, Kluwer.
9. H.Holden, T.Lindstrøm, B.Øksendal, J.Ubøe and T.-S.Zhang (1993). Stochastic boundary value problems. A white noise functional approach", *Probability Theory and Related Fields 95*, 391-419.
10. H.Holden, T.Lindstrøm, B.Øksendal, J.Ubøe and T.-S.Zhang (1995). *Stochastic partial differential equations - White Noise Functional Methods, Models and Applications*, book manuscript.
11. H-H.Kuo (1992). Lectures on white noise analysis, *Soochow J.Math. 18*, no.3, 229-300.
12. Y.G.Kondratiev, P.Leukert and L.Streit (1994). Wick Calculus in Gaussian analysis. Preprint nr. 637/6/94, University of Bielefeld, Germany.

13. T. Lindstrøm, B. Øksendal and J. Ubøe (1992). Wick multiplication and Ito-Skorohod stochastic differential equations. In S. Albeverio et al (editors): *Ideas and Methods in Mathematical Analysis, Stochastics, and Applications*. Cambridge Univ. Press, 183-206.
14. B.Øksendal (1992). *Stochastic Differential Equations*, 3 edn., Springer Verlag.
15. J.Potthoff and L.Streit (1991). A characterization of Hida distributions, *J.Func. Anal. 101*, 212-229.
16. W.Rudin (1980). *Function Theory in the Unit Ball of* $\mathbb{C}^n$, Springer.
17. L.Schwartz (1980). Semi-martingales sur des variétés, et martingales conformes sur des variétés analytiques complexes. *Lecture notes in Mathematics No.780*, Springer Verlag.
18. J.Ubøe (1987). Conformal martingales and analytic functions. *Math.Scand. 60*, 292-309.
19. J.Ubøe (1992). Riemann integration on complex brownian paths. *Stoch.Anal.. 10(3)*, 351-361.
20. J.Ubøe (1994). Complex valued, multiparameter stochastic integrals. To appear in *J.Theor.Prob.*.
21. M.Yor (1977). Étude de certains processus (stochastiquement) différentiables ou holomorphes. *Ann.Inst.Henri Poincare XIII No.1*, 1-25.
22. K.Yosida (1978). *Functional Analysis*, Springer.
23. T.-S.Zhang (1992). Characterizations of white noise test functions and Hida distributions, *Stochastics* 41, 71-87.

# Two classes of stochastic Dirichlet equations which admit explicit solution formulas

Jon Gjerde
Department of Mathematics
University of Oslo
Box 1053 Blindern, N-0316 Oslo
Norway

## Abstract

In this paper we look at stochastic Dirichlet equations of the type

$$\mathcal{A}u = (\sum_{i=1}^{m} c_i \cdot \mathrm{Exp}\{W_{\phi_x}^{(i)}\}) \diamond u - g$$

$$u|_{\partial D} = f$$

and

$$\mathrm{div}(\mathrm{Exp}\{W_{\phi_x}\} \diamond u) = \kappa \diamond u - g$$

$$u|_{\partial D} = f$$

where $\mathcal{A}$ is a uniformly elliptic second order differential operator and $\mathrm{Exp}\{W_{\phi_x}\}, \kappa, f$ and $g$ are elements in the space $(\mathcal{S})^{-1}$ of generalized white noise distributions. With suitable conditions on $\kappa, f$ and $g$ both classes of stochastic Dirichlet equations admit unique solution formulas in the space $(\mathcal{S})^{-1}$. These are used to give explicit solution formulas to the Scrödinger and wave equation when the boundary conditions are particularly simple.

Keywords : Generalized white noise distributions, Wick product, Hermite transform.

# §1 Introduction

It is well known that given a solution $u \in C(\bar{D}) \cap C^2(D)$ of the Dirichlet problem

$$\mathcal{A}u = \kappa \cdot u - g$$
$$u|_{\partial D} = f$$

(1)

where $D$ is an open, bounded domain, $\kappa$, $g$, $f$ suitable functions and $\mathcal{A}$ a uniformly elliptic second order differential operator, then $u$ has a stochastic representation given by

$$u(x) = E^x[f(X_{\tau_D}) \exp\{-\int_0^{\tau_D} \kappa(X_s)\, ds\} + \int_0^{\tau_D} g(X_t) \exp\{-\int_0^t \kappa(X_s)\, ds\}\, dt]$$

(2)

where $X_s$ is a certain stochastic process associated with $\mathcal{A}$.

If we would like to use the Dirichlet equation for physical modeling, then it would be natural to replace $\mathcal{A}$ and/or $\kappa$ by stochastic functionals. In the white noise setting, replacing $\kappa$ with $\text{Exp}\{W_{\phi_x}\}$ would seem to be an interesting choice, but where should we be looking for solutions? Although $\text{Exp}\{W_{\phi_x}\}$ is in $\mathcal{L}^2(\mu)$, it seems clear that this would not be the case for the solution candidate (2). The next logical step would be the space $(\mathcal{S})^*$ of Hida distributions, but since, given arbitrary real constants $K_1$ and $K_2$, there always exists $x \in \mathbb{R}$ such that

$$|S(\text{Exp}\{-\text{Exp}\{W_\phi\}\})((x + i\frac{\pi}{\|\phi\|^2})\phi)| = |\exp\{-\exp\{x\|\phi\|^2 + i\pi\}\}|$$

$$= \exp\{\exp\{x\|\phi\|^2\}\}$$

$$> K_1 \exp\{K_2|x + i\frac{\pi}{\|\phi\|^2}|^2\|\phi\|^2\}$$

it is clear from the characterization theorem in [HKPS] that $(\mathcal{S})^*$ is probably not the right space to look for solutions either. Fortunately, recently there have been constructed new spaces of generalized white noise distributions which will be adequate for our needs. These spaces will be described in the next section.

The methods used to solve (1) are generalizations of those used by Holden et al. in [HLØUZ] and [HLØUZ3]. In particular, theorem 3.1 may be seen as a generalization of theorem 10.2 in [HLØUZ] and theorem 4.1 generalizes theorem 3.1 in [HLØUZ3] ($\kappa \equiv 0$).

# §2 Preliminaries on multidimensional white noise

There are many problems of physical nature where the need for several independent white noise sources arises. For example, given $m$ independent positive white noise sources in a domain D, one wants to calculate the effect of these on a particle traveling in D. The result should intuitively be given by

$$\sum_{i=1}^{m} \mathrm{Exp}\{W_{\phi}^{(i)}\}$$

where $\{\mathrm{Exp}\{W_{\phi}^{(i)}\}\}_{i=1}^{m}$ are one dimensional independent positive white noise sources.

We will now give a short introduction of definitions and results from multidimensional Wick calculus, taken mostly from [Gj], [HLØUZ3], [HKPS] and [KLS].

In the following we will fix the parameter dimension $n$ and space dimension $m$.

Let

$$\mathcal{N} := \prod_{i=1}^{m} \mathcal{S}(\mathbb{R}^{n})$$

where $\mathcal{S}(\mathbb{R}^{n})$ is the Schwartz space of rapidly decreasing $C^{\infty}$-functions on $\mathbb{R}^{n}$, and

$$\mathcal{N}^{*} := (\prod_{i=1}^{m} \mathcal{S}(\mathbb{R}^{n}))^{*} \approx \prod_{i=1}^{m} \mathcal{S}'(\mathbb{R}^{n})$$

where $\mathcal{S}'(\mathbb{R}^{n})$ is the space of tempered distributions.

Let $\mathcal{B} := \mathcal{B}(\mathcal{N}^{*})$ denote the Borel $\sigma$-algebra on $\mathcal{N}^{*}$ equipped with the weak star topology and set

$$\mathcal{H} := \bigoplus_{i=1}^{m} \mathcal{L}^{2}(\mathbb{R}^{n})$$

where $\oplus$ denotes orthogonal sum.

Since $\mathcal{N}$ is a countably Hilbert nuclear space (cf. eg.[Gj]) we get, using Minlos' theorem, a unique probability measure $\nu$ on $(\mathcal{N}^*, \mathcal{B})$ such that

$$\int_{\mathcal{N}^*} e^{i\langle \omega, \phi \rangle} \, d\nu(\omega) = e^{-\frac{1}{2}\|\phi\|_{\mathcal{H}}^2} \quad \forall \phi \in \mathcal{N}$$

where $\|\phi\|_{\mathcal{H}}^2 = \sum_{i=1}^m \|\phi_i\|_{\mathcal{L}^2(\mathbb{R}^n)}^2$.

Note that if $m = 1$ then $\nu$ is usually denoted by $\mu$.

THEOREM 2.1 [Gj] We have the following

1. $\otimes_{i=1}^m \mathcal{B}(\mathcal{S}'(\mathbb{R}^n)) = \mathcal{B}(\prod_{i=1}^m \mathcal{S}'(\mathbb{R}^n))$

2. $\nu = \times_{i=1}^m \mu$

DEFINITION 2.2 [Gj] The triple

$$\left( \prod_{i=1}^m \mathcal{S}'(\mathbb{R}^n), \mathcal{B}, \nu \right)$$

is called the ($m$-dimensional) ($n$-parameter) white noise probability space.

For $k = 0, 1, 2, \ldots$ and $x \in \mathbb{R}$ let

$$h_k(x) := (-1)^k e^{\frac{x^2}{2}} \frac{d^k}{dx^k} \left( e^{-\frac{x^2}{2}} \right)$$

be the Hermite polynomials and

$$\xi_k(x) := \pi^{-\frac{1}{4}} ((k-1)!)^{-\frac{1}{2}} e^{-\frac{x^2}{2}} h_{k-1}(\sqrt{2}x) \; ; \; k \geq 1$$

the Hermite functions.

It is well known that the family $\{\tilde{e}_\alpha\} \subset \mathcal{S}(\mathbb{R}^n)$ of tensor products

$$\tilde{e}_\alpha := \xi_{\alpha_1} \otimes \cdots \otimes \xi_{\alpha_n}$$

forms an orthonormal basis for $\mathcal{L}^2(\mathbb{R}^n)$.

Give the family of all multi-indecies $\zeta = (\zeta_1, \dots, \zeta_n)$ a fixed ordering

$$(\zeta^{(1)}, \zeta^{(2)}, \dots, \zeta^{(k)}, \dots) \text{ where } \zeta^{(k)} = (\zeta_1^{(k)}, \dots, \zeta_n^{(k)})$$

and define $\tilde{e}_k := \tilde{e}_{\zeta^{(k)}}$.

Let $\{e_k\}_{k=1}^{\infty}$ be the orthonormal basis of $\mathcal{H}$ we get from the collection

$$\{(\overbrace{0, \dots, 0}^{i-1}, \tilde{e}_j, \overbrace{0, \dots, 0}^{m-i}) \in \mathcal{H} \ \ 1 \le i \le m, 1 \le j < \infty\}$$

and let $\gamma : \mathbf{N} \to \mathbf{N}$ be a function such that

$$e_k = (0, \dots, 0, \tilde{e}_{\zeta^{(\gamma(k))}}, 0, \dots, 0).$$

Finally , let $(\beta^{(1)}, \beta^{(2)}, \dots, \beta^{(k)}, \dots)$ with $\beta^{(k)} = (\beta_1^{(k)}, \dots, \beta_n^{(k)})$ be a sequence
such that $\beta^{(k)} = \zeta^{(\gamma(k))}$.

If $\alpha = (\alpha_1, \dots, \alpha_k)$ is a multi-index of non-negative integers we put

$$H_\alpha(\omega) := \prod_{i=1}^{k} h_{\alpha_i}(\langle \omega, e_i \rangle).$$

From theorem 2.1 in [HLØUZ] we know that the collection

$$\{H_\alpha(\cdot); \alpha \in \mathbf{N}_0^k; k = 0, 1, \dots\}$$

forms an orthogonal basis for $\mathcal{L}^2(\mathcal{N}^*, \mathcal{B}, \nu)$ with $\|H_\alpha\|_{\mathcal{L}^2(\nu)} = \alpha!$ where
$\alpha! = \prod_{i=1}^{k} \alpha_i!$.

This implies that any $f \in \mathcal{L}^2(\nu)$ has the unique representation

$$f(\omega) = \sum_\alpha c_\alpha H_\alpha(\omega)$$

where $c_\alpha \in \mathbb{R}$ for each multi-index $\alpha$ and

$$\|f\|_{\mathcal{L}^2(\nu)}^2 = \sum_\alpha \alpha! c_\alpha^2.$$

DEFINITION 2.3 [Gj] The $m$-dimensional white noise map is a map

$$W : \prod_{i=1}^{m} \mathcal{S}(\mathbb{R}^n) \times \prod_{i=1}^{m} \mathcal{S}'(\mathbb{R}^n) \to \mathbb{R}^m$$

given by

$$W^{(i)}(\phi, \omega) := \omega_i(\phi_i) \quad 1 \le i \le m$$

PROPOSITION 2.4 [Gj] The $m$-dimensional white noise map $W$ satisfies the following

1. $\{W^{(i)}(\phi, \cdot)\}_{i=1}^{m}$ is a family of independent normal random variables.

2. $W^{(i)}(\phi, \cdot) \in \mathcal{L}^2(\nu)$ for $1 \le i \le m$.

DEFINITION 2.5 [HLØUZ3] Let $0 \le \rho \le 1$.

• Let $(\mathcal{S}_n^m)^\rho$, the space of generalized white noise test functions, consist of all

$$f = \sum_\alpha H_\alpha \in \mathcal{L}^2(\nu)$$

such that

$$\|f\|_{\rho,k}^2 := \sum_\alpha c_\alpha^2 (\alpha!)^{1+\rho} (2N)^{\alpha k} < \infty \quad \forall k \in \mathbf{N}$$

• Let $(\mathcal{S}_n^m)^{-\rho}$, the space of generalized white noise distributions, consist of all formal expansions

$$F = \sum_\alpha b_\alpha H_\alpha$$

such that

$$\sum_\alpha b_\alpha^2 (\alpha!)^{1-\rho} (2N)^{-\alpha q} < \infty \text{ for some } q \in \mathbf{N}$$

where

$$(2N)^\alpha := \prod_{i=1}^{k} (2^n \beta_1^{(i)} \cdots \beta_n^{(i)})^{\alpha_i} \text{ if } \alpha = (\alpha_1, \ldots, \alpha_k).$$

We know that $(\mathcal{S}_n^m)^{-\rho}$ is the dual of $(\mathcal{S}_n^m)^\rho$ (when the later space has the topology given by the seminorms $\|\cdot\|_{\rho,k}$) and if $F = \sum b_\alpha H_\alpha \in (\mathcal{S}_n^m)^{-\rho}$ and $f = \sum c_\alpha H_\alpha \in (\mathcal{S}_n^m)^\rho$ then

$$\langle F, f \rangle = \sum_\alpha b_\alpha c_\alpha \alpha!.$$

It is obvious that we have the inclusions

$$(\mathcal{S}_n^m)^1 \subset (\mathcal{S}_n^m)^\rho \subset (\mathcal{S}_n^m)^{-\rho} \subset (\mathcal{S}_n^m)^{-1} \quad \rho \in [0, 1]$$

and in the remaining of this paper we will consider the larger space $(\mathcal{S}_n^m)^{-1}$.

DEFINITION 2.6 [HLØUZ3] The Wick product of two elements in $(\mathcal{S}_n^m)^{-1}$ given by

$$F = \sum_\alpha a_\alpha H_\alpha \ , \ \ G = \sum_\beta b_\beta H_\beta$$

is defined by

$$F \diamond G = \sum_\gamma c_\gamma H_\gamma$$

where

$$c_\gamma = \sum_{\alpha+\beta=\gamma} a_\alpha b_\beta$$

LEMMA 2.7 [HLØUZ3] We have the following

1. $F, G \in (\mathcal{S}_n^m)^{-1} \Rightarrow F \diamond G \in (\mathcal{S}_n^m)^{-1}$

2. $f, g \in (\mathcal{S}_n^m)^1 \Rightarrow f \diamond g \in (\mathcal{S}_n^m)^1$

DEFINITION 2.8 [HLØUZ3] Let $F = \sum b_\alpha H_\alpha$ be given. Then the Hermite transform of F, denoted by $\mathcal{H}F$, is defined to be (whenever convergent)

$$\mathcal{H}F := \sum_\alpha b_\alpha z^\alpha$$

where $z = (z_1, z_2, \cdots)$ and $z^\alpha = z_1^{\alpha_1} z_2^{\alpha_2} \cdots z_k^{\alpha_k}$ if $\alpha = (\alpha_1, \ldots, \alpha_k)$.

LEMMA 2.9 [HLØUZ3] If $F, G \in (\mathcal{S}_n^m)^{-1}$ then

$$\mathcal{H}(F \diamond G)(z) = \mathcal{H}F(z) \cdot \mathcal{H}G(z)$$

for all z such that $\mathcal{H}F(z)$ and $\mathcal{H}G(z)$ exists.

LEMMA 2.10 [HLØUZ3] Suppose $g(z_1, z_2, \cdots)$ is a bounded analytic function on $B_q(\delta)$ for some $\delta > 0, q < \infty$ where

$$B_q(\delta) := \{\zeta = (\zeta_1, \zeta_2, \cdots) \in \mathbb{C}_0^N; \sum_{\alpha \neq 0} |\zeta^\alpha|^2 (2N)^{\alpha q} < \delta^2\}.$$

Then there exists $X \in (S_n^m)^{-1}$ such that $\mathcal{H}X = g$.

LEMMA 2.11 [HLØUZ3] Suppose $X \in (S_n^m)^{-1}$ and that $f$ is an analytic function in a neighborhood of $\mathcal{H}X(0)$ in $\mathbb{C}$. Then there exists $Y \in (S_n^m)^{-1}$ such that $\mathcal{H}Y = f \circ \mathcal{H}X$.

THEOREM 2.12 [KLS] Let $(T, \Sigma, \tau)$ be a measure space and let $\Phi : T \rightarrow (S_n^m)^{-1}$ be such that there exists $q < \infty, \delta > 0$ such that

1. $\mathcal{H}\Phi_t(z) : T \rightarrow \mathbb{C}$ is measurable for all $z \in B_q(\delta)$

2. there exists $C \in \mathcal{L}^1(T, \tau)$ such that $|\mathcal{H}\Phi_t(z)| \leq C(t)$ for all $z \in B_q(\delta)$ and for $\tau$-almost all t.

Then $\int_T \Phi_t \, d\tau(t)$ exists as a Bochner integral in $(S_n^m)^{-1}$. In particular, $\langle \int_T \Phi_t \, d\tau(t), \phi \rangle = \int_T \langle \Phi_t, \phi \rangle \, d\tau(t)$ ; $\phi \in (S_n^m)^1$.

EXAMPLE 2.13 Define the x-shift of $\phi$, denoted by $\phi_x$, by $\phi_x(y) := \phi(y - x)$. Then

$$\mathrm{Exp}\{W_{\phi_x}^{(i)}\} \in (S_n^m)^{-1} \quad 1 \leq i \leq m, \forall x \in \mathbb{R}^n$$

which is an immediate consequence of proposition 2.4 and lemma 2.11.

# §3 The first class of stochastic Dirichlet equations

We will in this section let $X_t$ be the solution of the stochastic integral equation

$$X_t^x = x + \int_0^t b(X_\theta^x) \, d\theta + \int_0^t \sigma(X_\theta^x) \, db_\theta \tag{3}$$

under the assumptions that

- the coefficients $b_i(x)$, $\sigma_{i,k}(x) : \mathbb{R}^n \to \mathbb{R}$ are continuous and satisfy $\|b(x)\|_{\mathbb{R}^n}^2 + \|\sigma(x)\|_{\mathbb{R}^n}^2 \leq K(1 + \|x\|_{\mathbb{R}^n}^2)$ for all $x \in \mathbb{R}^n$, where K is a positive constant.

- the equation (3) has a weak solution $(X_t^x, b_t)$, $(\Omega, \mathcal{F}, \hat{P}^x)$, $\{\mathcal{F}_s\}$ for all $x \in \mathbb{R}^n$ and this solution is unique in the sense of probability law.

We will use the notation that

- $\hat{E}^x$ is expectation w.r.t. $\hat{P}^x$.

- $\tau_D = \tau_D^{X_t} := \inf\{t > 0 : X_t \in D^c\}$ is the first exit time from D for $X_t$.

THEOREM 3.1 Let D be an open, bounded domain in $\mathbb{R}^n$, $f \in C_b(\partial D)$, $g \in C^\alpha(\bar{D})$, $\mathcal{A}$ the differential operator on $C^2(\mathbb{R}^n)$, associated with $X_t$, given by

$$\mathcal{A}h(x) := \frac{1}{2} \sum_{i=1}^n \sum_{k=1}^n (\sigma\sigma^T)_{i,k}(x) \frac{\partial^2 h(x)}{\partial x_i \partial x_k} + \sum_{i=1}^n b_i(x) \frac{\partial h(x)}{\partial x_i} \; ; \; h \in C^2(D)$$

and suppose that we have the following

- $\mathcal{A}$ is uniformly elliptic in D.

- the functions $(\sigma\sigma^T)_{i,k}$, $b_i$ are Hölder continuous in D.

- every point $x \in \delta D$ has the exterior sphere property; i.e there exists a ball $B \ni x$ such that $\bar{B} \cap D = \emptyset$, $\bar{B} \cap \partial D = \{x\}$.

Then

$$u(x) = \hat{E}^x[f(X_{\tau_D})\text{Exp}\{-\int_0^{\tau_D} \mathcal{U}(X_s)\,ds\} + \int_0^{\tau_D} g(X_t)\text{Exp}\{-\int_0^t \mathcal{U}(X_s)\,ds\}\,dt]$$

is the unique $(\mathcal{S}_n^m)^{-1}$-valued process which solves the stochastic Dirichlet problem

$$\begin{aligned} \mathcal{A}u(x) &= \mathcal{U}(x) \diamond u(x) - g(x) \quad &x \in D \\ u(x) &= f(x) \quad &x \in \partial D \end{aligned} \tag{4}$$

where

$$\mathcal{U}(x) = \sum_{i=1}^{m} c_i \cdot \text{Exp}\{W_{\phi_x}^{(i)}\} \quad (c_i \in \mathbb{R}_+) \tag{5}$$

is the potential given by sources of independent positive white noise, $\hat{E}^x$, $\int_0^{\tau_D} \cdot \, ds$ and $\int_0^t \cdot \, ds$ are the Bochner integrals in $(\mathcal{S}_n^m)^{-1}$.

Remark : If $u(x) \in (\mathcal{S}_n^m)^{-1}$ and $\mathcal{A}(\mathcal{H}u(x)) \in A_b(B_q(\delta))$, where $A_b(B_q(\delta))$ is the space of all bounded analytic functions on $B_q(\delta)$ , for some $q \in \mathbb{N}, \delta > 0$, we will use the convention that $\mathcal{A}u(x) := \mathcal{H}^{-1}\mathcal{A}(\mathcal{H}u(x))$.

PROOF:

We will assume that $n = m = c = 1$ for simplicity.

We must find $\delta > 0, q < \infty$ such that $\tilde{u}(x, z) := \mathcal{H}(u(x))(z) \in A_b(B_q(\delta))$ solves the equation

$$\begin{aligned} \mathcal{A}\tilde{u}(x) &= \exp\{\tilde{W}_{\phi_x}\} \cdot \tilde{u}(x) - g(x) \quad x \in D \\ \tilde{u}(x) &= f(x) \qquad\qquad\qquad\qquad x \in \partial D \end{aligned} \tag{6}$$

when $z \in B_q(\delta)$.

Fix $\hat{q} \in \mathbb{N}$ and $\hat{\delta}$ with $0 < \hat{\delta} < \frac{\pi}{2\|\phi\|}$.

LEMMA 3.2 $\tilde{u}(x, z) \in A_b(B_{\hat{q}}(\hat{\delta})) \, \forall x \in D$.

PROOF:

Since

$$\begin{aligned} |\tilde{W}_{\phi_x}(z)|^2 &= |\sum_{k=0}^{\infty} (\phi_x, e_k)z_k|^2 \\ &\le \sum_{k=0}^{\infty} (\phi_x, e_k)^2 \cdot \sum_{k=0}^{\infty} |z_k|^2 \\ &\le \|\phi\|^2 \cdot \sum_{\alpha \neq 0} |z^\alpha|(2N)^{\alpha q} \\ &\le \hat{\delta}^2 \|\phi\|^2 \end{aligned}$$

when $z \in B_{\hat{q}}(\hat{\delta})$ it follows that

$$|\tilde{u}(x,z)| \le K_f \cdot \hat{E}^x [|\exp\{-\int_0^{\tau_D} \exp\{\tilde{W}_{\phi x_s}\}\,ds\}|]$$

$$+ K_g \cdot \hat{E}^x [\int_0^{\tau_D} |\exp\{-\int_0^t \exp\{\tilde{W}_{\phi x_s}\}\,ds\}|\,dt]$$

$$= K_f \cdot \hat{E}^x [|\exp\{-\int_0^{\tau_D} \exp\{\Re(\tilde{W}_{\phi x_s}) + i\Im(\tilde{W}_{\phi x_s})\}\,ds\}|]$$

$$+ K_g \cdot \hat{E}^x [\int_0^{\tau_D} |\exp\{-\int_0^t \exp\{\Re(\tilde{W}_{\phi x_s}) + i\Im(\tilde{W}_{\phi x_s})\}\,ds\}|\,dt]$$

$$= K_f \cdot \hat{E}^x [|\exp\{-\int_0^{\tau_D} \exp\{\Re(\tilde{W}_{\phi x_s})\}\cos\Im(\tilde{W}_{\phi x_s})\,ds\}|]$$

$$+ K_g \cdot \hat{E}^x [\int_0^{\tau_D} |\exp\{-\int_0^t \exp\{\Re(\tilde{W}_{\phi x_s})\}\cos\Im(\tilde{W}_{\phi x_s})\,ds\}|\,dt]$$

$$\le K_f + K_g \cdot \hat{E}^x [\int_0^{\tau_D} \exp\{-t\exp\{-\hat{\delta}\|\phi\|\}\cos(\hat{\delta}\|\phi\|)\}\,dt]$$

$$\le K_f + \frac{2K_g \cdot \exp\{\hat{\delta}\|\phi\|\}}{\cos(\hat{\delta}\|\phi\|)}$$

where

$$K_f := \sup_{x\in\partial D} |f(x)|, \quad K_g := \sup_{x\in\bar{D}} |g(x)|.$$

∎

LEMMA 3.3 $u(x)$ is well-defined as a Bochner integral in $(\mathcal{S}_n^m)^{-1}$.

PROOF:

Estimating as in lemma 3.2 , we get with

$$\Phi(x, z, \omega) := f(X_{\tau_D}) \exp\{- \int_0^{\tau_D} \exp\{\tilde{W}_{\phi x_s}\} \, ds\}$$

$$+ \int_0^{\tau_D} g(X_t) \exp\{- \int_0^t \exp\{\tilde{W}_{\phi x_s}\} \, ds\} \, dt$$

that

$$|\Phi(x, z, \omega)| \leq K_f + \frac{2K_g \cdot \exp\{\hat{\delta}\|\phi\|\}}{\cos(\hat{\delta}\|\phi\|)}$$

whenever $z \in B_{\hat{q}}(\hat{\delta})$, i.e it follows from theorem 2.12 that $u(x)$ is given as a Bochner integral in $(\mathcal{S}_n^m)^{-1}$. ∎

LEMMA 3.4 $\tilde{u}(x, z)$ is the unique function which solves equation (6) when $z \in B_{\hat{q}}(\hat{\delta})$.

PROOF:

Since

$$\tilde{W}_{\phi x}(z) = \sum_{k=0}^{\infty} (\phi_x, e_k) z_k$$

for all $z \in \mathbb{C}_0^N$, it follows that $\exp\{\tilde{W}_{\phi x}\} \geq 0 \; \forall x \in D$ whenever $z \in B_{\hat{q}}(\hat{\delta}) \cap \mathbb{R}_0^N$. In this case the claim now follows from proposition 7.2 and remark 7.5 in [KS], and by expanding the natural analytic extension of $\tilde{u}(x, z)$ into real and imaginary parts, the result follows for all $z \in B_{\hat{q}}(\hat{\delta})$.

∎

LEMMA 3.5 $\mathcal{A}u(x)$ is well-defined as an element in $(\mathcal{S}_n^m)^{-1} \; \forall x \in D$.

PROOF:

Since

$$\mathcal{A}\tilde{u} = \exp\{\tilde{W}_{\phi x}\} \cdot \tilde{u} - g$$

it follows from lemma 3.2 that

$$|\mathcal{A}\tilde{u}(x, z)| \leq \exp\{\hat{\delta}\|\phi\|\} \cdot (K_f + \frac{2K_g \cdot \exp\{\hat{\delta}\|\phi\|\}}{\cos(\hat{\delta}\|\phi\|)}) + K_g$$

when $z \in B_{\hat{q}}(\hat{\delta})$, i.e the claim follows. ∎

The theorem now follows from the previous lemmas. ∎

REMARK 3.6 It is easy to extend theorem 3.1 into more general situations.

Let $f(x) \in (S_n^m)^{-1}$ $\forall x \in \partial D$, $g(x) \in (S_n^m)^{-1}$ $\forall x \in \bar{D}$ and $\mathcal{U}(x) \in (S_n^m)^{-1}$ $\forall x \in \bar{D}$ be given and assume that there exits $\tilde{q} \in N$ and $\tilde{\delta} > 0$ such that the following holds:

- $\mathcal{H}f(x, z) \in C(\partial D)$ when $z \in B_{\tilde{q}}(\tilde{\delta})$ and $\exists K_f > 0$ (independent of $x, z$) such that
$$\sup_{x \in \partial D, z \in B_{\tilde{q}}(\tilde{\delta})} |\mathcal{H}f(x, z)| \leq K_f$$

- $\exists \alpha(z) > 0$ such that $\mathcal{H}g(x, z) \in C^{\alpha(z)}(\bar{D})$ when $z \in B_{\tilde{q}}(\tilde{\delta})$ and $\exists K_g > 0$ (independent of $x, z$) such that
$$\sup_{x \in \bar{D}, z \in B_{\tilde{q}}(\tilde{\delta})} |\mathcal{H}g(x, z)| \leq K_g$$

- $\exists \beta(z) > 0$ such that $\mathcal{H}\mathcal{U}(x, z) \in C^{\beta(z)}(\bar{D})$ when $z \in B_{\tilde{q}}(\tilde{\delta})$ and $\exists K_{\mathcal{U}}(x) > 0$ (independent of $z$) such that
$$\sup_{z \in B_{\tilde{q}}(\tilde{\delta})} |\mathcal{H}\mathcal{U}(x, z)| \leq K_{\mathcal{U}}(x) \quad \forall x \in \bar{D}$$

- $\Re(\mathcal{H}\mathcal{U}(x, z)) \geq 0$ $\forall x \in \bar{D}, z \in B_{\tilde{q}}(\tilde{\delta})$.

Then

$$u(x) = \hat{E}^x[f(X_{\tau_D}) \diamond \text{Exp}\{-\int_0^{\tau_D} \mathcal{U}(X_s) \, ds\} + \int_0^{\tau_D} g(X_t) \diamond \text{Exp}\{-\int_0^t \mathcal{U}(X_s) \, ds\} \, dt]$$

is the unique $(S_n^m)^{-1}$-valued process which solves the (modified) stochastic Dirichlet problem given by (4).

COROLLARY 3.7 (The wave equation)

Assume that the assumptions of theorem 3.1 are valid.
Then

$$\Psi(t, x) = \hat{E}^x[\sinh(t)f(b_{\tau_D})\text{Exp}\{-\int_0^{\tau_D} (\mathcal{U}(b_s) \wedge c)^{\diamond-1} \, ds\}]$$

is an $(\mathcal{S}_n^m)^{-1}$-valued process which solves the stochastic wave equation

$$\frac{\partial^2 \Psi}{\partial t^2}(t,x) = (\mathcal{U}(x) \wedge c) \diamond \Delta_x \Psi(t,x) \quad (t,x) \in \mathbb{R}_+ \times D$$

$$\Psi(0,x) = 0 \qquad\qquad\qquad x \in \bar{D}$$

$$\frac{\partial \Psi}{\partial t}(0,x) = f(x) \qquad\qquad\qquad x \in \partial D$$

where $c \in \mathbb{R}_+ \cup \{\infty\}$ and $\mathcal{U}(x)$ is the potential given by (5).

PROOF:

It is clear that the boundary conditions are satisfied and that

$$\frac{\partial^2 \Psi}{\partial t^2}(t,x) = \Psi(t,x)$$

i.e we must show that the following equation is satisfied

$$\Psi(t,x) = (\mathcal{U}(x) \wedge c) \diamond \Delta_x \Psi(t,x)$$

or equivalent

$$\Delta_x \Psi(t,x) = (\mathcal{U}(x) \wedge c)^{\diamond -1} \diamond \Psi(t,x)$$

but this is nothing but the (modified) Dirichlet equation of remark 3.6, i.e the result follows. ∎

COROLLARY 3.8 (The Schrödinger equation)

Assume that the assumptions of theorem 3.1 are valid.
Then

$$\Psi(t,x) = \hat{E}^x[\exp\{i\frac{\hbar}{2\bar{m}}t\}f(b_{\tau_D})\text{Exp}\{-\int_0^{\tau_D} (1 + \frac{2\bar{m}}{\hbar^2}\mathcal{U}(b_s))\,ds\}]$$

is an $(\mathcal{S}_n^m)_c^{-1}$-valued process which solves the stochastic Schrödinger equation

$$-\frac{\hbar^2}{2\bar{m}}\Delta_x \Psi(t,x) + \mathcal{U}(x) \diamond \Psi(t,x) = i\hbar\frac{\partial \Psi}{\partial t}(t,x) \quad (t,x) \in \mathbb{R}_+ \times D$$

$$\Psi(0,x) = f(x) \qquad\qquad x \in \partial D$$

where $\mathcal{U}(x)$ is the potential given by (5), $i = \sqrt{-1}$ is the imaginary unit, $\hbar$ is Planck's constant divided by $2\pi$ and $\bar{m}$ is the mass of the particle in study.

PROOF:

It is clear that the boundary condition is satisfied and that

$$\frac{\partial \Psi}{\partial t}(t, x) = \frac{\hbar}{2\bar{m}} i \Psi(t, x)$$

i.e we must show that the following equation is satisfied

$$\Delta_x \Psi(t, x) = (1 + \frac{2\bar{m}}{\hbar} \mathcal{U}(x)) \diamond \Psi(t, x)$$

but this is again nothing but the (modified) stochastic Dirichlet equation of remark 3.6, i.e the result follows. $\blacksquare$

# §4 The second class of stochastic Dirichlet equations

We will in this section assume that $m = 1$ and use the notation

- $\hat{E}^x$ is expectation w.r.t. the measure $\hat{P}^x$.

- $\tau_D = \tau_D^{b_t} := \inf\{t > 0 : b_t \in D^c\}$ is the first exit time from $D$ for $b_t$.

where $(b_t(\tilde{\omega}), \hat{P}^x)$ is a Brownian motion in $\mathbb{R}^n$.

THEOREM 4.1 Let $D$ be an open, bounded domain in $\mathbb{R}^n$ such that every point $x \in \partial D$ has the exterior sphere property; i.e. there exists a ball $B \ni {}_\wedge$ such that $\bar{B} \cap D = \emptyset$, $\bar{B} \cap \partial D = \{x\}$.

Assume further that we are given functions $\bar{D} \ni x \mapsto \kappa(x) \in (\mathcal{S})^{-1}$, $\partial D \ni x \mapsto f(x) \in (\mathcal{S})^{-1}$ and $\bar{D} \ni x \mapsto g(x) \in (\mathcal{S})^{-1}$ such that

- $\exists (q_f \in \mathbb{N}, \delta_f > 0, K_f > 0)$ such that

    1. $\sup_{x \in \partial D, z \in B_{q_f}(\delta_f)} |\mathcal{H}f(x, z)| \leq K_f$
    2. $x \mapsto \mathcal{H}f(x, z) \in C(\partial D)$ whenever $z \in B_{q_f}(\delta_f)$.

- $\exists(q_g \in \mathbb{N}, \delta_g > 0, K_g > 0)$ such that

    1. $\sup_{x \in \bar{D}, z \in B_{q_g}(\delta_g)} |\mathcal{H}g(x,z)| \le K_g$
    2. $\exists(\alpha(z) > 0 \ \forall z \in B_{q_g}(\delta_g))$ such that $x \mapsto \mathcal{H}g(x,z) \in C^{\alpha(z)}(\bar{D})$.

- $\exists(q_\kappa \in \mathbb{N}, \delta_\kappa > 0, 0 < \epsilon_\kappa < \frac{\pi}{2}, K_\kappa(x) > 0 \ \forall x \in \bar{D})$ such that

    1. $\sup_{z \in B_{q_\kappa}(\delta_\kappa)} |\mathcal{H}\kappa(x,z)| \le K_\kappa(x)$
    2. $\exists(\beta(z) > 0 \ \forall z \in B_{q_\kappa}(\delta_\kappa))$ such that $x \mapsto \mathcal{H}\kappa(x,z) \in C^{\beta(z)}(\bar{D})$.
    3. $\sup_{x \in \bar{D}, z \in B_{q_\kappa}(\delta_\kappa)} |\arg(\mathcal{H}\kappa(x,z))| \le \frac{\pi}{2} - \epsilon_\kappa$.
    4. $\mathcal{H}\kappa(x,z) \ge 0 \ \ \forall x \in \bar{D}$ whenever $z \in B_{q_\kappa}(\delta_\kappa) \cap \mathbb{R}_0^N$.

Then

$$u(x) = \mathrm{Exp}\{W_{-\frac{1}{2}\phi_x}\}$$

$$\diamond \hat{E}^x[f(b_{\tau_D}) \diamond \mathrm{Exp}\{-\frac{1}{2}\int_0^{\tau_D} \kappa(b_s) \diamond \mathrm{Exp}\{W_{-\phi_{b_s}}\} \, ds\} \diamond \mathcal{J}(\tau_D, x)]$$

$$+ \frac{1}{2}\mathrm{Exp}\{W_{-\frac{1}{2}\phi_x}\}$$

$$\diamond \hat{E}^x[\int_0^{\tau_D} \hat{g}(b_t) \diamond \mathrm{Exp}\{-\frac{1}{2}\int_0^t \kappa(b_s) \diamond \{W_{-\phi_{b_s}}\} \, ds\} \diamond \mathcal{J}(t, x) \, dt]$$

where

$$\mathcal{J}(t, x) := \mathrm{Exp}\{\frac{1}{2}W_{\phi_{b_t}} - \frac{1}{4}\int_0^t [\frac{1}{2}(\nabla W_{\phi_x})^{\diamond 2} + \Delta W_{\phi_x}]_{x=b_s} \, ds\} \tag{7}$$

and

$$\hat{g}(x) := g(x) \diamond \mathrm{Exp}\{W_{-\phi_x}\}$$

is the unique $(\mathcal{S})^{-1}$-valued process which solves

$$\mathrm{div}(\mathrm{Exp}\{W_{\phi_x}\} \diamond \nabla u) = \kappa(x) \diamond u(x) - g(x) \quad x \in D$$
$$u(x) = f(x) \quad\quad\quad\quad x \in \partial D$$

where $\hat{E}^x$, $\int_0^t \cdot \, ds$ and $\int_0^{\tau_D} \cdot \, ds$ are Bochner integrals in $(\mathcal{S})^{-1}$.

REMARK 4.2 : We use the convention that arg is defined to be the function
arg : $\mathbb{C} \to < -\pi, \pi]$ given by the relation $z = |z| \exp\{i \cdot \arg(z)\}$.

REMARK 4.3 : If $u(x) \in (\mathcal{S})^{-1}$ and $\operatorname{div}(\exp\{\mathcal{H}W_{\phi_x}\} \cdot \mathcal{H}u(x)) \in A_b(B_q(\delta))$ for some
$q \in \mathbb{N}, \delta > 0$, we will use the convention that

$$\operatorname{div}(\operatorname{Exp}\{W_{\phi_x}\} \diamond \nabla u) := \mathcal{H}^{-1}(\operatorname{div}(\exp\{\mathcal{H}W_{\phi_x}\} \cdot \mathcal{H}u(x))).$$

PROOF:

We must find $\hat{\delta} > 0$, $\hat{q} \in \mathbb{N}$ such that $\tilde{u}(x, z) := \mathcal{H}(u(x))(z) \in A_b(B_{\hat{q}}(\hat{\delta}))$ solves the equation

$$\begin{aligned} \operatorname{div}(\exp\{\tilde{W}_{\phi_x}\} \cdot \nabla \tilde{u}) &= \tilde{\kappa}(x) \cdot \tilde{u}(x) - \tilde{g}(x) & x \in D \\ \tilde{u}(x) &= \tilde{f}(x) & x \in \partial D \end{aligned} \tag{8}$$

when $z \in B_{\hat{q}}(\hat{\delta})$.

LEMMA 4.4 $\exists (\hat{\delta} > 0, \hat{q} \in \mathbb{N})$ such that $z \mapsto \tilde{u}(x, z) \in A_b(B_{\hat{q}}(\hat{\delta})) \, \forall x \in D$.

PROOF:

It is clear that

$$|\exp\{\tilde{W}_{-\frac{1}{2}\phi_x}\}| \leq \exp\{\frac{\delta}{2}\|\phi\|\}$$

and

$$|\exp\{\frac{1}{2}\tilde{W}_{\phi_{b_{\tau_D}}}\}| \leq \exp\{\frac{\delta}{2}\|\phi_{b_{\tau_D}}\|\} = \exp\{\frac{\delta}{2}\|\phi\|\}$$

when $z \in B_q(\delta)$, since $\forall (q \in \mathbb{N}, \delta > 0)$

$$|\tilde{W}_{\phi_x}(z)|^2 \leq \delta^2 \|\phi\|^2.$$

Using this last estimate on

$$w(x, z) := \exp\{-\frac{1}{2} \int\limits_0^{\tau_D} \tilde{\kappa}(b_s) \exp\{\tilde{W}_{-\phi_{b_s}}\} ds\}$$

we get

$$|w| = |\exp\{-\frac{1}{2}\int_0^{\tau_D} |\tilde{\kappa}(b_s)|\exp\{i\arg(\tilde{\kappa}(b_s))\}\exp\{\Re(\tilde{W}_{-\phi_{b_s}}) + i\Im(\tilde{W}_{-\phi_{b_s}})\}\,ds\}|$$

$$= |\exp\{-\frac{1}{2}\int_0^{\tau_D} |\tilde{\kappa}(b_s)|\exp\{\Re(\tilde{W}_{-\phi_{b_s}})\}\exp\{i(\arg(\tilde{\kappa}(b_s)) + \Im(\tilde{W}_{-\phi_{b_s}}))\}\,ds\}|$$

$$= \exp\{-\frac{1}{2}\int_0^{\tau_D} |\tilde{\kappa}(b_s)|\exp\{\Re(\tilde{W}_{-\phi_{b_s}})\}\cos(\arg(\tilde{\kappa}(b_s)) + \Im(\tilde{W}_{-\phi_{b_s}}))\,ds\}$$

$$\leq 1$$

whenever $q \leq \min\{q_f, q_g, q_\kappa\}$ and $0 < \delta < \min\{\delta_f, \delta_g, \delta_\kappa, \frac{\epsilon_\kappa}{\|\phi\|}\}$.

Applying these estimates on $\tilde{u}(x,z)$, still assuming $q \leq \min\{q_f, q_g, q_\kappa\}$ and
$0 < \delta < \min\{\delta_f, \delta_g, \delta_\kappa, \frac{\epsilon_\kappa}{\|\phi\|}\}$, we get

$$|\tilde{u}| \leq K_f \exp\{\delta\|\phi\|\}\hat{E}^x[|\exp\{-\frac{1}{4}\int_0^{\tau_D}[\frac{1}{2}(\nabla\tilde{W}_{\phi_x})^2 + \Delta\tilde{W}_{\phi_x}]_{x=b_s}\,ds\}|]$$

$$+ \frac{1}{2}K_g\exp\{2\delta\|\phi\|\}\hat{E}^x[\int_0^{\tau_D}|\exp\{-\frac{1}{4}\int_0^t[\frac{1}{2}(\nabla\tilde{W}_{\phi_x})^2 + \Delta\tilde{W}_{\phi_x}]_{x=b_s}\,ds\}|\,dt]$$

$$\leq (K_f\exp\{\delta\|\phi\|\} + \frac{1}{2}K_g\frac{1}{c(\delta)}\exp\{2\delta\|\phi\|\})\hat{E}^x[\exp\{c(\delta)\tau_D\}]$$

where

$$c(\delta) := \frac{1}{4}(\frac{\delta^2}{2}\sum_{i=1}^n \|\frac{\partial\phi}{\partial y_i}\|^2 + \delta\sum_{i=1}^n \|\frac{\partial^2\phi}{\partial y_i^2}\|)$$

since

$$|(\nabla \tilde{W}_{\phi_x})^2| = |\sum_{i=1}^{n} (\frac{\partial}{\partial x_i} \tilde{W}_{\phi_x})^2|$$

$$\leq \sum_{i=1}^{n} |\frac{\partial}{\partial x_i} \tilde{W}_{\phi_x}|^2$$

$$= \sum_{i=1}^{n} |\tilde{W}_{(\frac{\partial \phi}{\partial y_i})_x}|^2$$

$$\leq \delta^2 \sum_{i=1}^{n} \|\frac{\partial \phi}{\partial y_i}\|^2$$

and

$$|\Delta \tilde{W}_{\phi_x}| = |\sum_{i=1}^{n} \frac{\partial^2}{\partial x_i^2} \tilde{W}_{\phi_x}|$$

$$\leq \sum_{i=1}^{n} |\frac{\partial^2}{\partial x_i^2} \tilde{W}_{\phi_x}|$$

$$= \sum_{i=1}^{n} |\tilde{W}_{(\frac{\partial^2 \phi}{\partial y_i^2})_x}|$$

$$\leq \delta \sum_{i=1}^{n} \|\frac{\partial^2 \phi}{\partial y_i^2}\|.$$

We know from [DUR] that there exists $\rho > 0$ such that

$$\hat{E}^x[\exp\{\rho \tau_D\}] < \infty \quad \forall x \in D$$

i.e if we choose a $0 < \hat{\delta} < \min\{\delta_f, \delta_g, \delta_\kappa, \frac{\epsilon_\kappa}{\|\phi\|}\}$ such that

$$\frac{1}{4}(\frac{\delta^2}{2} \sum_{i=1}^{n} \|\frac{\partial \phi}{\partial y_i}\|^2 + \delta \sum_{i=1}^{n} \|\frac{\partial^2 \phi}{\partial y_i^2}\|) \leq \rho$$

and a $\hat{q} \leq \min\{q_f, q_g, q_\kappa\}$, then the claim follows. ∎

LEMMA 4.5 The Bochner integrals in the expression for $u(x)$ are well-defined.

PROOF:

This is obvious from the estimates in lemma 4.4. ∎

LEMMA 4.6 $\tilde{u}(x,z)$ is the unique function which solves equation (8) when $z \in B_{\hat{q}}(\hat{\delta})$.

PROOF:

Since

$$\mathrm{div}(\exp\{\tilde{W}_{\phi_x}\} \cdot \nabla \tilde{u}) = \sum_{i=1}^{n} \frac{\partial}{\partial x_i}(\exp\{\tilde{W}_{\phi_x}\}) \frac{\partial \tilde{u}}{\partial x_i} + \sum_{i=1}^{n} \exp\{\tilde{W}_{\phi_x}\} \frac{\partial^2 \tilde{u}}{\partial x_i^2}$$

$$= \sum_{i=1}^{n} \frac{\partial}{\partial x_i}(\tilde{W}_{\phi_x}) \exp\{\tilde{W}_{\phi_x}\} \frac{\partial \tilde{u}}{\partial x_i} + \sum_{i=1}^{n} \exp\{\tilde{W}_{\phi_x}\} \frac{\partial^2 \tilde{u}}{\partial x_i^2}$$

our problem (8) may be written as

$$\begin{aligned} \mathcal{A}^z \tilde{u} &= \frac{1}{2} \exp\{\tilde{W}_{-\phi_x}\} \tilde{\kappa}(x) \tilde{u} - \frac{1}{2} \exp\{\tilde{W}_{-\phi_x}\} \tilde{g}(x) \quad & x \in D \\ \tilde{u}(x) &= \tilde{f}(x) & x \in \partial D \end{aligned} \tag{9}$$

where $\mathcal{A}^z$ is the second order differential operator given by

$$\mathcal{A}^z h = \sum_{i=1}^{n} \frac{1}{2} \frac{\partial^2 h}{\partial x_i^2} + \sum_{i=1}^{n} \frac{1}{2} \frac{\partial}{\partial x_i}(\tilde{W}_{\phi_x}) \frac{\partial h}{\partial x_i} \; ; \; h \in C^2(D).$$

Assume now that $z = \zeta \in B_{\hat{q}}(\hat{\delta}) \cap \mathbb{R}_0^N$.

Then this operator is clearly uniformly elliptic in D and since the drift coefficient satisfies the linear growth condition

$$|(\frac{\partial}{\partial x_i} \tilde{W}_{\phi_{h_2}}) - (\frac{\partial}{\partial x_i} \tilde{W}_{\phi_{h_1}})|(\zeta) = |\tilde{W}_{(\frac{\partial \phi}{\partial y_i})_{h_2}}(\zeta) - \tilde{W}_{(\frac{\partial \phi}{\partial y_i})_{h_1}}(\zeta)|$$

$$= |\sum_{k=0}^{\infty} ((\frac{\partial \phi}{\partial y_i})_{h_2} - (\frac{\partial \phi}{\partial y_i})_{h_1}, e_k) \zeta_k|$$

$$\leq \sum_{k=0}^{\infty} |((\frac{\partial \phi}{\partial y_i})_{h_2} - (\frac{\partial \phi}{\partial y_i})_{h_1}, e_k)||\zeta_k|$$

$$\leq (M \sum_{k=0}^{\infty} \int_{\mathbb{R}^n} |e_k(x)| \, dx |\zeta_k|)|h_2 - h_1|$$

where

$$M := \max_{1 \leq i \leq n} \{\sup_{x \in \mathbb{R}^n} |\frac{\partial^2 \phi}{\partial x_i^2}|\}$$

the process given by

$$dX_t = \frac{1}{2}\nabla\tilde{W}_{\phi_x}(\zeta)\,dt + db_t \ ; \ X_0 = x.$$

exists with $\mathcal{A}^\zeta$ as the generator.

We know that

$$\tilde{W}_{\phi_x} = \sum_{k=0}^{\infty}(\phi_x, e_k)\zeta_k$$

so it follows that $\exp\{\tilde{W}_{\phi_x}(\zeta)\} \geq 0 \ \forall x \in D$ and from [KS] we know that the solution of (9) is given uniquely by

$$\tilde{u}(x, \zeta) = \hat{E}^x[\tilde{f}(X_{\tau_D^{X_t}})\exp\{-\frac{1}{2}\int_0^{\tau_D^{X_t}}\tilde{\kappa}(X_s)\exp\{\tilde{W}_{-\phi_{x_s}}\}\,ds\}]$$

$$+ \frac{1}{2}\hat{E}^x[\int_0^{\tau_D^{X_t}}\tilde{g}(X_t)\exp\{\tilde{W}_{-\phi_{x_t}}\}\exp\{-\int_0^t\tilde{\kappa}(X_s)\exp\{\tilde{W}_{-\phi_{x_s}}\}\,ds\}\,dt].$$

By a change of measure this may be written as

$$\tilde{u}(x, \zeta) = \hat{E}^x[\tilde{f}(b_{\tau_D})\exp\{-\frac{1}{2}\int_0^{\tau_D}\tilde{\kappa}(b_s)\exp\{\tilde{W}_{-\phi_{b_s}}\}\,ds\}\mathcal{M}(\tau_D)]$$

$$+ \frac{1}{2}\hat{E}^x[\int_0^{\tau_D}\tilde{g}(b_t)\exp\{\tilde{W}_{-\phi_{b_t}}\}\exp\{-\int_0^t\tilde{\kappa}(b_s)\exp\{\tilde{W}_{-\phi_{b_s}}\}\,ds\}\mathcal{M}(t)\,dt]$$

where

$$\mathcal{M}(t) := \exp\{\frac{1}{2}\int_0^t(\nabla\tilde{W}_{\phi_x})_{x=b_s}\,db_s - \frac{1}{8}\int_0^t(\nabla\tilde{W}_{\phi_x})_{x=b_s}^2\,ds\} \qquad (10)$$

and by applying the Ito-formula we know that

$$\frac{1}{2}\int_0^t[\nabla\tilde{W}_{\phi_x}]_{x=b_s}\,db_s = -\frac{1}{4}\int_0^t[\Delta\tilde{W}_{\phi_x}]_{x=b_s}\,ds + \frac{1}{2}\tilde{W}_{\phi_{b_t}} - \frac{1}{2}\tilde{W}_{\phi_x}$$

so finally, by substituting this expression into (10), we obtain equation (8).

This expression is easily seen to have an analytic extension to all $z \in B_q(\hat{\delta})$ and by applying the generator of $b_t$ on both the real and imaginary part of $\tilde{u}(x, z)$ we see that equation (8) also holds in this case. ∎

**LEMMA 4.7** The differential operator $\text{div}(\text{Exp}\{W_{\phi_x}\} \diamond \nabla u)$ is well-defined as an element in $(\mathcal{S})^{-1}$ $\forall x \in D$.

PROOF:

This is an obvious consequence of lemma 4.4 since we have shown that

$$\text{div}(\exp\{\tilde{W}_{\phi_x}\} \diamond \nabla \tilde{u}) = \tilde{\kappa}(x)\tilde{u}(x) - \tilde{g}(x)$$

in lemma 4.6. ∎

The theorem now follows from the previous lemmas. ∎

**COROLLARY 4.8** (The wave equation in an isotropic stochastic medium)

Assume that the assumptions of theorem 4.1 are valid.

Then

$$\Psi(t, x) = \text{Exp}\{W_{-\frac{1}{2}\phi_x}\}$$

$$\diamond \hat{E}^x[\sinh(t)f(b_{\tau_D}) \diamond \text{Exp}\{-\frac{1}{2}\int_0^{\tau_D} \text{Exp}\{W_{-\phi_{b_s}}\}\,ds\} \diamond \mathcal{J}(\tau_D, x)]$$

where $\mathcal{J}(t, x)$ is given in (7) is an $(\mathcal{S})^{-1}$-valued process which solves the wave equation in an isotropic medium given by

$$\frac{\partial^2 \Psi}{\partial t^2}(t, x) = \text{div}(\text{Exp}\{W_{\phi_x}\} \diamond \nabla_x \Psi) \quad x \in D$$

$$\Psi(0, x) = 0 \quad x \in \bar{D}$$

$$\frac{\partial \Psi}{\partial t}(0, x) = f(x) \quad x \in \partial D.$$

PROOF:

It is clear that the boundary conditions are satisfied and that

$$\frac{\partial^2 \Psi}{\partial t^2}(t, x) = \Psi(t, x)$$

i.e we must show that

$$\Psi(t, x) = \mathrm{div}(\mathrm{Exp}\{W_{\phi_x}\} \diamond \nabla_x \Psi)$$

but this is nothing but the Dirichlet equation of theorem 4.1.　■

Acknowledgments: I would like to thank my supervisor Bernt Øksendal for always taking time to answer my questions and introducing me to interesting problems.

This work is supported by Vista, a research cooperation between The Norwegian Academy of Science and Letters and Den Norske Stats Oljeselskap A.S (Statoil).

# References

[BØ]　B. Øksendal: Stochastic Differential Equations (third edition). Springer Verlag 1992.

[BØ2]　B. Øksendal: A stochastic approach to moving boundary problems. To appear in M.Pinsky (editor): Diffusion Processes and Related Problems in Analysis. Birkhäuser. 1990

[BØ3]　B. Øksendal: Stochastic partial differential equations ; A mathematical connection between macrocosmos and microcosmos. Preprint, University of Oslo 1993.

[BØ4]　B. Øksendal: Stochastic partial differential equations and applications to hydrodynamics. Preprint,University of Oslo 1994.

[DUR]　R. Durret: Brownian motion and martingales in analysis. Wadsworth Inc. 1984.

[GBF]　G. B. Folland: Real Analysis. J.Wiley & Sons 1984.

[Gj]　J. Gjerde: Multidimensional white noise with applications: Cand. Scient. thesis, University of Oslo 1993.

[GjHØUZ]  J. Gjerde, H. Holden, B. Øksendal, J. Ubøe and T. -S. Zhang: An equation modelling transport of a substance in a stochastic medium. Manuscript, University of Oslo 1993.

[GS2]  I. M. Gelfand and G. E. Shilov: Generalized Functions ,Vol.2 : Spaces of fundamental and generalized functions. Academic Press 1968 (English translation).

[GV4]  I. M. Gelfand and N. Y. Vilenkin: Generalized Functions ,Vol.4: Application of Harmonic Analysis. Academic Press 1964 (English translation).

[H]  T. Hida: Brownian Motion. Springer Verlag 1980.

[HKPS]  T. Hida, H. -H. Kuo,J.Potthoff and L.Streit: White Noise Analysis. Kluwer 1993.

[HLØUZ]  H. Holden, T. Lindstrøm, B. Øksendal, J. Ubøe and T. -S. Zhang: Stochastic boundary value problems. A white noise functional approach. Preprint, University of Oslo 1991

[HLØUZ2]  H. Holden, T. Lindstrøm, B. Øksendal, J. Ubøe and T. -S. Zhang: The Burgers equation with a noise force. Manuscript, University of Oslo 1992.

[HLØUZ3]  H. Holden, T. Lindstrøm, B. Øksendal, J. Ubøe and T. -S. Zhang: The pressure equation for fluid flow in a stochastic medium. Manuscript, University of Oslo 1994.

[HP]  E. Hille and R. S. Phillips: Functional analysis and semigroups. Amer. Math. Soc. Colloq. Publ.31.1957

[ITO]  K. Ito: Multiple Wiener integral. J.Math.Soc .Japan 3 (1951),157-169

[KLS]  Y. G. Kondratiev, P. Leukert and L. Streit: Wick calculus in Gaussian analysis. Manuscript. University of Bielefeld 1994.

[KS]  I. Karatzas and S. E. Shreve: Brownian motion and stochastic calculus. Springer Verlag 1988

[LØU1]  T. Lindstrøm, B. Øksendal and J. Ubøe: Stochastic differential equations involving positive noise. In M.Barlow and N.Bingham (editors): Stochastic Analysis. Cambridge Univ. Press 1991, pp 261-303

[LØU2]   T. Lindstøm, B. Øksendal and J. Ubøe: Wick multiplication and
         Ito-Skorohod stochastic differential equations. In S.Albeverio et al
         (editors): Ideas and Methods in Mathematical Analysis, Stochastics
         and Applications. Cambridge Univ. Press 1992, pp.183-206.

[LØU3]   T. Lindstrøm, B. Øksendal and J. Ubøe: Stochastic modelling of
         fluid flow in porous media. Preprint University of Oslo 1991.

[NZ]     D. Nualart and M. Zakai: Generalized stochastic integrals and the
         Malliavin calculus. Probab. Th. Rel. Fields 73 (1986), pp 255-280.

[RAM]    R. A. Minlos: Generalized random processes and their extension to
         a measure.
         Selected translations in mathematical statistics and probability. No.
         3, Am. Math. Soc., Providence, Rhode Island,1963.

[WICK]   H. Gjessing, H. Holden, T. Lindstrøm, B. Øksendal, J. Ubøe and
         T. -S. Zhang: The Wick Product. Preprint University of Oslo 1992

[Z]      T. -S. Zhang: Characterizations of white noise test functions and
         Hida distributions. In Stochastics and stochastics reports, Vol. 41,
         pp 71-87.

[ØZ]     B. Øksendal and T. -S. Zhang: The stochastic Volterra equation.
         Preprint, University of Oslo 1992.

# SEMI-IMPLICIT EULER-MARUYAMA SCHEME FOR STIFF STOCHASTIC EQUATIONS

*Yaozhong* $Hu^*$

Department of Mathematics
University of Oslo
P.O. Box 1053, Blindern
N-0316, Oslo, NORWAY
e-mail: huyao@ma-mail.uio.no
fax: (47) 22 85 43 49

**ABSTRACT**   We discuss a semi-implicit time discretization scheme to approximate the solution of a kind of stiff stochastic differential equations. Roughly speaking, by stiffness for an SDE we mean that the drift coefficient of the considered equtaions satisfies the so-called one-side Lipschitz condition and the diffusion coefficients have the bounded derivatives. We prove that the semi-implicit scheme has the convergence rate 0.5 (as for the explicit Euler scheme for nonstiff problems). The proof of this result needs some results on moment estimation of the solution of the original stiff stochastic differential equations, which is also done in this paper.

**KEYWORDS**   stochastic differential equations, stiff problem, one-side Lipschitz condition, semi-implicit time discretization schemes, convergence rate.

---

*   The author also holds a position in the Young Scientist Laboratory of Mathemtical Physics, Wuhan Institute of Mathematical Sciences, Chinese Academy of Sciences, Wuhan 430071, China.

# §1 Introduction

Let $\mathbb{R}^d$ be the $d$-dimensional Euclidean space. Let $a_i : \mathbb{R}^d \to \mathbb{R}^d, i = 1, \cdots, m$ and $b : \mathbb{R}^d \to \mathbb{R}^d$ be $m+1$ continuous functions (vector fields on $\mathbb{R}^d$). Let $w_t^i, t \geq 0, i = 1, \cdots, m$ be $m$ independent Wiener processes on the canonical Wiener space $(\Omega, \mathcal{F}, (\mathcal{F}_t)_{t \geq 0}, \mathbb{P})$. We are concerned with the following stochastic differential equation on the fixed time interval $[0, T]$:

$$x_t = x + \int_0^t b(x_s)ds + \sum_{i=1}^m \int_0^t a_i(x_s)dw_s, \quad 0 \leq t \leq T. \tag{1.1}$$

When the coefficients $b$ and $a_i$'s are globally Lipschitz continuous, the above equation has a unique solution. Under some more conditions such as the coefficients have bounded derivatives till certain orders, one can use the time discretization schemes to obtain explicit but approximate solutions for (1.1), see, e.g. [PK] (also [Hu]) for details. In [CH] an exact convergence rate problem is considered for the Euler-Maruyama scheme. All these results require that all the coefficients satisfy the global Lipschitz conditions.

But in the deterministic case, as explained for instance in [HW], the stiff equations occur frenquently in the applications. Recently there is an extensive study on the numerical methods (in particular the time discretization) for the stiff *ordinary* equations, i.e., equation (1.1) in the case when all $a_i = 0$. We refer to [HW] and the references therein for details.

If a stiff ordinary system is perturbed by a noise, we are led naturally to the consideration of the stiff stochastic differential equations ( Stiff SDE).

There is no common mathematical definition on the stiffness, see, e.g. p.1 of chapter IV of [HW] and R.C. Aiken ed. Stiff computation, Oxford Univ. Press, 1985, pp. 360-363.

We do not intend to discuss the most general problem. The stiffness of the stochastic differential equations in this paper is precised by the following assumptions.

(A) The diffusion coefficients $a_i$ are continuously differentiable with bounded derivatives,. i.e.,

$$\left\|\frac{\partial a_i}{\partial x}(x)\right\|_\infty \le \mu, \quad x \in \mathbb{R}^d, \quad i = 1, \cdots, m, \tag{1.2}$$

where $\| \cdot \|_\infty$ means the operator norm for an operator from $\mathbb{R}^d$ to $\mathbb{R}^d$ (equipped with the Euclidean norm).

(B) The drift coefficient $b$ is continuously differentiable and satisfies the *one-side Lipschitz condition*: There is a constant $\lambda \ge 0$ such that for any $x, y \in \mathbb{R}^d$,

$$< b(x) - b(y), x - y > \le \lambda |x - y|^2, \tag{1.3}$$

where $< \cdot, \cdot >$ means the Euclidean scalar product and $| \cdot |$ means the Euclidean norm.

(B)' Without loss of generality, we assume furthermore:

$$< b(x), x > \le \lambda |x|^2. \tag{1.4}$$

REMARK In the left hand side of (1.3) and (1.4), there is no absolute sign. This is the reason one called it the one-side Lipschitz condition.

The assumption (B)' can be made really without loss of generality. In fact, introduce $\bar{b}(x) = b(x) - b(0)$. Since $\bar{b}(0) = 0$, (1.4) is a particular case of (1.3) (when $y = 0$). Now by the Itô formula, one can prove easily that $x_t$ is the solution of equation (1.1) iff $\tilde{x}_t \equiv e^{tb(0)} x_t$ is the solution of equation (1.1) with $b$ replaced by $\bar{b}$.

Assumptions (A) and (B)' are enough for the existence and uniqueness of the solution of equation (1.1), see, e.g. the following theorem 3.1. But author is not able to use only these three conditions (A), (B) and (B)' to prove the convegence theorem. We need to make a technical assumption about the solution of equation (1.1):

(S): There is a constant $C$ dependent only on the coefficients and $T$ such that for any $0 \leq s \leq t \leq T$

$$\mathbb{E}\left| \int_s^t [b(x_t) - b(x_u)] du \right|^p \leq C|t - s|^{3p/2}. \qquad (1.5)$$

This assumption might not be automatically satisfied but it is satisfied when we have the following

(B)" there are constants $C > 0$, $\alpha \geq 0$ and $\beta \geq 0$ such that

$$|b(x) - b(y)| \leq C[\exp\{\alpha|x|^\beta\} + \exp\{\alpha|y|^\beta\}]|x - y| \qquad (1.6)$$

This is a local Lipschitz condition and the local Lipschitz constant satisfies a kind of exponential growth. The global Lipschitz condition correponds to the case $\alpha = \beta = 0$.

Condition (B)" implies (S) will be proved in section 3.

Now we describe our semi-implicit Euler-Maruyama time discretization scheme. Let $\pi : 0 = t_0 < t_1 < \cdots t_{n-1} < t_n = T$ be a partition of $[0, T]$. Let $\delta = \sup_{0 \leq k \leq n-1}(t_{k+1} - t_k)$. On $[t_k, t_{k+1}]$, we use the following formula to approximate the solution of (1.1)

$$y_{k+1}^\pi = y_k^\pi + b(y_{k+1}^\pi)\Delta_k + \sum_{i=1}^m a_i(y_k^\pi)\Delta w_k^i, \qquad (1.7)$$

where

$$\Delta_k = t_{k+1} - t_k, \qquad \Delta w_k^i = w_{t_{k+1}}^i - w_{t_k}^i.$$

The main result of this paper is to prove that when the partition becomes finer and finer, $y_n^\pi$ converges to the original solution $x_T$ at point $T$ with the (strong) convergence rate 0.5: There is a constant independent of the partition $\pi$ such that

$$\mathbb{E}|y_n - x_T|^2 \leq C \delta. \qquad (1.8)$$

Example: Let $d = 1$ and $b(x) = b_{2n+1}x^{2n+1} + \cdots + b_1 x + b_0$ be a polynomial with $b_{2n+1} < 0$. Since $b'(x) \le \lambda$ for some $\lambda < \infty$, one can check easily that $b(x)$ satisfies assumptions (B), (B)' and (B)".

## §2 The convergence of the scheme

Let $b : \mathbb{R}^d \to \mathbb{R}^d$ be a continuously differentiable function. Let

$$J_b(x) \equiv \frac{\partial b}{\partial x} \equiv (\frac{\partial b_i}{\partial x_j})_{1 \le i,j \le d}. \tag{2.1}$$

be the Jacobian of (the vector field) $b$. $J_b$ is a continuous function from $\mathbb{R}^d$ to the set of $d \times d$ matrices. We want to express the one-side Lipschitz condition on $b$ by a condition on its Jacobian $J_b(x)$. The following lemma is taken from excercise 6 of section 1.10 of [HNW]. We give a brief proof.

**LEMMA** 2.1 $b$ *satisfying the one-side Lipschitz condition (B) iff*

$$< x, J_b(y)x > \le \lambda |x|^2. \tag{2.2}$$

*for any* $x, y \in \mathbb{R}^d$.

**PROOF** By the mean value theorem, it is obvious that condition (2.2) implies (B). Now suppose that the one-side Lipschitz condition (B) is satisfied. Let $x, y \in \mathbb{R}^d$. Denote $\tilde{x} = x/|x|$, i.e. $|\tilde{x}| = 1$. By the one-side Lipschitz condition we have for any $\delta > 0$,

$$< 2\delta\tilde{x}, b(y + \delta\tilde{x}) - b(y - \delta\tilde{x}) > \le 4\lambda\delta^2 |\tilde{x}|^2.$$

By the mean value theorem, there is a $-1 \le \theta \le 1$ such that

$$< 2\delta\tilde{x}, 2\delta J_b(y + \theta\delta\tilde{x})\tilde{x} > \le 4\lambda\delta^2.$$

Hence

$$< x, J_b(y + \theta\delta\tilde{x})x > \le \lambda |x|^2.$$

Fixing $x$ and $y$ and letting $\delta \to 0$ we obtain

$$< x, J_b(y)x > \le \lambda |x|^2.$$

This has proved the lemma. ∎

**LEMMA 2.2** *If $J$ is a $d \times d$ square matrix such that*

$$< x, Jx > \le \lambda |x|^2. \tag{2.3}$$

*Then when $\lambda t < 1$, $(I - tJ)$ is invertible and*

$$\|(I - tJ)^{-1}\|_\infty \le (1 - \lambda t)^{-1}, \tag{2.4}$$

*where $\| \cdot \|_\infty$ means the operator norm as an operator from $\mathbb{R}^d$ to $\mathbb{R}^d$ (equipped with the Euclidean norm).*

**PROOF** Let $y = (I - tJ)x$ for $x \in \mathbb{R}^d$, i.e., $< y, x > = < (I - tJ)x, x >$. When $\lambda t < 1$,

$$|y||x| \ge | < y, x > | = | < x - tJx, x > |$$
$$= \|x\|^2 - t < Jx, x > | \ge |x|^2 - t\lambda|x|^2$$
$$= (1 - \lambda t)|x|^2.$$

From this we see that when $t\lambda < 1$

$$|x| \le (1 - \lambda t)^{-1}|y|. \tag{2.5}$$

This implies that $I - tJ$ is injective. So it is also surjective. Hence $I - tJ$ is inversible and also by (2.5) we see that (2.4) is true. ∎

Combining lemma 2.1 with lemma 2.2 we obtain

**COROLLARY 2.3** *If $b$ satisfies the one-side Lipschitz condition (B) and is continuously differentiable, then when $t\lambda < 1$, $I - tJ_b(x)$ is invertible (for any $x \in \mathbb{R}^d$) and*

$$\|(I - tJ_b(x))^{-1}\|_\infty \le (1 - \lambda t)^{-1}. \tag{2.6}$$

After this preparation, we return to the stochastic differential equation (1.1):

$$x_t = x + \int_0^t b(x_s)ds + \sum_{i=1}^m \int_0^t a_i(x_s)dw_s^i, \quad 0 \le t \le T. \tag{2.7}$$

Let $\pi : 0 = t_0 < t_1 < \cdots < t_{n-1} < t_n = T$ be a partition of the time interval $[0, T]$. Denote $\Delta_k = t_{k+1} - t_k$, $\Delta w_k^i = w_{t_{k+1}}^i - w_{t_k}^i$ and $\delta = \sup_{0 \le k \le n-1} \Delta_k$. Let $y_k^\pi$ be the sequence given by the semi-implicit scheme as follows:

$$y_t^\pi = y_k^\pi + b(y_t^\pi)(t - t_k) + \sum_{i=1}^m a(y_t^\pi)(w_t^i - w_{t_k}^i), \quad y_{k+1}^\pi = y_{t_{k+1}}^\pi, \quad (2.8)$$

where $t_k \le t \le t_{k+1}$, $k = 0, 1 \cdots, n$, $y_0^\pi = x$. Since $y_0^\pi = x$ is given, we can use formula (2.8) to compute $y_1^\pi$ (when $t = t_1$) $^*$ and then $y_2^\pi, \cdots, y_n^\pi$ etc. So formula (2.8) gives a recursive formula to compute, which can be done by computer, an approximate solution of the solution of (2.7). The main object of this section is to prove the following

**THEOREM** 2.4 *Under assumptions (A), (B), (B)' and (S), there is a constant $C$ independent of the partition $\pi$ (but dependent on $\lambda$, $\mu$, $T$ etc) such that*

$$\mathbb{E}|y_T^\pi - x_T|^2 \le C\delta. \qquad (2.9)$$

*and*

$$\int_0^T \mathbb{E}|y_t^\pi - x_t|^2 dt \le C\delta. \qquad (2.10)$$

**PROOF** We will prove (2.9). (2.10) follows from (2.9) easily. Equation (2.7) can be written as

$$x_t = x_{t_k} + \int_{t_k}^t b(x_s)ds + \sum_{i=1}^m \int_{t_k}^t a_i(x_s)dw_s^i$$

$$= x_{t_k} + b(x_t)(t - t_k) + \sum_{i=1}^m a(x_{t_k})(w_t^i - w_{t_k}^i)$$

---

$^*$ Professor E. Hairer kindly communicated to me that solving the nonlinear Runge-Kutta equation $y = x + hb(y)$ when $h$ small (which appeared in 2.8), the most appropriate way is to apply "simplified Newton iterations" as described in their book, Section IV.8. They have a good experience with their code RADU5, where this technique is implented, see [HW], Appendix, Fortran codes, p. 547.

$$+ R_k(t) + H_k(t), \tag{2.11}$$

where $t_k \le t \le t_{k+1}$,

$$R_k(t) = \int_{t_k}^{t} [b(x_s) - b(x_t)]ds \tag{2.12}$$

and

$$H_k(t) = \sum_{i=1}^{m} \int_{t_k}^{t} [a_i(x_s) - a_i(x_{t_k})]dw_s^i. \tag{2.13}$$

We also denote $R_k \equiv R_k(t_{k+1})$ and $H_k \equiv H_k(t_{k+1})$.

For simplicity we omit the explicit dependence of $y_t \equiv y_t^\pi$ on the partition $\pi$. Let $z_t = x_t - y_t$ and $z_k = z_{t_k}$. Substracting (2.8) from (2.11), we get for $t_k \le t \le t_{k+1}$

$$z_t = z_k + (t - t_k)\{b(x_t) - b(y_t)\} + R_k(t) + H_k(t) + \tilde{H}_k(t),$$

where

$$\tilde{H}_k(t) = \sum_{i=1}^{m} [a_i(x_{t_k}) - a_i(y_{t_k})](w_t^i - w_{t_k}^i). \tag{2.14}$$

We can rewrite the above formula as

$$z_t - (t - t_k)\{b(x_t) - b(y_t)\} = z_k + R_k(t) + H_k(t) + \tilde{H}_k(t).$$

i.e.

$$z_t - (t - t_k)\{\int_0^1 J_b(ux_t + (1-u)y_t)du\}z_t = z_k + R_k(t) + H_k(t) + \tilde{H}_k(t).$$

Denote when $t_k \le t \le t_{k+1}$,

$$A_t = \{I - (t - t_k)\int_0^1 J_b(ux_t + (1-u)y_t)du\}^{-1} \tag{2.15}$$

and also $A_{k+1} = A_{t_{k+1}}$. Similar to corollary 2.3, it is easy to see the existence of $A_t$ when $\lambda\delta < 1$ and

$$\|A_t\|_\infty \leq (1 - \lambda(t - t_k))^{-1}. \tag{2.16}$$

Therefore

$$z_t = A_t z_k + A_t R_k(t) + A_t H_k(t) + A_t \tilde{H}_k(t). \tag{2.17}$$

In particular

$$z_{k+1} = A_{k+1} z_k + A_{k+1} R_k + A_{k+1} H_k + A_{k+1} \tilde{H}_k. \tag{2.18}$$

Iterating (2.18), we have

$$z_t = \sum_{j=0}^{k-1} A_k \cdots A_{j+1} R_j + A_t R_k(t)$$

$$+ \sum_{j=0}^{k-1} A_k \cdots A_{j+1} H_j + A_t H_k(t)$$

$$+ \sum_{j=0}^{k-1} A_k \cdots A_{j+1} \tilde{H}_j + A_t \tilde{H}_k(t)$$

$$\equiv I_1(t) + I_2(t) + I_3(t), \tag{2.19}$$

where $I_i$ represents the sum in the $i$−th line of (2.19). First we treat $I_1$. Let $p'$ satisfy $1/p + 1/p' = 1$.

$$\mathbb{E} \sup_{0 \leq t \leq T} |I_1(t)|^p \leq \mathbb{E}\{\sum_{j=0}^{n-1} |A_k \cdots A_{j+1} R_j|\}^p$$

$$\leq \mathbb{E}\{\sum_{j=0}^{n-1} \|A_k \cdots A_{j+1}\|_\infty |R_j|\}^p$$

$$\leq \mathbb{E}\{\sum_{j=0}^{n-1} (1 - \lambda\Delta_{n-1})^{-1} \cdots (1 - \lambda\Delta_j)^{-1} |R_j|\}^p$$

$$\leq (1 - \lambda\Delta_{n-1})^{-1} \cdots (1 - \lambda\Delta_0)^{-1} \mathbb{E}\{\sum_{j=0}^{n-1} |R_j|\}^p$$

But when $\lambda\delta < 1/2$, $|\log(1 - \lambda\Delta_k)^{-1}| \leq C\Delta_k$. So

$$(1 - \lambda\Delta_{n-1})^{-1} \cdots (1 - \lambda\Delta_0)^{-1}$$

$$\leq \prod_{j=0}^{n-1} \exp\{|\log(1 - \lambda\Delta_k)^{-1}|\}$$

$$\leq \prod_{j=0}^{n-1} \exp\{C\Delta_k\} \leq C. \qquad (2.20)$$

Thus

$$\mathbb{E} \sup_{0 \leq t \leq T} |I_1(t)|^p \leq C\mathbb{E}\{\sum_{j=0}^{n-1} |R_j|\}^p.$$

Using the Hölder inequality,

$$\left(\sum_{j=0}^{n-1} |R_j|\right)^p \leq \left(\sum_{j=0}^{n-1} [|R_j|/\Delta_j^{1/p'}][\Delta_j^{1/p'}]\right)^p$$

$$\leq \sum_{j=0}^{n-1} [|R_j|^p \Delta_j^{-p/p'}]\{\sum_{j=0}^{n-1} \Delta_j\}^{p/p'}$$

$$\leq C \sum_{j=0}^{n-1} |R_j|^p \Delta_j^{-p/p'}.$$

So by assumption (S), i.e. $\mathbb{E}|R_j|^p \leq C|\Delta_j|^{3p/2}$, we have

$$\mathbb{E} \sup_{0 \leq t \leq T} |I_1(t)|^p \leq C \sum_{j=0}^{n-1} \Delta_j^{3p/2} \Delta_j^{-p/p'}$$

$$\leq C \sum_{j=0}^{n-1} \Delta_j^{3p/2 - p/p'}$$

$$\leq C' \sup_{0 \leq j \leq n-1} \Delta_j^{3p/2 - p/p' - 1} \leq C\delta^{p/2}.$$

In particular, we have

$$\mathbb{E}|I_1(t_k)|^2 \leq C\delta, \qquad (2.21)$$

where the constant $C$ is independent of $k$ and the partition $\pi$.

Now we are going to study the remaining two terms in formula (2.19). The main difficulty is the nonadaptness of the coefficients $A_k \cdots A_{j+1} \notin \mathcal{F}_{t_j}$. Let

$$\tilde{z}_k = I_2(t_k) + I_3(t_k).$$

Then $\tilde{z}_k$ satisfies equation (2.18) with $R_k \equiv 0$, i.e.

$$\tilde{z}_{k+1} = A_{k+1}\{\tilde{z}_k + H_k + \tilde{H}_k\}.$$

Since $\tilde{Z}_k$ and $H_k$, $\tilde{H}_k$ are orthogonal,

$$
\begin{aligned}
\mathbb{E}|\tilde{z}_{k+1}|^2 &= \mathbb{E}|A_{k+1}\{\tilde{z}_k + H_k + \tilde{H}_k\}|^2 \\
&\leq \mathbb{E}\|A_{k+1}\|_\infty^2 |\tilde{z}_k + H_k + \tilde{H}_k|^2 \\
&\leq (1 - \lambda\Delta_k)^{-2}\mathbb{E}|\tilde{z}_k + H_k + \tilde{H}_k|^2 \\
&= (1 - \lambda\Delta_k)^{-2}\left(\mathbb{E}|\tilde{z}_k|^2 + \mathbb{E}|H_k + \tilde{H}_k|^2\right) \\
&\leq (1 - \lambda\Delta_k)^{-2}\left(\mathbb{E}|\tilde{z}_k|^2 + 4\mathbb{E}|H_k|^2 + 4\mathbb{E}|\tilde{H}_k|^2\right). \quad (2.22)
\end{aligned}
$$

By (2.13) and lemma 3.4 (which is proved in next section),

$$
\begin{aligned}
\mathbb{E}|H_k|^2 &= \sum_{i=1}^m \int_{t_k}^{t_{k+1}} \mathbb{E}[a_i(x_s) - a_i(x_{t_k})]^2 ds \\
&\leq C \int_{t_k}^{t_{k+1}} \mathbb{E}|x_s - x_{t_k}|^2 ds \\
&\leq C \int_{t_k}^{t_{k+1}} |s - t_{t_k}| ds = C\Delta_k^2. \quad (2.23)
\end{aligned}
$$

By (2.14)

$$
\begin{aligned}
\mathbb{E}|\tilde{H}_k|^2 &= \sum_{i=1}^m \mathbb{E}|a_i(x_k) - a_i(y_k)|^2 \Delta_k \\
&\leq C\Delta_k \mathbb{E}|x_k - y_k|^2 = C\Delta_k \mathbb{E}|z_k|^2. \quad (2.24)
\end{aligned}
$$

But by the definition of $\tilde{Z}_k$ and estimate (2.21),

$$\mathbb{E}|z_k|^2 = \mathbb{E}|\tilde{z}_k + I_1(t_k)|^2$$

$$\leq 4\mathbb{E}|\tilde{z}_k|^2 + 4\delta. \tag{2.25}$$

Combining (2.21), (2.22), (2.23), (2.24) with (2.25), we have

$$\mathbb{E}|\tilde{z}_{k+1}|^2 \leq (1 - \lambda\Delta_k)^{-2}\{(1 + C\Delta_k)\mathbb{E}|\tilde{z}_k|^2 + C_1\delta\Delta_k + C_2\Delta_k^2\}. \tag{2.26}$$

When $\Delta_k$ is sufficiently small, we can prove easily, using for instance the L'Hopital rule, that there is a constant $C$ (independent of $\pi$) such that

$$(1 - \lambda\Delta_k)^{-2}(1 + C_1\Delta_k) \leq 1 + C\Delta_k.$$

We see also that when $\Delta_k$ small, $(1 - \lambda\Delta_k)^{-2}$ is bounded by a constant $C$. Denote $B_k = 1 + \Delta_k$. Then (2.26) becomes

$$\mathbb{E}|\tilde{z}_{k+1}|^2 \leq B_k\mathbb{E}|\tilde{z}_k|^2 + C\delta\Delta_k$$

Iterating the above formula we get $(t_n = T)$

$$\mathbb{E}|\tilde{z}_n|^2 = C\sum_{j=0}^{n-1} B_{n-1}\cdots B_j\delta\Delta_j. \tag{2.27}$$

Similar to (2.20), we have

$$B_{n-1}\cdots B_j \leq B_{n-1}\cdots B_0 \leq C < \infty.$$

By (2.27) we have

$$\mathbb{E}|\tilde{z}_n|^2 = C\sum_{j=0}^{n-1} \delta\Delta_j \leq C\delta.$$

Combining this formula with (2.21) we obtain

$$\mathbb{E}|y_T - x_T|^2 = \mathbb{E}|\tilde{z}_n|^2 \leq C\delta.$$

Therefore we have proved (2.9). The proof of (2.10) follows easily. ∎

## §3 The properties of the solution

We shall discuss some conditions on the coefficients $b$ and $a_i$ such that the property (S) is satisfied. First let us discuss the existence and uniqueness of the solution.

**THEOREM** 3.1 *Under the conditions (A) and (B), the stochastic differential equation (1.1) has a unique solution (no explosion) $x_t$ ($0 \leq t < \infty$).*

**PROOF**  We can prove the following fact by imitating word by word the proof of theorem 10.2.2 of [SV79], p.255:

If $b$ and $a_i$ ($i = 1, \cdots, m$) are continuously differentiable, if there are $b_n$ and $a_{i,n}$ such that

$$b_n(x) = b(x), \quad a_{i,n}(x) = a_i(x) \qquad (3.1)$$

when $|x| \leq n$ ($n = 1, \cdots, \infty$) and $b_n$, $a_{i,n}$ are globally Lipschitz continuous, and furthermore if there is constant $C$ such that

$$|a_i(x)| \leq C(1 + |x|^2), \quad < b(x), x > \leq C(1 + |x|^2), \qquad (3.2)$$

then there is a unique strong solution $x_t$ ($0 \leq t < \infty$) (without explosion) for (1.1).

The first inequality of (3.2) is implied by (A). The second inequality of (3.2) is implied by assumption (B)'. To prove theorem 3.1, it suffices then to construct a sequence $b_n$ and $a_{i,n}$ satisfying (3.1). Let $\rho_n(x)$ be a $C^\infty$ function on $\mathbb{R}^d$ such that $\rho_n$ has compact support and $\rho_n(x) = 1$ when $|x| \leq n$. Put

$$b_n(x) = b(x)\rho_n(x), \quad a_{i,n}(x) = a_i(x)\rho_n(x).$$

We can check that these $b_n$ and $a_{i,n}$ satisfies (3.1) easily.  ∎

We turn to study some properties of the solution of (1.1).

**LEMMA** 3.2 *There is a constant C (dependent on $\lambda, \mu, x, T$) such that*

$$x_t^2 \le C + \sum_{i=1}^{m} \int_0^t \sigma_i(u) dw_u^i, \tag{3.3}$$

*where $\sigma_i(u)$ satisfy*

$$\mathbb{E}|\sigma_i(u)|^2 \le C\mathbb{E}|x_u|^2. \tag{3.4}$$

**PROOF** Applying the Itô formula to $f(x_t)$ with $f(x) = |x|^2$, we get

$$|x_t|^2 = |x|^2 + 2\sum_{i=1}^{m} \int_0^t <a_i(x_u), x_u> dw_u^i + 2\int_0^t <b(x_u), x_u> du$$
$$+ \sum_{i=1}^{m} \int_0^t a_i(x_u)^2 du.$$

From assumptions (A) and (B)', we have

$$x_t^2 \le |x|^2 + m\mu^2 t + \lambda \int_0^t |x_u|^2 du + 2\sum_{i=1}^{m} \int_0^t <a_i(x_u), x_u> dw_u^i.$$

Let

$$\sigma_i(u) = 2 <a_i(x_u), x_u> .$$

Then by assumption (A), $\sigma_i$ satisfies (3.4). Therefore

$$x_t^2 \le C + \lambda \int_0^t |x_u|^2 du + \sum_{i=1}^{m} \int_0^t \sigma_i(u) dw_u^i.$$

By Gronwall lemma we show this lemma. ∎

**LEMMA** 3.3 *Let $x_t$ be the solution of (1.1) and let assumptions (A) and (B)' be satisfied. Then for any $\alpha > 0$ and $0 < \beta < 1$,*

$$\mathbb{E}\exp\{\alpha|x_t|^\beta\} \le C_{\alpha\beta} < \infty. \tag{3.5}$$

**PROOF**   Lemma 3.2 says that there is constant $C_1$ such that

$$x_t^2 \le C_1 + \sum_{i=1}^{m} \int_0^t \sigma_i(u) dw_u^i.$$

So for any $n \ge 1$,

$$\mathbb{E}x_t^{2n} \le (m+1)^n \{C_1^n + \sum_{i=1}^{m} \mathbb{E}|\int_0^t \sigma_i(u) dw_u^i|^n\}.$$

By Burkholder-Davis-Gundy's inequality (see for example [HWY], Theorem 10.36, or [Pr], chapter IV, Theorem 54),

$$\mathbb{E}|\sup_{0 \le s \le t} \int_0^t \sigma_i(u) dw_u^i|^n \le n^{n+1} \mathbb{E}(\int_0^t \sigma_i(u)^2 du)^{n/2}$$

$$\le n^{n+1} \mu^n \mathbb{E}(\int_0^t |x_u|^2 du)^{n/2}$$

$$\le n^{n+1} \mu^n T^n \mathbb{E}(\int_0^t |x_u|^n du)$$

$$\le n^{n+1} \mu^n T^n \mathbb{E}(\int_0^t \sup_{0 \le s \le u} |x_s|^n du)$$

$$\le n^{n+1} \mu^n T^n \int_0^t \{\mathbb{E} \sup_{0 \le s \le u} |x_s|^{2n}\}^{1/2} du.$$

Let $\xi_t = \mathbb{E} \sup_{0 \le s \le t} |x_s|^{2n}$. Then $\xi_t$ is positive and increasing as a function of $t$ and satisfies the following inequality

$$\xi_t \le C_1^n + C_2 n^n \int_0^t \xi_u^{1/2} du$$

which can be rewritten as

$$\xi_t - C_2 n^n \int_0^t \xi_u^{1/2} du \le C_1^n$$

or equivalently

$$\int_0^t [\dot{\xi}_u - C_2 n^n \xi_u^{1/2}] du \le C_1^n.$$

Setting $\xi_t = (\eta_t + \frac{1}{2} C_2 n^n t)^2$, the above inequality is equivalent to

$$\int_0^t \eta_u \dot{\eta}_u du + \frac{1}{2} C_2 n^n \int_0^t u \dot{\eta}_u du \le C_1.$$

By definition, $\dot{\eta}_u \ge -\frac{1}{2} C_2 n^n$, so we have

$$\frac{1}{2}\{\eta_t^2 - \eta_0^2\} - \frac{T^2}{8} C_2^2 n^{2n} \le C_1^2,$$

i.e.,

$$\eta_t^2 \le C_1^{2n} + C_2 n^{2n}.$$

Consequently

$$\xi_t \le C_1^{2n} + C_2 n^{2n},$$

where $C_1$ and $C_2$ are constants independent of $\pi$ which may differ from line to line. In other word, we have

$$\mathbb{E} \sup_{0 \le s \le t} |x_s|^{2n} \le C_1^n + C_2 n^{2n},$$

or equivalently

$$\mathbb{E} \sup_{0 \le s \le t} |x_s|^n \le C_1^n + C_2 n^n \le C^n n^n. \tag{3.6}$$

Note that the above $n$ can be any positive real number.

So when $0 \le \beta < 1$,

$$\mathbb{E} \exp\{\alpha \sup_{0 \le t \le T} |x_t|\} = \sum_{n=0}^{\infty} \frac{\alpha^n}{n!} \mathbb{E} \big( \sup_{0 \le t \le T} |x_t| \big)^{\beta n}$$

$$\le \sum_{n=0}^{\infty} \frac{\alpha^n C^n}{n!} n^{\beta n}.$$

By the Stirling formula the above sum is convergent when $0 \leq \beta < 1$. This proves the lemma. ∎

Now we suppose the growth condition on $b$:

(B)": Assume there are constants $C$, $\alpha$ and $\beta$ such that

$$|b(x) - b(y)| \leq C[\exp\{\alpha|x|^\beta\} + \exp\{\alpha|y|^\beta\}]|x - y| \qquad (3.7)$$

A direct consequence of assumption (3.7) is that (when one takes $y = 0$):

$$|b(x)| \leq C \exp\{\alpha|x|^\beta\} \qquad (3.8)$$

for some constants $C$, $\alpha$ and $\beta$.

**LEMMA 3.4** *Let $x_t$ be the solution of (1.1). Let assumptions (A), (B) and (B)" be true. Then there is a constant $C$ dependent only on the coefficients $a_i$, $b$ and $T$ such that*

$$\mathbb{E} \sup_{s \leq r \leq t} |x_r - x_s|^p \leq C(t - s)^{p/2}. \qquad (3.9)$$

**PROOF** We can write (1.1) as

$$x_r - x_s = \int_s^r b(x_u)du + \sum_{i=1}^m \int_s^r \sigma_i(x_u)dw_u^i.$$

Therefore

$$\mathbb{E} \sup_{s \leq r \leq t} |x_r - x_s|^p$$

$$\leq (m+1)^p \{\mathbb{E} \sup_{s \leq r \leq t} |\int_s^r b(x_u)du|^p + \sum_{i=1}^m \mathbb{E} \sup_{s \leq r \leq t} |\int_s^r \sigma_i(x_u)dw_u^i|^p\}$$

$$\leq C\mathbb{E}(\int_s^t |b(x_u)|du)^p + C \sum_{i=1}^m \mathbb{E}|\int_s^t |\sigma_i(x_u)|^2 du|^{p/2}. \qquad (3.10)$$

Using the Hölder inequality we have

$$\left(\int_s^t |b(x_u)|du\right)^p \leq \int_s^t |b(x_u)|^p du(t-s)^{p/p'}. \tag{3.11}$$

By (3.5) and (3.8)

$$\mathbb{E}|b(x_u)|^p \leq C\mathbb{E}\exp\{\alpha|x_u|^\beta\} \leq C < \infty. \tag{3.12}$$

From (3.11) and (3.12) above, we see that the first term of (3.10) can be dominated by

$$\mathbb{E}\left(\int_s^t |b(x_u)|du\right)^p \leq C(t-s)^{1+p/p'} \leq C(t-s)^p. \tag{3.13}$$

By the same argument (easier than what we did), we see that each term in the sum of the second term of (3.10) is less than $(t-s)^{p/2}$. This proves the lemma. ∎

Now we can prove our main theorem of this section:

**THEOREM** 3.5 *Let $x_t$ be the solution of (1.1). Let assumptions (A), (B) and (B)" be satisfied. Then there is a constant $C$ dependent only on the coefficients $a_i$, $b$ and $T$ such that*

$$\mathbb{E}\sup_{s\leq r\leq t} |b(x_r) - b(x_s)|^p \leq C(t-s)^{p/2}. \tag{3.14}$$

**PROOF** By (B)", we see that there is some $\alpha > 0$ and $C$

$$|b(x_r) - b(x_s)|^p \leq C[\exp\{\alpha|x_r|^\beta\} + \exp\{\alpha|x_s|^\beta\}]|x_r - x_s|^p$$

So

$$\mathbb{E}\sup_{s\leq r\leq t} |b(x_r) - b(x_s)|^p$$
$$\leq C\mathbb{E}\exp\{\alpha \sup_{0\leq r\leq T} |x_r|^\beta\}\{\mathbb{E}|x_r - x_s|^{2p}\}^{1/2}$$

By lemma 3.3 and lemma 3.4 we have proved the theorem. ∎

Theorem 3.5 implies easily assumption (S):

**COROLLARY** 3.6 *Let $x_t$ be the solution of (1.1). Let assumptions (A), (B) and (B)" be true. Then there is a constant $C$ dependent only on the coefficients $a_i$, $b$ and $T$ such that*

$$\mathbb{E} \int_s^t |b(x_t) - b(x_u)|^p du \le C(t-s)^{p/2}. \tag{3.15}$$

**REMARK** To finish the paper we make some remarks. In the deterministic case, the control of derivatives of the solution of the original solution can be done, for example, as in the following way: Firstly we note that when all $a_i \equiv 0$, and $b$ satisfies the assumption (B), and (B)', then by lemma 12.1 of [HW], p.191,

$$|x_t| \le |x_0| e^{\lambda t},$$

where $x_t$ is the solution of the original solution of the equation $\dot{x}_t = b(x_t)$. So the solution $x_t$ is bounded on any finite interval $[0,T]$. So if we assume that $b$ has continuous derivatives till some orders, then by the fact that a continuous function is bounded on the compact set, we see that the solution $x_t$ has also the bounded derivatives till certain orders.

There are still much to do in this direction: For example, to obtain the higher convergence schemes; to prove the $L^p$ convergence etc. There are also a lot of equations which do not satisfies the one-side Lipschitz condition (for instance the following Lorentz system): $d = 3$ and $b$ is given by

$$b(x,y,z) = (a_1 x + b_1 y + c_1 z, a_2 x + b_2 + c_2 z - \lambda xz, a_3 x + b_3 + c_3 z + \mu xy).$$

where $a_i, b_i, c_i, \lambda, \mu, i = 1,2,3$ are constants.

**ACKNOWLEDGEMENTS** The author holds an NAVF research scholarship of Norwegian Research Council during this work.

**REFERENCES**

[CH] S. Cambanis & Y.Z. Hu, Exact convergence rate of Euler-Maruyama scheme and application to sample design, Preprint 1994.

[Da] G. Dalquist, Positive functions and some applications to stability questions for numerical methods, in Recent Advances in Numerical Analysis.

[He] P. Henrici, Discrete variable methods in ordinary differential equations, Wiley, New York, 1962.

[HWY] S. W. He, J. G. Wang & J. A. Yan, Semimartingale theory and stochastic calculus, Science Press and CRC Press Inc. 1992.

[HS] W.H. Hundsdorfer & B.I. Steininger, Convergence of linear multistep and one-leg methods for stiff nonlinear initial valued problems, BIT (1991), 124-143.

[HNW] E. Hairer, S.P. Nørsett & G. Wanner, Solving ordinary differential equations I, nonstiff problems, 2nd edition, Springer, 1993.

[HW] E. Hairer & G. Wanner, Solving ordinary differential equations II, stiff and differential-algebraic problems, Springer, 1991.

[Hu] Y. Z. Hu, Strong and weak convergence rate of time discretization schemes, to appear in Sem. de Probabilités, Lect. Notes in Mathematics, Springer, 1995.

[Mc] H.P. Mckean, Jr., Stochastic integrals, Academic Press, 1969.

[Pr] P. Protter, Stochastic integration and differential equations, A new approach, Springer, 1990.

[SV] D. W. Stroock & S.R.S. Varadhan, Multidimensional diffusion processes, Springer, 1979.

# WICK APPROXIMATION OF QUASILINEAR STOCHASTIC DIFFERENTIAL EQUATIONS

*Yaozhong Hu    and Bernt Øksendal*

Department of Mathematics
University of Oslo
N-0316, Oslo, NORWAY

**Abstract:** For a $\varepsilon > 0$ let $\{W_s^\varepsilon\}_{s \geq 0}$ be a smooth approximation to 1-dimensional Brownian motion $\{W_s\}_{s \geq 0}$. We consider the equation

$$1) \quad X_t^\varepsilon = \eta + \int_0^t \sigma(s, X_s^\varepsilon) \diamond W_s^\varepsilon ds + \int_0^t b(s, X_s^\varepsilon) ds; \quad 0 \leq s < \infty$$

where $\diamond$ denotes the Wick product. It is conjectured that (with reasonable conditions on $b$ and $\sigma$) a unique strong solution $X_t^\varepsilon$ exists for all $\varepsilon$ and that $X_t^\varepsilon \to X_t$ as $\varepsilon \to 0$ (i.e., $W_s^\varepsilon \to W_s$), where $X_t$ is the solution of the Itô differential equation

$$2) \quad X_t = \eta + \int_0^t \sigma(s, X_s) dW_s + \int_0^t b(s, X_s) ds; \quad 0 \leq s < \infty$$

We prove the conjecture in the quasilinear case, i.e., where $\sigma(s, x) = \sigma_s x$, where $\sigma_s$ is independent of $x$.

The conjecture should be compared to the Wong-Zakai theorem, which says that if we let $Y_t^\varepsilon$ be the solution of the stochastic equation (with $\diamond$ replaced by ordinary product)

$$1)' \quad Y_t^\varepsilon = \eta + \int_0^t \sigma(s, Y_s^\varepsilon) W_s^\varepsilon ds + \int_0^t b(s, Y_s^\varepsilon) ds; \quad 0 \leq s < \infty$$

then $Y_t^\varepsilon \to Y_t$, where $Y_t$ is the solution of the <u>Stratonovitch</u> differential equation

$$2)' \quad Y_t = \eta + \int_0^t \sigma(s, Y_s) \circ dW_s + \int_0^t b(s, Y_s) ds; \quad 0 \leq s < \infty$$

## §1. Introduction

Let $(\Omega, H, P)$ be the classical canonical Wiener space, where $\Omega = \{\omega : \mathbb{R}_+ \to \mathbb{R}; \omega \text{ is continuous}, \omega(0) = 0\}$ is the space continuous functions starting at 0, $H = \{h \in \Omega \ h \text{ is absolutely continuous and } |h|_H^2 := \int_0^\infty |\dot{h}|^2 dt < \infty\}$ is the Cameron-Martin Hilbert space and $P$ is the Wiener measure. Let $W : \mathbb{R}_+ \times \Omega \to \mathbb{R}$ defined by $W_t(\omega) = \omega(t)$ be the canonical Wiener process on the time interval $\mathbb{R}_+$ starting at the origin.

Let $\rho \geq 0$ be a smooth $(C^\infty)$ function on the real line $\mathbb{R}$ with support $\mathrm{supp}(\rho) \subset [0, 1]$ and such that

$$\int_0^1 \rho(s)ds = 1. \tag{1.1}$$

Define $\tilde{\rho}(s) := \rho(-s)$ and for $k = 1, 2, \cdots$ put

$$\rho^{(k)}(s) = k\rho(ks), \quad \tilde{\rho}^{(k)}(s) = k\rho(-ks). \tag{1.2}$$

Define

$$W_t^{(k)}(\omega) := (W_\cdot(\omega) * \tilde{\rho}^{(k)})(t) \tag{1.3}$$

$$= \int_{\mathbb{R}} W_s(\omega)\tilde{\rho}^{(k)}(t-s)ds = \int_t^{1+t} W_s(\omega)\rho^{(k)}(s-t)ds.$$

In other words, $W_t^{(k)}$ is a $t$-smoothed version of $W_t$, obtained by taking the convolution of (each component of) $W_t$ with $\tilde{\rho}^{(k)}$.

Now we consider an $n$-dimensional stochastic differential equation of the form

$$dX_t = b(X_t)dt + \sigma(X_t)dW_t; \quad X_0 = x \in \mathbb{R}^n, \tag{1.4}$$

where the coefficients $b : \mathbb{R}^n \to \mathbb{R}^n$, $\sigma : \mathbb{R}^n \to \mathbb{R}^{n \times m}$ satisfy the usual Lipschitz requirement

$$|b(x_1) - b(x_2)| + |\sigma(x_1) - \sigma(x_2)| \leq C|x_1 - x_2|; \quad x_1, x_2 \in \mathbb{R}^n, \tag{1.5}$$

where $C$ is a constant. The first version of the following well-known result was proved by Wong and Zakai [WZ65] (see e.g. [IW89], Theorem 7.2 and the references there):

<u>The Wong-Zakai theorem</u>: If we replace $W_t$ by the smoothed version $W_t^{(k)}$ and for each fixed $\omega \in \Omega$ let $Y_t^{(k)} = Y_t^{(k)}(\omega)$ be the solution of the corresponding ordinary equation

$$\frac{dY_t^{(k)}}{dt} = b(Y_t^{(k)}) + \sigma(Y_t^{(k)})\dot{W}_t^{(k)}(\omega); \quad Y_0^{(k)} = x, \qquad (1.6)$$

(where $\dot{W}_t^{(k)} = \frac{dW_t^{(k)}}{dt}$), then $Y_t^{(k)} \to Y_t$ as $k \to \infty$, where $Y_t$ is the solution of the Stratonovitch stochastic differential equation

$$dY_t = b(Y_t)dt + \sigma(Y_t) \circ dW_t; \quad Y_0 = x. \qquad (1.7)$$

The convergence is uniform on bounded $t$-intervals, for a.e. $\omega$. In fact, we have

$$\mathbb{E}[\sup_{0 \le t \le T} |Y_t(\cdot) - Y_t^{(k)}(\cdot)|^2] \to 0 \quad \text{as} \quad k \to \infty \qquad (1.8)$$

for all $T < \infty$, where $\mathbb{E}$ denotes the expectation with respect to the probability law for $W_t$.

At first glance it is rather surprising that the limit of $Y_t^{(k)}$ is $Y_t$ and not the solution $X_t$ of the Itô equation (1.4). It is natural to ask if there is an analogous approximation procedure which converges to $X_t$. We propose the following procedure: Replace the ordinary, $\omega$-wise (matrix) product in (1.6) by the *Wick product* $\diamond$, i.e., let $X_t^{(k)}$ be the solution of the stochastic equation

$$\frac{dX_t^{(k)}}{dt} = b(X_t^{(k)}) + \sigma(X_t^{(k)}) \diamond \dot{W}_t^{(k)}; \quad X_0^{(k)} = x. \qquad (1.9)$$

Then we conjecture that (possibly under some extra conditions) $X_t^{(k)}$ converges to $X_t$ as $k \to \infty$, in the same sense as above, i.e.,

$$\mathbb{E}[\sup_{0 \le t \le T} |X_t(\cdot) - X_t^{(k)}(\cdot)|^2] \to 0 \quad \text{as} \quad k \to \infty \qquad (1.10)$$

for all $T < \infty$.

The motivation for this conjecture comes from the following remarkable connection between Itô/Skorohod integrals and Wick products (see e.g. [LØU92], [HKPS93, ch.8] and [B93] for proofs and discussions): If $Y_s(\omega)$ is a Skorohod integrable stochastic process then

$$\int_0^t Y_s \delta W_s = \int_0^t Y_s \diamond N_s ds \quad \text{for all} \quad t \geq 0,$$

where $\delta W_s$ denotes Skorohod integration and $N_s$ is white noise. In particular, if $Y_s(\cdot)$ is also adapted with respect to the filtration $\{\mathcal{F}_t\}_{t\geq 0}$ generated by $\{W_t(\cdot)\}_{t\geq 0}$, then

$$\int_0^t Y_s dW_s = \int_0^t Y_s \diamond N_s ds. \tag{1.12}$$

The smoothed version of this is that

$$\int_{\mathbb{R}} (\rho^{(k)} * Y)_t \delta W_t = \int_{\mathbb{R}} Y_t \diamond W_t^{(k)} dt \tag{1.13}$$

for all $k = 1, 2, \cdots$, where

$$(\rho^{(k)} * Y)(t) = \int_{\mathbb{R}} Y_s \rho^{(k)}(t - s) ds.$$

It seems natural to guess that a similar connection between Itô/Skorohod integrals and Wick products also holds in the setting of Itô stochastic differential equation, in the sense of our conjecture (1.9)-(1.10) above.

The purpose of this paper is to prove this conjecture for a class of 1-dimensional quasilinear stochastic differential equations of the form

$$dX_t = b(t, X_t, \omega)dt + \sigma(t, \omega)X_t \delta W_t; \quad X_0 = \eta,$$

where $b : \mathbb{R} \times \mathbb{R} \times \Omega \to \mathbb{R}$ and $\sigma : \mathbb{R} \times \Omega \to \mathbb{R}$, $\eta : \Omega \to \mathbb{R}$ are given functions, possibly anticipating (see exact conditions below). This is done by reducing the equation to a quasilinear (hyperbolic) partial differential equation in infinitely many variables and then applying a characteristic curve method (see §3).

We emphasize however, that our results in this paper are far from a solution to the general conjecture. In general it is not even clear that the Wick approximate equation (1.9) has a solution (unique or not) for each $k$, nor is it clear that this solution -if it exists- converges in any sense and to what.

Now we explain the main results of our paper in details:

Let $\sigma$ and $b : \mathbb{R}_+ \times \mathbb{R} \times \Omega \to \mathbb{R}$ satisfy the following conditions:

(A) For any fixed $x \in \mathbb{R}$, $\sigma(t, x, \omega)$ and $b(t, x, \omega)$ are (real valued) stochastic processes (not necessarily adapted);

B) For a.e. $(t, \omega) \in \mathbb{R}_+ \times \Omega$, $\sigma(t, x, \omega)$ and $b(t, x, \omega)$ are continuously differentiable with respect to $x$.

We will impose more conditions as we go on. Let $\eta$ be a random variable on $(\Omega, \mathcal{F}, P)$. Consider the following (anticipating) stochastic differential equation:

$$X_t = \eta + \int_0^t \sigma(s, X_s, \omega)dW_s + \int_0^t b(s, X_s, \omega)ds, \quad 0 \le t < \infty, \quad (1.15)$$

Here and in the following $dW_s$ is the Skorohod integral and we use $\sigma(s, X_s)$ or sometimes $\sigma_s(X_s)$ to represent $\sigma(s, X_s, \omega)$. When $\sigma$ and $b$ are adapted, the above equation has been studied since decades. When $\sigma$ and $b$ are not adapted (anticipating), several special cases have been studied recently. Let us point out two particular cases:

I) When $\sigma$ and $b$ are linear and deterministic and $\eta$ is not necessarily in $\mathcal{F}_0$, this equation was studied by Shiota [Sh86], Ustunel [Us88] etc. A global solution (a solution for all times $t \in \mathbb{R}_+$) was proved to exist.

II) When $\sigma$ and $b$ are linear but possibly anticipating, this equation was studied by Buckdahn [91] and Pardoux [Pa90] etc. Under some mild conditions, a global solution exists in this case too.

III) When $\sigma$ and $b$ are nonlinear but deterministic and time independent, i.e., $\sigma(t, x, \omega) = \sigma(x)$ and $b(t, x, \omega) = b(x)$, a local solution exists up to some stopping time. Under some more restrictive conditions, a global solution exists. See, e.g. [Bu92] etc.

In this paper, we study the approximation of Equation (1.1) by Wick product equations.

Let now $\{\phi_\varepsilon; \varepsilon > 0\}$ be a family of piecewise differentiable functions on $\mathbb{R}_+ \times \mathbb{R}_+$ such that

$$\phi_\varepsilon(s, t) \to \delta(s - t) \quad (1.16)$$

as $\varepsilon \to 0$, where $\delta$ is the Dirac delta function on $\mathbb{R}_+$, i.e., $\int_{\mathbb{R}_+} \delta(t - s)f(s)ds = f(t)$ for all $C_0^\infty$ functions ($C^\infty$ functions with compact support) $f$ on $\mathbb{R}_+$. Let

$$W_s^\varepsilon = \int_{\mathbb{R}_+} \phi_\varepsilon(s,t) dW_t \qquad (1.17)$$

be an approximation of the (one dimensional) Brownian motion. We consider the following stochastic equation:

$$X_t^\varepsilon = \eta + \int_0^t \sigma(s, X_s^\varepsilon) \diamond W_s^\varepsilon ds + \int_0^t b(s, X_s^\varepsilon) ds, \quad 0 \le t < \infty, \quad (1.18)$$

where $\diamond$ means the Wick product, which we will recall in Section 2, see e.g. [GHLØUZ92], [Me93] etc. As expained above, it is natural to guess that $X_t^\varepsilon \to X_s$ as $\varepsilon \to 0$. We will prove that it turns out be true in the linear case and also in the quasilinear case.

Unlike the Wong-Zakai approximation [WZ65] etc, in which case, the approximate equations are well-posed, our approximate equation (1.18) is not a familiar one. So we should first study this equation, in particular, the problems such as the existence and the uniqueness of the solution.

Our strategy to solve Equation (1.18) is to reduce it to a first order quasilinear (hyperbolic) partial differential equation in infinitely many variables. The latter in the case of finitely many variables is also hard and has been studied by many people. It is well-known that a strong solution fails to exist generally. There is a classical method, called the *characteristic curve method* [Jo78] for solving the above reduced equation in finite dimensional case. We will adopt it for our purpose. The characteristic curve method is particularly powerful for the linear case and for what we would like to call the quasilinear case. In these two cases, we can find the "explicit" solution of (1.18) using this approach. Formally, Equation (1.15) can be considered as the limiting case ($\varepsilon \to 0$) of (1.18). In the linear case, this characteristic method concides with the so-called (anticipating) Girsanov transformation method.

We feel that the approximate equation of type (1.18) is easier than Equation (1.15) because we have, for instance, an "energy integral" inequality for it. We state it in Section 4 and there we also use it to prove the uniqueness.

We will also give a direct approach to a particular quasilinear case in Section 6 and prove a strong convergence there.

Here is a summary of each section:

In Section 2 we introduce some notations and study some properties of the Wick product. In particular, we state two formulas which express the Wick product through ordinary (Wiener) product and vice versa, see (2.3) and (2.4) below. From this formula we deduce a formula (2.6) obtained recently by Benth and Gjessing [BG94]. We prove a special case which is needed in this paper. We also prove an integration by parts formula which is of useful in the following.

In Section 3 we reduce Equation (1.18) to a first order (hyperbolic) equation in infinitely many variables and then we explain the characteristic curve method used to solve the equation. We also write formally a corresponding equation for (1.15).

In Section 4 we deal with the linear case. We first obtain the explicit solution of (1.18) by solving the characteristic equations. We also prove that the density of the nonlinear transformation induced by one characteristic equation solves the equation. We obtain in this way an "explicit" expression of the density. This also means that our method coincides with the Girsanov transformation method in the limit. We establish an analogue of the energy inequality and use it to prove the uniqueness of the solution of Equation (1.18) (Section 4.2).

In Section 5 we deal with a quasilinear case. We prove that the characteristic curve method works here too. We prove the existence and solve the approximation problem.

In Section 6 we discuss a particular case (the adapted case) of the above quasilinear equation. In this case we can solve the equation using a simpler method (called the reduction method). A stronger approximation result is also obtained here.

## 2. The Wick product

Starting with the family of $L^p$ spaces $\mathcal{L}^p$ over the Wiener space $(\Omega, H, P)$ and the basic differential operators $\nabla$ (the gradient), $\nabla^*$ (the dual of $\nabla$, the divergence) and $L = -\nabla^*\nabla$ (the Ornstein-Uhlenbeck operator), a family of Sobolev spaces $\mathbb{D}_\alpha^p$, $1 \leq p \leq \infty, \alpha \in \mathbb{R}$, can be introduced as

$$\mathbb{D}_\alpha^p = (I - L)^{-\alpha/2}(\mathcal{L}^p)$$

with the norm

$$\|F\|_{\alpha,p} = \|(I - L)^{\alpha/2}F\|_p, \quad F \in \mathbb{D}_\alpha^p,$$

where $\|\cdot\|_p$ is the $L_p$ norm on $\mathcal{L}^p$. Then obviously,

$$\mathbb{D}_0^p = \mathcal{L}^p, \quad \mathbb{D}_\alpha^p \subset \mathbb{D}_{\alpha'}^{p'} \quad \text{if} \quad p > p' \text{ and } \alpha > \alpha'$$

and

$$(\mathbb{D}_\alpha^p)' = \mathbb{D}_{-\alpha}^{p/(p-1)}.$$

More generally, when we consider $E-$valued functionals, $E$ being a separable real Hilbert space, the corresponding Sobolev spaces are denoted by $\mathbb{D}_\alpha^p[E]$. We will also use the notation $D_t F$ defined by

$$\int_{\mathbb{R}_+} D_t F \dot{h}_t dt = \nabla_h F \quad \text{for any} \quad h \in H.$$

This means that if we let $\mathcal{H}_t$ be the Heaviside function

$$\mathcal{H}_t(s) = \begin{cases} 1 & s > t \\ 0 & s < r, \end{cases}$$

then $D_t F = \nabla_{\mathcal{H}_t} F$. Let $f_k : \mathbb{R}_+^k \to \mathbb{R}$ be such that $f_k \in L^2(\mathbb{R}_+^{\otimes k})$. Then the multiple Itô-Wiener integral

$$I_k(f_k) = \int_{\mathbb{R}_+^k} f_k(t_1, \cdots, t_k) dW_{t_1} \cdots dW_{t_k}$$

is well-defined. As usual, we identify $\nabla^n F$ as a mapping from $\Omega \to H^{\otimes n}$. It is easy to see that if $F = I_k(f_k)$ with $f_k \in L^2(\mathbb{R}_+^{\otimes k})$, then $\nabla^n F$ exists and

$$\nabla^n F = \begin{cases} (k!/n!) \int_{\mathbb{R}_+^{k-n}} f_k(t_1, \cdots, t_{k-n}; \cdot, \cdots, \cdot) dW_{t_1} \cdots dW_{t_{k-n}} & \text{if } n \le k \\ 0 & \text{otherwise.} \end{cases}$$

**DEFINITION** Let $F = I_k(f_k)$ and $G = I_l(g_l)$. We define the Wick product of $F$ and $G$ by

$$F \diamond G = I_{k+l}(f_k \otimes g_l), \tag{2.1}$$

where $f_k \otimes g_l$ is a function of $k+l$ variables, which is the symmetrization of $f_k(s_1, \cdots, s_k) g_l(s_{k+1}, \cdots, s_{k+l})$. Then we define the Wick product of two arbitrary random variables by linearity.

Given two $E$–valued Wiener functionals $F$ and $G$ we define their Wick scalar product by

$$< F \diamond G >_E = \sum_{n=1}^{\infty} < F, e_n > \diamond < G, e_n > \qquad (2.2)$$

if each $< F, e_n > \diamond < G, e_n >$ exists and the above limit exists, where $\{e_n, n = 1, 2, \cdots\}$ is an orthonormal basis of $E$.

We are going to deduce two formulas formally.

For $f, g \in L^2(\mathbb{R}_+; dt)$ we define $\tilde{f} = \int_0^{\infty} f_s dW_s$, $< f, g >= \int_0^{\infty} f_s g_s ds$ and $\|f\| =< f, f >^{1/2}$.

Set $F = \exp\{s\tilde{f} - s^2\|f\|^2/2\}$ and $G = \exp\{t\tilde{g} - t^2\|g\|^2/2\}$. Then according to the definition of Wick product, $F \diamond G$ exists and

$$
\begin{aligned}
F \diamond G &= \exp\{s\tilde{f} + t\tilde{g} - \|sf + tg\|^2/2\} \\
&= \exp\{s\tilde{f}g - s^2\|f\|^2/2\} \exp\{t\tilde{g} - t^2\|g\|^2/2\} \exp\{st < f, g >\} \\
&= \sum_{m,n=1}^{\infty} s^m t^n \sum_{p \le m \wedge n} \frac{(-1)^p I_{m-p}(f^{\otimes m-p}) I_{n-p}(g^{\otimes n-p}) < f, g >^p}{p!(n-p)!(m-p)!}.
\end{aligned}
$$

On other hand, we have

$$F \diamond G = \sum_{n,m=1}^{\infty} s^m t^n I_m(f^{\otimes m}) I_n(g^{\otimes n})/m!n! .$$

Comparing the coefficients of $s^m t^n$, we can write the above formula as

$$I_m(f^{\otimes m}) \diamond I_n(g^{\otimes n}) = \sum_{p \le m \wedge n} \frac{(-1)^p < f, g >^p}{p!(n-p)!(m-p)!} I_{m-p}(f^{\otimes m-p}) I_{n-p}(g^{\otimes n-p}).$$

Using the notation of derivative, we obtain

$$I_m(f^{\otimes m}) \diamond I_n(g^{\otimes n}) = \sum_{p \le m \wedge n} \frac{(-1)^p}{p!} < \nabla^p I_m(f^{\otimes m}), \nabla^p I_n(g^{\otimes n} >_{H^{\otimes p}},$$

where $\nabla^p F$ is identified as a mapping from $\Omega$ to $H^{\otimes p}$. By polarization, we have

$$I_m(f_m) \diamond I_n(g^{\otimes n}) = \sum_{p \leq m \wedge n} \frac{(-1)^p}{p!} < \nabla^p I_m(f_m), \nabla^p I_n(g_n >_{H^{\otimes p}} .$$

This is a formula for the single chaos. For the general functionals, we have

$$F \diamond G = \sum_{p=0}^{\infty} \frac{(-1)^p}{p!} < \nabla^p F, \nabla^p G >_{H^{\otimes p}} . \tag{2.3}$$

This general formula represents the Wick product by the ordinary product. We can also obtain similarly a formula to represent the ordinary product by the Wick product, namely

$$FG = \sum_{p=0}^{\infty} \frac{1}{p!} < \nabla^p F \diamond \nabla^p G >_{H^{\otimes p}} . \tag{2.4}$$

Summarizing the above formal results, we have

*Under some conditions on $F$ and $G$ and a suitable sense of convergence of the concerned series, the formulas (2.3) and (2.4) are true.*

There is no unique way to give a rigorous justification of the above formulas (2.3) and (2.4), which involves to identify the spaces where $FG$ and $F \diamond G$ belong. In this paper we only need the formula (2.3) in the case $F \in \mathbb{D}_1^2$ and $G = \int_0^\infty h_t dW_t$ for $h \in L^2(\mathbb{R}_+; dt)$. In this case we can state a rigorous result as

**LEMMA** 2.1 *If $F \in \mathbb{D}_1^p$, $p > 1$ and $H = \int_0^\infty h_t dW_t$ for $h \in L^2(\mathbb{R}_+; dt)$. Then $F \diamond H$ is well-defined as an element of $L^q(\Omega; P)$ for any $q < p$ and*

$$F \diamond H = FH - \nabla_h F = FH - \int_0^\infty h_t D_t F dt. \tag{2.5}$$

**PROOF** When $F = \exp\{\tilde{f} - \|f\|^2/2\}$, the formula is true by preceding computation. But the set of finite linear combinations of such $F$'s is dense in $\mathbb{D}_1^p$. So using a limit argument, we see (2.5) is true for $F \in \mathbb{D}_1^p$.
∎

Another particular case of (2.3) is a formula obtained recently by Benth and Gjessing [BG94], which we will deduce as follows

Let $F = \exp\{\int_0^\infty f_s dW_s - \frac{1}{2}|f_s|^2 ds\}$, $f \in L^2$. Then it is easy to see that $\nabla^p F = F f^{\otimes p}$. According (2.3), we have

$$F \diamond G = \sum_{p=0}^\infty \frac{(-1)^p}{p!} < \nabla^p F, \nabla^p G >_{H^{\otimes p}} = F \sum_{p=0}^\infty \frac{(-1)^p}{p!} < f^{\otimes p}, \nabla^p G >_{H^{\otimes p}}$$

$$= F \sum_{p=0}^\infty \frac{(-1)^p}{p!} \nabla_f^p G = FG(\cdot - \int_0^\cdot f_s ds). \tag{2.6}$$

Formula (2.6) appeared in [BG94].

From Lemma 2.1, we can deduce the following "integration by parts" formula which will be useful in Section 4.

**LEMMA 2.2** *If* $F \in \mathbb{D}_1^p$, $p > 1$ *and* $H = \int_0^\infty h_t dW_t$ *for* $h \in L^2(\mathbb{R}_+; dt)$. *Let* $G \in \mathbb{D}_1^q$, $q > 1$. *Then*

$$\mathbb{E}\{(F \diamond H)G\} = \mathbb{E}\{F\nabla_h G\}. \tag{2.7}$$

**PROOF**   By Lemma 2.1,

$$\mathbb{E}\{(F \diamond H)G\} = \mathbb{E}\{[FH - \nabla_h F]G\} = \mathbb{E}[FHG] - \mathbb{E}[\nabla_h FG]$$
$$= \mathbb{E}[FHG] - \mathbb{E}[F\nabla^*(h(\cdot)G)].$$

From the well-known fact $\nabla^*(h(\cdot)G) = HG - \nabla_h G$, we prove the lemma.
∎

**REMARK** H. Gjessing communicated us that formulas (2.3) and (2.4) are known, see e.g. (3.8) and (3.9) of [NZ93], pp.137-138.

## §3. Reduction to a hyperbolic equation

For simplicity we will omit the index $\epsilon$ in this subsection. Let $\phi(t, s)$ be a (deterministic) piecewise differentiable function and let $W_\phi(t) = \int_0^\infty \phi(t, s)dW_s$. Let $\sigma$ and $b$: $\mathbb{R}_+ \times R \times \Omega \to R$ satisfy the conditions (A) and (B) stated in the introduction. Consider the following equation

$$\begin{cases} \dot{X}_t = \sigma(t, X_t) \diamond W_\phi(t) + b(t, X_t), & 0 \le t < \infty \\ X_0 = \eta. \end{cases} \tag{3.1}$$

For any fixed $t \in \mathbb{R}_+$, $\hat{\phi}_t(\cdot) = \int_0 \phi(t,s)ds$ is an element of $H$. Using Lemma 2.1, Equation (3.1) is equivalent to

$$
\begin{aligned}
\dot{X}_t &= \sigma(t, X_t) W_\phi(t) + b(t, X_t) - \nabla_{\hat{\phi}_t} \sigma(t, X_t) \\
&= \sigma(t, X_t) W_\phi(t) + b(t, X_t) - (\nabla_{\hat{\phi}_t} \sigma)(t, X_t) - \sigma'(t, X_t) \nabla_{\hat{\phi}_t} X_t \\
&= F(t, X_t) + \sigma'(t, X_t) \int_0^\infty \phi(t,r) D_r X_t dr = F(t, X_t) + \nabla_{\hat{G}(t,X_t)} X_t
\end{aligned}
\tag{3.2}
$$

where

$$
F(t, x, \omega) = b(t, x, \omega) + \sigma(t, x, \omega) W_\phi(t) - \nabla_{\hat{\phi}_t} \sigma(t, x, \omega) \tag{3.3}
$$

$$
\equiv b(t, x, \omega) + \sigma(t, x, \omega) W_\phi(t) - \int_0^\infty \phi(t,r) D_r \sigma_t dr
$$

$$
G(t, x, \omega, \cdot) = -\sigma'(t, x, \omega) \phi(t, \cdot) \tag{3.4}
$$

$$
\hat{G}(t, \cdot; \omega) = \int_0^\cdot G(t, x, \omega, r) dr = -\sigma'(t, x, \omega) \hat{\phi}_t . \tag{3.5}
$$

For almost every $(t, \omega) \in \mathbb{R}_+ \times \Omega$, $\hat{G}(t, \cdot; \omega)$ can be considered an element of $H \subset \Omega$. We also write it as $\hat{G}(t, \omega)$. So $\hat{G}$ is a mapping from $\mathbb{R}_+ \times \mathbb{R} \times \Omega$ to $H \subset \Omega$.

Now let $\sigma(t, x) \in \mathbb{D}_1^\infty$. If $X_t \in \mathbb{D}_1^p$ for some $p > 1$, then it is easy to see that $\sigma(t, X_t) \in \mathbb{D}_1^p$. According to Lemma 2.1, $\sigma(t, X_t) \diamond W_\phi(t)$ is well-defined.

**DEFINITION** A (not necessarily adapted) stochastic process $X_t$, $0 \le t < \infty$ is called a strong solution of Equation (3.1) if $X_t \in \mathbb{D}_1^p$ for some $p > 1$ and $X_t$ is differentiable with respect to $t$ such that $X_0 = \eta$ a.s. and (3.1) holds almost surely on $\mathbb{R}_+ \times \Omega$.

From the above computation, we have

**LEMMA** 3.1 *$X_t$ is a strong solution of Equation (3.1) if and only if it satisfies (3.2) with the data given by (3.3)-(3.5).*

Equation (3.2) is a first order quasilinear (hyperbolic) partial differential equation of variable coefficients in infinitely many variables. To see this clearly, let $\{e_n, n = 1, 2, \cdots\}$ be an orthonormal basis of $L^2(\mathbb{R}_+; dt)$. Then formally we can write

$$
X_t = X(t, \tilde{e}_1, \cdots, \tilde{e}_n, \cdots), \quad F_t = F(t, x, \tilde{e}_1, \cdots, \tilde{e}_n, \cdots)
$$

and $G(t, x) = \sum_{n=1}^{\infty} G_n(t, x)e_n(t)$ with

$$G_n = \int_0^{\infty} e_n(t)G(t, x)dt = G_n(\tilde{e}_1, \cdots, \tilde{e}_n, \cdots).$$

Then Equation (3.2) becomes

$$\frac{\partial X}{\partial t} = F(t, X) + \sum_{n=1}^{\infty} G_n(t, X_t)\frac{\partial X}{\partial \tilde{e}_n}, \tag{3.6}$$

where we consider $\tilde{e}_1, \cdots, \tilde{e}_n, \cdots$ as independent variables.

We will establish some rules of differentiation adapted to our situation.

**LEMMA 3.2** *Let $g : \mathbb{R}_+ \times \Omega \to E$ be a continuous mapping, $E$ being an arbitrary Banach space. If $K : \mathbb{R}_+ \to \Omega$ is continuous and differentiable such that $\frac{d}{dt}K(t) \in H$ and $g \in \mathbb{D}_1^p$ for some $p \geq 1$. Then $g(t, K(t))$ is differentiable with respect to $t$ and we have the following chain rule:*

$$\frac{d}{dt}g(t, K(t)) = \frac{\partial}{\partial t}g(t, K(t)) + \nabla g(t, K(t))(\frac{d}{dt}K(t)), \tag{3.7}$$

*where $\nabla g(\omega)$, a mapping from $H$ to $E$, is the derivative of $g$ with respect to $\omega$.*

**PROOF** Easy. ∎

We will explain how to use the so-called characteristic curve method to solve the above equation.

Let $\Gamma_t : \Omega \to \Omega$, $0 \leq t < \infty$ and $v_t : \Omega \to \mathbb{R}$, $0 \leq t < \infty$ satisfy the following ordinary differential equations on the Banach space $\Omega \times \mathbb{R}$:

$$\Gamma_t = \omega - \int_0^t G(s, v_s, \Gamma_s)\hat{\phi}_s ds, \tag{3.8}$$

$$v_t = \eta + \int_0^t F(s, v_s, \Gamma_s)ds. \tag{3.9}$$

**THEOREM 3.3** *Suppose Equations (3.8)-(3.9) have a unique solution $(\Gamma_t, v_t)$, $0 \leq t < \infty$ such that for any $t \geq 0$, $\Gamma_t : \Omega \to \Omega$ is invertible, i.e., there is a $\Lambda_t : \Omega \to \Omega$ such that for a.e. $\omega \in \Omega$, $\Gamma_t(\Lambda_t(\omega)) = \omega$ and $\Lambda_t(\Gamma_t(\omega)) = \omega$ and $\Lambda_{\cdot}(\omega)$ is differentiable for a.e. $\omega \in \Omega$. Then*

$$X(t, \omega) = v_t(\Lambda_t(\omega)) \tag{3.10}$$

*is a solution of (3.2), i.e., a solution of (3.1).*

**PROOF**   Differentiating the identity $\Gamma(t, \Lambda_t(\omega)) = \omega$ with respect to $t$, we get

$$\frac{\partial \Gamma_t}{\partial t}(\Lambda_t) + (\nabla \Gamma_t)(\Lambda_t)\frac{\partial \Lambda_t}{\partial t} = 0,$$

i.e.,

$$\frac{\partial \Lambda_t}{\partial t} = -\{(\nabla \Gamma_t)(\Lambda_t)^{-1}\frac{\partial \Gamma_t}{\partial t}\}(\Lambda_t).$$

Differentiating the identity $\Gamma(t, \Lambda_t(\omega)) = \omega$ with respect to $\omega$, we get

$$[(\nabla \Gamma_t)(\Lambda_t)]^{-1} = \nabla \Lambda_t. \tag{3.11}$$

By Lemma 3.2, we get

$$
\begin{aligned}
\dot{X}_t &= \frac{\partial v}{\partial t}(t, \Lambda_t) + (\nabla v)(t, \Lambda_t)\frac{d}{dt}\Lambda_t \\
&= \frac{\partial v}{\partial t}(t, \Lambda_t) - (\nabla v)(t, \Lambda_t)\{(\nabla \Gamma_t)(\Lambda_t)^{-1}\frac{\partial \Gamma_t}{\partial t}\}(\Lambda_t) \\
&= \frac{\partial v}{\partial t}(t, \Lambda_t) - (\nabla v)(t, \Lambda_t)\{\nabla \Lambda_t \frac{\partial \Gamma_t}{\partial t}\}(\Lambda_t) = \frac{\partial v}{\partial t}(t, \Lambda_t) - <\nabla X_t, \hat{G}>_H \\
&= F(t, v_t, \Gamma_t)|_{\omega = \Lambda_t} + \nabla_{\hat{G}}X_t = F(t, X_t) + \nabla_{\hat{G}}X_t.
\end{aligned}
$$

This proves the theorem.   ∎

**REMARK** The main difficulty in applying this theorem is to prove that the solution $\Gamma_t$ has inverse, i.e., $\Gamma_t : \Omega \to \Gamma$ is one-to-one for any $t \in \mathbb{R}_+$, since the characteristic equations are in general coupled. In the linear and quasilinear case, it turns out there is no coupling at all. So we can solve Equation (3.8) and then (3.9) separately. This is why we can handle these two cases easily.

   In the finite dimensional case, there is a necessary and sufficient condition [ML85] for $\Gamma_t$ to be invertible. But it seems too hard to apply it in our case.

**REMARK** It is natural to think that the stochastic differential equation (1.15) is a particular case of the equation of Wick type with the Dirac function. The characteristic curve method makes sense in this singular case also.

## §4. The linear case

In this section we will solve the characteristic curve Equations (3.8) and (3.9) and find the explicit solution of (3.1) for linear case. Then we prove an energy integral inequality which implies the uniqueness easily. We also establish the relation between the solution and the Girsanov density which permits us to use the existing theory.

### 4.1 Existence

We use the same notation as in Section 3. For the simplicity of notation we will omit the index $\varepsilon$ in this subsection.

Let $\sigma = \sigma_s(\omega)$ and $b = b_s(\omega)$ be two anticipating processes and let $\eta$ be a random variable. Consider the following linear anticipating stochastic differential equation:

$$X_t = \eta + \int_0^t \sigma_s X_s dW_s + \int_0^t b_s X_s ds, \quad 0 \le t < \infty \qquad (4.1)$$

and its approximate equation

$$\dot{Y}_t = (\sigma_s Y_s) \diamond W_\phi(t) + b_s Y_s, \quad 0 \le t < \infty, \quad Y_0 = \eta. \qquad (4.2)$$

Equation (4.2) can be reduced as in Section 3 to the following one

$$\dot{Y}_t = F_t Y_t + \nabla_{\hat{G}_t} Y_t, \quad Y_0 = \eta,$$

where (we omit the explicit dependence on $\varepsilon$)

$$F_t = b_t + \sigma_t W_\phi(t) - \nabla_{\hat{\phi}_t} \sigma_t \qquad (4.3)$$

$$\hat{G}(t, \cdot; \omega) = \int_0^{\cdot} G(t, r; \omega) dr = -\sigma_t(\omega)\hat{\phi}_t . \qquad (4.4)$$

Equation (4.2) is a **linear** first order (hyperbolic) partial differential equation of variable coefficients in infinitely many variables.

We are going to solve the characteristic equations to find the explicit solution. On the Banach space $\Omega$, endowed with the sup norm, Equation (3.8) becomes

$$\Gamma_t = \omega - \int_0^t \hat{G}(s, \Gamma_s) ds = \omega + \int_0^t \sigma_s(\Gamma_s)\hat{\phi}_s ds, \quad 0 \le t < \infty. \qquad (4.5)$$

There is no $v_t$ in this equation and we can solve it in a usual way.

**LEMMA 4.1** *Let $\sigma \in \mathbb{D}_1^\infty$ uniformly on $t \in \mathbb{R}_+$, i.e.,*

$$\sup_{0 \leq t \leq T} \ ess \ sup_{\omega \in \Omega} |\nabla \sigma_t(\omega)|_H < \infty.$$

*Then Equation (4.5) has a unique solution $\Gamma : \mathbb{R}_+ \times \Omega \to \Omega$. Moreover, for any fixed $t \in \mathbb{R}_+$, $\Gamma(t, \cdot)$ is invertible as a mapping from $\Omega$ to $\Omega$. Denote its inverse by $\Lambda(t, \omega)$, i.e.,*

$$\Gamma(t, \Lambda(t, \omega)) = \omega, \qquad \Lambda(t, \Gamma(t, \omega)) = \omega. \qquad (4.6)$$

*Furthermore, $\Lambda(t, \omega)$ is given by $\Lambda(t, \omega) = \Lambda(0, t; \omega)$, where $\Lambda(s, t; \omega)$ satisfies the following equation:*

$$\Lambda(s, t; \omega) = \omega + \int_s^t \hat{G}(r, \Lambda(r, t; \omega)) dr = \omega - \int_s^t \sigma_r(\Lambda(r, t; \omega)) \hat{\phi}_r dr, \quad 0 \leq s \leq t.$$
$$(4.7)$$

**PROOF** The proof is a routine one. We will not provide the details here, see for instance, [Bu91], the proof of Lemma 3.1 there. ∎

Now substituting $\Gamma_t$ into (3.9), we get

$$\frac{\partial v(t, \omega)}{\partial t} = F(t, \Gamma(t, \omega)) v(t, \omega).$$

The explicit solution of this equation is given by

$$v = v(0, \omega) \exp\{ \int_0^\infty F(s, \Gamma(s, \Lambda(t, \omega))) ds \}.$$

To determine $v(0, \omega)$ we shall use the initial condition $X_0 = \eta$. Because $\Lambda(0, \omega) = \omega$, we have

$$\eta(\omega) = Y(0, \omega) = v(0, \Lambda(0, \omega)) = v(0, \omega). \qquad (4.8)$$

Hence we obtain

**THEOREM 4.2** *If $\sigma$ has bounded Malliavin derivative and $F(s, \Lambda(s, \omega))$ is well-defined and is Bochner integrable, then the solution of (4.2) exists and is given by*

$$Y(t, \omega) = \eta(\Lambda(t, \omega)) \exp\{ \int_0^\infty F(s, \Gamma(s, \Lambda(t, \omega))) ds \}. \qquad (4.9)$$

To obtain the expression for the solution of Equation (4.2) from the above formula, we have to compute the expression $W_\phi(s)$ (which is a functional of $\omega$) when we replace $\omega$ by $\Gamma(s, \Lambda(t, \omega))$, i.e., we want to compute

$$W_\phi(s)|_{\omega=\Gamma(s,\Lambda(t,\omega))}.$$

We shall make use of the following lemma whose proof is obvious.

**LEMMA 4.3** *If* $\tilde{\omega} = \omega + \int_0^{\cdot} h_r(\omega)dr$, *then*

$$W_\phi(s)|_{\omega=\tilde{\omega}} = W_\phi(s) + \int_0^\infty \phi(s,r)h_r(\omega)dr.$$

From (4.5), we have

$$W_\phi(s)|_{\omega=\Gamma(s,\omega)} = W_\phi(s) + \int_0^\infty \int_0^s \phi(s,r)\phi(u,r)\sigma_u(\Gamma(u,\omega))dudr.$$
$$(4.10)$$

Using (4.7) and Lemma 4.3, we then have

$$W_\phi(s)|_{\omega=\Gamma(s,\Lambda(t,\omega))} = \left(W_\phi(s)|_{\omega=\Gamma(s,\omega)}\right)|_{\omega=\Lambda(t,\omega)}$$

$$= W_\phi(s)|_{\omega=\Lambda(t,\omega)} + \int_0^\infty \int_0^s \phi(s,r)\phi(u,r)\sigma_u(\Gamma(u,\Lambda(t,\omega)))dudr$$

$$= W_\phi(s) - \int_0^\infty \int_0^\infty \phi(s,r)\phi(v,r)\sigma_v(\Lambda(v,t;\omega))dvdr$$

$$+ \int_0^\infty \int_0^s \phi(s,r)\phi(u,r)\sigma_u(\Gamma(u,\Lambda(t,\omega)))dudr \quad (4.11)$$

From the above computation we get

**THEOREM 4.4** *The explicit solution of Equation (4.2) is given by*

$$X(t,\omega) = \eta(\Lambda(t,\omega))\exp\{\int_0^t b(s,\Gamma(s,\Lambda(t,\omega)))ds\} \quad (4.12)$$

$$\times \exp\{\int_0^t \sigma(s,\Gamma(s,\Lambda(t,\omega)))W_\phi(s)ds - \int_0^t (\nabla_{\dot{\phi}_s}\sigma)(s,\Gamma(s,\Lambda(t,\omega)))ds\}$$

$$\times \exp\{-\int_0^t \int_0^T \int_0^t \sigma(s,\Gamma(s,\Lambda(t,\omega)))\sigma(v,\Gamma(v,\Lambda(v,t;\omega)))\phi(s,r)\phi(v,r)dsdvdr\}$$

$$\times \exp\{\int_0^t \int_0^T \int_0^s \sigma(s,\Gamma(s,\Lambda(t,\omega)))\sigma(u,\Gamma(u,\Lambda(t,\omega)))\phi(u,r)\phi(s,r)dsdudr\},$$

*where $\Gamma_t$ and $\Lambda$ are given respectively by (4.5) and (4.7).*

## 4.2. Energy integral and uniqueness

Let $W_h = \int_0^\infty h_s dW_s$ for some $h \in H$. Let $X_t$ be the solution of the following equation:

$$\dot{X}_t = b_t X_t + (\sigma_t X_t) \diamond W_h, \quad 0 \le t \le T. \tag{4.13}$$

We want to deduce an energy integral inequality of the following type, see for instance [Fr56].

**THEOREM 4.5** *Let $p \ge 1$ be an integer. Assume that there is a constant $0 < K < \infty$ such that*

$$\sup_{0 \le t < \infty} |\nabla_h(\sigma_t)| \le K, \quad \sup_{0 \le t < \infty} |W_h \sigma_t)| \le K, \quad \sup_{0 \le t < \infty} |b_t| \le K. \tag{4.14}$$

*Then*

$$\mathbb{E}|X_t|^{2p} \le \exp\{3(2p-1)Kt\}\mathbb{E}|X_0|^{2p}. \tag{4.15}$$

**PROOF** Differentiating $\mathbb{E}|X_t|^{2p}$ with respect to $t$, we have

$$\frac{d}{dt}\mathbb{E}|X_t|^{2p} = 2p\mathbb{E}\{X_t^{2p-1}\dot{X}_t\} = 2p\mathbb{E}\{X_t^{2p-1}b_tX_t\}+2p\mathbb{E}\{X_t^{2p-1}[(\sigma_t,X_t)\diamond W_h]\}.$$

But by the integration by parts formula (2.7), we have

$$2p\mathbb{E}\{X_t^{2p-1}[(\sigma_t,X_t)\diamond W_h]\} = 2p\mathbb{E}\{\sigma_t X_t \nabla_h(X_t^{2p-1}\}$$
$$= 2p(2p-1)\mathbb{E}\{\sigma_t X_t^{2p-1}\nabla_h X_t\} = (2p-1)\mathbb{E}\{\sigma_t \nabla_h(X_t^{2p})\}$$
$$= (2p-1)\mathbb{E}\{(\nabla_h^*\sigma_t)(X_t^{2p})\} = (2p-1)\mathbb{E}\{(W_h\sigma_t - \nabla_h\sigma_t)(X_t^{2p})\} \tag{4.16}$$

Thus

$$\frac{d}{dt}\mathbb{E}|X_t|^{2p} = (2p-1)\mathbb{E}\{(W_h\sigma_t - \nabla_h\sigma_t)(X_t^{2p})\}.$$

If the conditions of the theorem are satisfied, then

$$\frac{d}{dt}\mathbb{E}|X_t|^{2p} \le 3(2p-1)K\mathbb{E}|X_t|^{2p}.$$

This implies the theorem by the Grownwall lemma.　∎

From (4.15), we deduce

**COROLLARY** 4.6 *Let the coefficients satisfy the conditions of Theorem 4.5. Then Equation (4.13) has a unique solution once $X_0$ is given in some $L^2$ space.*

## 4.3 Approximation

The solution of (4.1) is already known and is given by (see [Bu], Theorem 4.1)

$$X_t = \eta(A(t,\omega)) \exp\{\int_0^\infty b(s, T(s, A(t,\omega)))ds\} L_t , \qquad (4.17)$$

where $T_t$ and $A_t = A_{0,t}$ are given by the following ordinary differential equations in $\Omega$:

$$T_t\omega = \omega + \int_0^{t\wedge\cdot} \sigma_s(T_s\omega)ds, \quad 0 \le t \le T$$

$$A_s\omega = \omega - \int_{s\wedge\cdot}^{t\wedge\cdot} \sigma_s(A_{r,t}\omega)dr, \quad 0 \le t \le T$$

and $L_t$ is the density of $T_t$. The explicit expression for $L_t$ can be obtained as

$$L_t(\omega) = [\mathcal{L}_t(A_t(\omega))]^{-1}$$

with

$$\mathcal{L}_t(\omega) = \frac{dP \circ A_t^{-1}}{dP}(\omega) = \exp\{-\int_0^t \sigma_s(T_s)dW_s$$
$$-\frac{1}{2}\int_0^t \sigma_s(T_s)^2 ds - \int_0^t \int_0^s (D_r\sigma_s)(T_s)(D_s[\sigma_r(T_r)]drds\}.$$

Using (4.10) to compute $\int_0^\infty \sigma_s(T_s\omega)dW_s|_{\omega=A_t(\omega)}$, we obtain from the above formula $L_t(\omega) = \exp\left\{I_1 + I_2 + I_3 + I_4\right\}$, where

$$I_1 = \int_0^\infty \sigma_s(T_s A_t\omega)dW_s \qquad (4.18)$$

$$I_2 = -\int_0^t \sigma_s(T_s A_t \omega)\sigma_s(T_s A_{s,t}\omega)ds \qquad (4.19)$$

$$I_3 = \frac{1}{2}\int_0^\infty [\sigma_s(T_s A_t \omega)]^2 ds \qquad (4.20)$$

$$I_4 = \int_0^t \int_0^s (D_r\sigma_s)(T_s A_t \omega)([D_s\sigma_r(T_r)](A_t\omega)drds. \qquad (4.21)$$

We write this formula to point out that it seems that there is a mild error in the remark of p.55 of [Bu94], where the term (4.19) is missing. **REMARK** Note the apparent relation between the above Girsanov transformation method and the characteristic curve method.

We want to prove that the solution $X^\varepsilon := Y$ (We ressume the dependence on $\varepsilon$) of (4.2) is also the density $L_t^\varepsilon$ of the transformation of $\Gamma$ given by (4.5):

$$\int_\Omega L_t^\varepsilon(\omega)G(\omega)P(d\omega) = \int_\Omega G(\Gamma_t^\varepsilon(\omega))P(d\omega). \qquad (4.22)$$

**LEMMA** 4.6 $X_t^\varepsilon$ *is the solution of (4.2) with $\eta = 1$ iff $X_t^\varepsilon = L_t^\varepsilon$.*
**PROOF** Let $L_t^\varepsilon$ satisfy (4.22). Then by Lemma 2.2, for any $G \in \mathbb{D}_1^\infty$,

$$\mathbb{E}\int_0^t \{[(\sigma_s L_s^\varepsilon) \diamond W_\phi(s)]G\}ds = \int_0^t \mathbb{E}[\sigma_s L_s^\varepsilon \nabla_{\phi_s} G]ds$$
$$= \int_0^t \mathbb{E}[\sigma_s(\Gamma_s)(\nabla_{\phi_s} G)(\Gamma_s^\varepsilon)]ds.$$

Now

$$\frac{d}{ds}G(\Gamma_s^\varepsilon) = (\nabla_{\frac{d}{ds}\Gamma_s^\varepsilon} G)(\Gamma_s^\varepsilon) = \sigma_s(\Gamma_s^\varepsilon)(\nabla_{\phi_s} G)(\Gamma_s^\varepsilon). \qquad (4.23)$$

Thus

$$\mathbb{E}\int_0^t (\sigma_s L_s^\varepsilon) \diamond W_\phi(s)Gds = \mathbb{E}\int_0^t \frac{d}{ds}G(\Gamma_s^\varepsilon)ds$$
$$= \mathbb{E}G(\Gamma_t^\varepsilon) - \mathbb{E}G = \mathbb{E}\{[L_t^\varepsilon - 1]G\}.$$

Since the above formula is true for any $G$, we prove that

$$\int_0^t (\sigma_s L_s^\varepsilon) \diamond W_\phi(s)ds = L_t^\varepsilon - 1.$$

This proves that $L_t^\varepsilon$ is the solution of (4.2). By the uniqueness of the solution of (4.2), the lemma is proved. ∎

Then by the results of [Bu91], under some mild conditions on the coefficients $b$ and $\sigma$, $L_t^\varepsilon$ converges to $L_t$ in the topology $\sigma(L^1, L^\infty)$. We will not repeat this work here.

## §5. The quasilinear equation I

Let $b : \mathbb{R}_+ \times \mathbb{R} \times \Omega \to \mathbb{R}$ and $\sigma : \mathbb{R}_+ \times \Omega \to \mathbb{R}$ satisfy the conditions (A) and (B) in the introduction. Consider the following quasilinear stochastic differential equation

$$X_t = \eta + \int_0^t b(s, X_s, \omega)ds + \int_0^t \sigma(s, \omega)X_s dW_s, \quad 0 \le t < \infty \quad (5.1)$$

and its approximation equation

$$\frac{dX_t^\varepsilon}{dt} = b(t, \omega, X_t^\varepsilon) + (\sigma_t X_t^\varepsilon) \diamond W_{\phi^\varepsilon}(t), \quad X_0 = \eta. \quad (5.2)$$

We will discuss Equation (5.2) first. By (3.2)-(3.5), the solution of (5.2) is given by the solution of the following equation

$$\frac{\partial X_t^\varepsilon}{\partial t} = F(t, X_t^\varepsilon, \omega) + \nabla_{\hat{G}(t, X_t^\varepsilon, \omega)} X_t^\varepsilon, \quad (5.3)$$

where

$$F(t, \omega, x) = b(t, x, \omega) + \sigma(t, \omega)W_\phi(t) - \nabla_{\phi_t^\varepsilon} \sigma(t, \omega). \quad (5.4)$$

As in the linear case we can solve this equation by the "characteristic curve method".

Let $\Lambda(t, \omega)$ and $\Gamma(t, \omega)$ be constructed as in preceding section and let the solution of (5.3) have the form (4.10), i.e.,

$$X^\varepsilon(t, \omega) = g(t, \Lambda(t, \omega)) \quad (5.5)$$

for some $g : \mathbb{R}_+ \times \Omega \to \mathbb{R}$. Then

$$\frac{\partial g(t, \Lambda(t, \omega))}{\partial t} = F(t, g(t, \Lambda(t, \omega)), \omega).$$

Making the transform $\omega \to \Gamma(t,\omega)$, we get

$$\frac{\partial g(t,\omega)}{\partial t} = F(t,g(t,\omega),\Gamma(t,\omega)).$$

From the explicit expression of $F$ we have

$$\frac{\partial g(t,\omega)}{\partial t} = b(t,g(t,\omega),\Gamma(t,\omega))g(t,\omega) + h_t(\omega)g(t,\omega), \qquad (5.6)$$

where

$$h_t(\omega) = \{\sigma(t,\omega)W_{\phi^\epsilon}(t) - \nabla_{\phi_t^\epsilon}\sigma(t,\omega)\}|_{\omega=\Gamma(t,\omega)}. \qquad (5.7)$$

Introduce $Y_t(\omega) = g(t,\omega)e^{-\int_0^t h_s\,ds}$ to simplify (5.6) to

$$\frac{dY_t}{dt} = e^{-\int_0^t h_s\,ds}b(t,e^{\int_0^t h_s\,ds}Y_t,\Gamma(t,\omega)). \qquad (5.8)$$

(5.8) is an ordinary equation with $\omega$ as a parameter. It can be solved in the usual way. Let us summarize the above computation as

**PROPOSITION** 5.1 *Let $h$ be given by (5.7) and make the above transformation. Then (5.2) becomes (5.8).*

Let $b = 0$ and $\eta = 1$. Then $Y = 1$. So it is easy to see that

$$L_t^\epsilon = e^{\int_0^t h_s\,ds|_{\omega=\Lambda(t,\omega)}} \qquad (5.9)$$

is the solution of the equation $X_t = (\sigma_t X_t) \diamond W_\phi(t)$. Equation (5.8) becomes

$$\frac{dY_t}{dt} = (L_t^\epsilon)^{-1}|_{\omega=\Gamma_t(\omega)}b(t,L_t^\epsilon|_{\omega=\Gamma_t(\omega)}Y_t,\Gamma(t,\omega)). \qquad (5.10)$$

Summarizing the above results, we have

**THEOREM** 5.2 *Let $Y_t(\omega)$ be given by the unique solution of (5.10). Put*

$$X^\epsilon(t,\omega) = Y(t,\Lambda(t,\omega))L_t^\epsilon. \qquad (5.11)$$

*Then $X_t^\epsilon$ is a solution of (5.2).*

Formally letting $\varepsilon \to 0$, we can obtain the solution of (5.1) this way. Let $T_t$ and $A_{s,t}$ be defined as in Section 4.3. Let $L_t$ be the density of the transformation of $T$ given by (4.5), i.e.,

$$\int_\Omega L_t F(\omega) P(d\omega) = \int_\Omega F(T_t(\omega)) P(d\omega).$$

**THEOREM** 5.3 *Let $Y_t(\omega)$ be given by the unique solution of the following ordinary differential equation*

$$\frac{dY_t}{dt} = (L_t(A_t))^{-1} b(t, L_t(A_t) Y_t, T_t). \tag{5.12}$$

*Put*

$$X(t, \omega) = Y(t, A(t, \omega)) L_t. \tag{5.13}$$

*Then $X_t$ is a solution of (5.1).*

**PROOF**    We will use the same idea as in Lemma 4.6. For any $G \in \mathbb{D}_1^\infty$,

$$
\begin{aligned}
\mathbb{E}[(\int_0^t \sigma_s X_s dW_s) G] ds &= \int_0^t \mathbb{E}\{\sigma_s X_s D_s G\} ds = \int_0^t \mathbb{E}\{\sigma_s (Y_s(T_s) L_s) D_s G\} ds \\
&= \int_0^t \mathbb{E}\{\sigma_s(A_s) Y_s(D_s G)(A_s)\} = \int_0^t \mathbb{E}\{Y_s \frac{d}{ds}[G(A_s)]\} \\
&= \mathbb{E}\{Y_t G(A_t) - \eta G\} - \int_0^t \mathbb{E}[G(A_s) \frac{d}{ds} Y_s] \\
&= \mathbb{E}\{Y_t G(A_t) - \eta G\} - \int_0^t \mathbb{E}[G(A_s)(L_s(A_s))^{-1} b(s, L_s(A_s) Y_s, T_s)] \\
&= \mathbb{E}\{Y_t G(A_t) - \eta G\} - \int_0^t \mathbb{E}[G(L_s)^{-1} b(s, L_s Y_s(T_s) L_s)] \\
&= \mathbb{E}\{Y_t(T_t) L_t G - \eta G\} - \int_0^t \mathbb{E}[Gb(s, L_s Y_s(T_s))] = \mathbb{E}\{X_t G - \eta G\} - \int_0^t \mathbb{E}[Gb(s, X_s)].
\end{aligned}
$$

Since $G$ is arbitrary, this proves the theorem.    ∎

It is easy to see that the solution of (5.8) tends to the solution of (5.12) in probability. So we can prove the convergence of $X_t^\varepsilon$ to $X_t$ in probability.

## §6. The quasilinear equation II: A simple approach

### 6.1. Reduction of the equation

Consider the nonlinear stochastic equation involving the Wick product

$$\frac{dX_t}{dt} = b(t, X_t, \omega) + \sigma_t X_t \diamond W_\phi(t). \tag{6.1}$$

Suppose now that for any $x \in \mathbb{R}$, $b(t, x, \omega)$ is an adapted process and $\sigma_t$ is deterministic. Define

$$Y_t = X_t \diamond J_\phi(t), \tag{6.2}$$

where

$$J_\phi(t) = \exp^\diamond\{-\int_0^t \sigma_s W_\phi(s)ds\} = \exp\{-\int_{\mathbb{R}_+} \psi_s^t dW_s - \frac{1}{2}\int_{\mathbb{R}_+} |\psi_s^t|^2 ds\} \tag{6.3}$$

with

$$\psi_s^t = \int_0^t \sigma_s \phi(t-s)ds. \tag{6.4}$$

With this notation, (6.1) can be written as

$$\frac{dY_t}{dt} = \frac{dX_t}{dt} \diamond J_\phi(t) - \sigma_t J_\phi(t) \diamond X_t \diamond W_\phi(t) = J_\phi(t) \diamond b(t, \omega, X_t). \tag{6.5}$$

By formula (5) in [BG94] (see also (2.6) of this paper), if we define

$$\tau_\eta f(\omega) = f(\omega - \int_0^{\cdot} \eta_s ds),$$

then we have

$$J_\phi \diamond F = J_\phi \cdot (\tau_{-\psi_t} F).$$

Therefore from (6.3)

$$\frac{dY_t}{dt} = J_\phi(t)\tau_{-\psi_t} b(t, \tau_{-\psi_t} X_t). \tag{6.6}$$

On the other hand

$$\tau_{-\psi_t} X_t(t, \omega) = \tau_{-\psi_t} [J_\phi(t) \diamond Y_t(t, \omega)$$
$$= \tau_{-\psi_t} [J_\phi(t)\tau_{\psi_t} Y_t] = \tau_{-\psi_t} J_\phi(t) Y_t$$

Since $\tau_{-\psi_t} J_\phi(t) = [J_\phi(t)]^{-1}$, we obtain the equivalent equation of $Y_t$ for (6.1):

$$\frac{dY_t}{dt} = J_\phi(t)b(t, \omega - \hat{\psi}_t, [J_\phi(t)]^{-1} Y_t). \tag{6.7}$$

This equation can be used to prove the existence and approximation theorem.

In fact, the equation for $Y_t$ can also be obtained from the characteristic curve method. To see this, note that when $\sigma_t$ is deterministic we have

$$\Gamma_t = \omega. + \int_0^t \sigma_s \hat{\delta}_\epsilon(s - \cdot)ds$$

and

$$\Lambda_t = \omega. - \int_0^t \sigma_s \hat{\delta}_\epsilon(s - \cdot)ds.$$

These equations are easy to solve and we can also check that $L_t|_{\omega = \Gamma_t(\omega)} = (J_\phi(t))^{-1}$, i.e., (5.10) gives (6.7).

## 6.2 Approximation

Now suppose that $b$ is adapted and satisfies

$$|b(t, \omega_1, x) - b(t, \omega_2, x)| \le C|\omega_1 - \omega_2|_H.$$

We will sketch how to obtain a stronger convergence theorem.

First note that the stochastic differential equation

$$dX_t = b(t, \omega, X_t)dt + \sigma_t X_t dW_t \tag{6.8}$$

has a unique strong solution. Let

$$K_t = \exp\{-\int_0^t \sigma_s dW_s - \frac{1}{2}\int_0^t |\sigma_s|^2 ds\}$$

and

$$Y_t = K_t X_t(\omega - \int_0^{t\wedge\cdot} \sigma_s ds).$$

First of all, $\tilde{X}_t := X_t(\omega - \int_0^{t\wedge\cdot} \sigma_s ds)$ satisfies

$$d\tilde{X}_t = b(t, \omega - \int_0^{t\wedge\cdot} \sigma_s ds, \tilde{X}_t)dt + \sigma_t \tilde{X}_t dW_t - \sigma_t^2 \tilde{X}_t dt.$$

Applying the Itô formula, we obtain

$$\frac{dY_t}{dt} = K_t b(t, \omega - \int_0^{t\wedge\cdot} \sigma_s ds, K_t^{-1} Y_t). \tag{6.9}$$

Thus

$$
\begin{aligned}
Y_t^\varepsilon - Y_t &= Y_0^\varepsilon - Y_0 + \int_0^t [K_s^\varepsilon b(t, \omega - \int_0^{t\wedge\cdot} \phi_s^\varepsilon ds, (K_s^\varepsilon)^{-1} Y_s^\varepsilon) \\
&\quad - K_s b(s, \omega - \int_0^{s\wedge\cdot} \sigma_u du, K_s^{-1} Y_s)] ds \\
&= Y_0^\varepsilon - Y_0 + \int_0^t [K_s^\varepsilon - K_s] b(t, \omega - \int_0^{t\wedge\cdot} \psi_s^\varepsilon ds, (K_s^\varepsilon)^{-1} Y_s^\varepsilon) \\
&\quad + \int_0^t K_s [b(t, \omega - \int_0^{t\wedge\cdot} \psi_s^\varepsilon ds, (K_s^\varepsilon)^{-1} Y_s^\varepsilon) - b(t, \omega - \int_0^{t\wedge\cdot} \phi_s ds, (K_s^\varepsilon)^{-1} Y_s^\varepsilon)] ds \\
&\quad + \int_0^t K_s [b(t, \omega - \int_0^{t\wedge\cdot} \phi_s ds, (K_s^\varepsilon)^{-1} Y_s^\varepsilon) - b(t, \omega - \int_0^{t\wedge\cdot} \phi_s ds, K_s^{-1} Y_s)] ds \\
&\leq C_1(\varepsilon) + C_2 \int_0^t K_s |(K_s^\varepsilon)^{-1} Y_s^\varepsilon - K_s^{-1} Y_s| ds \\
&\leq C_1(\varepsilon) + C_2 \int_0^t K_s (K_s^\varepsilon)^{-1} |Y_s^\varepsilon - Y_s| ds,
\end{aligned}
$$

where $C_1(\varepsilon)$ is a random variable which converges to 0 in any $L^p$ norm and $C_2$ is a constant (with respect to $\varepsilon$) which may differ from line to line. So

$$\sup_{0 \leq t \leq T} |Y_t^\varepsilon - Y_t| \leq C_1(\varepsilon) \exp\{C_2 \int_0^T K_t (K_t^\varepsilon)^{-1} dt\}.$$

This implies that

$$\lim_{\varepsilon \to 0} \mathbb{E} \sup_{0 \leq t \leq T} |Y_t^\varepsilon - Y_t|^p = 0.$$

Therefore, we obtain

**THEOREM** 6.1 *Let* $b(\cdot, \cdot, x)$ *is adapted and globally Lipschitz continuous, bounded and H continuous on* $\omega$. *Then for any* $p > 1$,

$$\lim_{\varepsilon \to 0} \mathbb{E} \sup_{0 \leq t \leq T} |Y_t^\varepsilon - Y_t|^p = 0.$$

**ACKNOWLEDGEMENTS:** Y. Hu is supported by NAVF research scholarship and also holds a position at Institute of Mathematical Sciences, Chinese Academy of Sciences, Wuhan, China.

B. Øksendal is supported by VISTA, a research cooperation between the Norwegian Academy of Science and Letters and Den Norske Stats Oljeselskap A.S. (Statoil).

**REFERENCES**

[Be93] F.E. Benth, Integrals in the Hida distribution space $(\mathcal{S})^*$, in T. Lindstrøm, B. Øksendal and A.S. Ustunel (editors): Stochastic Analysis and Related Topics, Gordon and Breach 1993, 89-99.

[BG94] F.E. Benth and H.K. Gjessing, A nonlinear parabolic equation with noise. A reduction method, Preprint, University of Mannheim 1994.

[BN94] R. Buckdahn and Nualart, Linear stochastic differential equations and Wick products, Probab. Th. Rel. Fields 99 (1994), 501-524.

[Bu91] R. Buckdahn, Linear Skorohod stochastic differential equations, Probab. Th. Rel. Fields 90 (1991), 223-240.

[Bu92] R. Buckdahn, Skorohod stochastic differential equations of diffusion type, Probab. Th. Rel. Fields 92 (1991), 297-323.

[Fr56] K. O. Friedrichs, Symmetric hyperbolic linear differential equations, Comm. Pure Appl. Math. 7 (1954), 345-392.

[GHLØUZ92] H. Gjessing, H. Holden, T. Lindstrøm, B. Øksendal, J. Ubøe and T.S. Zhang, The Wick product, in H. Niemi et al (editors):

Frontiers in Pure and Applied Probability, vol. 1, TVP Science Pubilishers, Moscow 1993, 29-67.

[HKPS93] T. Hida, H.-H. Kuo, J. Potthoff and L. Streit, White Noise, Kluwer 1993.

[Jo78] F. John, Partial differential equations, 2nd edition, Springer, 1978.

[Ku82] S. Kusuoka, The nonlinear transformation of Gaussian measure on Banach space and its absolute continuity, J. Fac. Sci. Univ. Tokyo, Sect. 1, A 29 (1982), 567-592.

[LØU] Lindstrøm, B. Øksendal and J. Ubøe, Wick-multiplication and Itô-Skorohod stochastic differential equations, in S. Albeverio et al. (editors): Ideas and Methods in Mathematical Analysis, Stochastics and Applications, Cambridge Univ. Press 1992, 183-206.

[Me93] P. A. Meyer, Quantum Probability for Probabilists, Lecture Notes in Mathematics 1538, Springer, 1993.

[ML85] T. Mikio and T.T. Li, Globally classical solutions for nonlinear equations of first order, Comm. in Partial Diff. equ. 10 (1985), 1451-1463.

[NP88] D. Nualart and Pardoux, Stochastic calculus with anticipating integrands, Probab. Th. Rel. Fields 78 (1988), 535-581.

[NZ93] D. Nualart and M. Zakai, Positive and strongly positive Wiener functionals, in D. Nualart and M. Sanz Solé (editors): Barcelona Seminar on Stochastic Analysis, Birkhauser 1993, 132-146.

[Pa90] E. Pardoux, Application of anticipating stochastic calculus to stochastic differential equations, in H. Korezlioglu and A.S. Ustunel (editors): Stochastic Analysis and Related Topics II, Springer Lect. Notes in Math. 1444 (1990), 63-105.

[Ra74] R. Ramer, On non-linear transformation of Gaussian measures, J. Funct. Anal. 15 (1974), 166-187.

[Sh86] Y. Shiota, A linear stochastic integral equation containing the extended Itô integral, Math. Rep. Toyama Univ. 9 (1986), 43-65.

[Us88] A.S. Ustunel, Some comments on the filtering of diffusions and the Malliavin calculus, Springer Lect. Notes in Math. 1316 (1988), 247-266.

[UZ92] A.S. Ustunel and M. Zakai, Transformation of Wiener measure under anticipative flows, Probab. Th. Rel. Fields 93 (1992), 91-136.

[WZ65] E. Wong and M. Zakai, On the relation between ordinary and stochastic differential equations, Intern. J. Engng. Sci. 3 (1965), 213-229.

# THE CIRCLE AS A FERMIONIC DISTRIBUTION

R. Leandre

Let us consider the free loop space of a manifold. It is very well known that the equivariant cohomology of the free loop spce is related to the index theorem for a finite dimensional Dirac operator. The reader can see [Bis85], [Bis86], [GJP90], [JP90] for instance.

Let us consider the case of the Brownian loop, in order to give an ana lytical meaning to the theory of [GJP90] : a stochastic cohomology of the free loop space is studied in [Leac] , [Leaa] in relation with stochastic Chen forms ([JL91]) and with the Hochschild cohomology of the manifold. But the infinitesimal action of the circle is not studied in these papers. The purpose of this paper is to give a preliminary definition of the action of the circle. (For the same type of considerations, the reader can see [HKPS93] in the flat case.)

Let $L_\infty(M)$ be the space of continuous applications $\gamma$ from the circle $S_1$ into the compact Riemannian manifold $M$: the Laplace-Beltrami operator is denoted by $\Delta$, the associated heat-kernel by $p_t(x, y)$ and the law of the Brownian bridge by $dP_{1,x}$. We consider the B.H.K measure $\frac{p_1(x,x)dx \otimes dP_{1,x}}{\int p_1(x,x)dx}$ (See [HK74], [Bis85], [Bis86]. Let us denote by $\tau_t$ the holonomy from $\gamma_0$ to $\gamma_t$ : it is almost surely defined. The tangent space of a loop is the space of continuous sections $X_s$ over $\gamma_s$ of the shape $X_s = \tau_s H_s$ with the boundary condition $X_0 = X_1$ ( See [Bis84] ). There exists two Hilbert structures invariant by rotation over the free loop space:

-) The first one is defined in [JL91]:

$$||X||^2_\infty = \int_0^1 < X_s, X_s > ds + \int_0^1 < DX_s, DX_s > ds \qquad (1)$$

where $DX_s = \tau_s \dot{H}_s$.

-) The second one is defined in [JL] [Leab] and works only when the holonomy checks $||\tau_1 - I||^2 < \delta < 1$ ($||\cdot||$ is the Hilbert-Schimdt norm).

We split the tangent space $T_g$ of a loop space, after complexification, into $\sum T_\gamma^n$: $T_\gamma^n$ is the space of vectors of the type $\tau_t \left( \frac{\exp(2i\pi nt)\exp(-t\log\tau_1)}{\sqrt{cn^2+1}} \right) X_0$ The decomposition of $\tau_\gamma$ into $\sum T_\gamma^n$ is orthogonal. If we don't take the complexified tangent space, we can split with respect of $\cos(2\pi nt)$ and $\sin(2\pi nt)$: minor difficulties arise. Over each $T_\gamma^n$, we take the morm given by $||X_0||^2$. We get another Hilbert structure $|| \cdot ||_\infty$, which works only for small loops. We choose this Hilbert structure because it is easier to get an orthonormal basis, without taking the Cram-Schmidt procedure.

Let us come back for one moment to the smooth loop case. Let us study the component of $\frac{d\gamma_t}{dt} = X_\infty(\gamma)_t$ on the tangent space $T_\gamma$. We use the first Hilbert structure. We choose a suitable partition of unity in order that

$$\frac{\tau_t \exp(2i\pi nt) \exp[-t \log(\tau_1)] X_i(\gamma)}{\sqrt{cn^2 + 1}} = X_{n,i}(t)$$

is an orthonormal basis for $T_{\gamma_0}$ for the first Hilbert-Schmidt procedure, we have to compute:

$$\int_0^1 < X_{n,i}(t), \frac{d\gamma_t}{dt} > dt + \int_0^1 < DX_{n,i}(t), D\frac{d\gamma_t}{dt} > dt. \tag{2}$$

Let us write

$$\tau_t^{-1} \frac{d\gamma_t}{dt} = \frac{dB_t}{dt} \cdot D\frac{d\gamma_t}{dt} = \tau_t \frac{d^2 B_t}{dt^2}.$$

we have to compute

$$\int_0^1 < DX_{n,i}(t), \tau_t \frac{d^2 B_t}{dt^2} > . \tag{3}$$

We integrate by part using the fact we integrate over the circle, we find that this integral is equal to $\int_0^1 < \tau_t, \tilde{H}_{n,i}(t), d\gamma_t >$ where in $\tilde{H}_n(t)$ we take at most two derivatives of $\exp(2i\pi nt)$.

Moreover, on $T_\gamma^n$, the two Hilbert structures are uniformly equivalent if $\delta$ is small enough. Let $X_n(e_i)$ be an orthonormal basis for the second Hilbert structure over $T_\gamma^n$. We deduce from the previous previous considerations that:

$$||i_{X_\infty} X_n(e_i)||_{L^p} \leq c(|n| + 2). \tag{4}$$

Let us enumerate the $X_n(e_i)$ by lexicographic order. Let $A(i_1, n_1, \ldots, i_r, n_r)$ be component of $X_{n_1}(e_{i_1}) \wedge \ldots \wedge X_{n_r}(e_{i_r})$ after performing a local partition of unity, in order to get an orthornormal basis of the exterior algebra complexified. We choose a space of test sections of forms in order to construct the interior product by the killing vector field $X_\infty(\gamma)$. The space $L^p(k)$ of forms is defined as following:

Let us take a form

$$\sigma = \sum A(i_1, n_1, \ldots, i_r, n_r) X_{n_1}(e_{i_1}) \wedge \ldots \wedge X_{n_r}(e_{i_r})$$

$$\|\sigma\|_{L^p(k)} = \left[ \mathbf{E}\sqrt{\sum A^2(i_1, n_1, \ldots, i_r, n_r)(|n_{i_1}| + 2)^k \ldots (|n_{i_r}| + 2)^k)}\,^p \right]^{\frac{1}{p}} \quad (5)$$

the expecatation is taken over small loops. This choice of norm is invariant by rotation, because the metric is invariant by rotation, and the decomposition of $T_\gamma$ into $\sum T_\gamma^n$ is also invariant under rotations. The choice of $(|n_i| + 2)^k$ is done in order to simplify the computations.

**Theorem 0.1** $i_{X_\infty}$ *is continuous over the topological space defined by the families of norms* $L^p(k)$.

**Proof.** The component of $i_{X_\infty}\sigma$ over $X_{n_1}(e_{i_1}) \wedge \ldots \wedge X_{n_r}(e_{i_r})$ is

$$\begin{aligned}
&B(i_1, n_1, \ldots, i_r, n_r) \\
&= \sum_{j,l,l'} A(i_1, n_{i_1}, \ldots, i_{j-1}, n_{i_j}-1, i_\ell, \ell', j_j, n_{i_j}, \ldots, i_r, n_{i_r}) B(\ell, \ell')
\end{aligned} \quad (6)$$

where

$$\|B(\ell, \ell')\|_{L^p} \le c(|\ell'| + 2). \quad (7)$$

Therefore

$$\sum B^2(i_1, n_1, \ldots, i_r, n_{i_r})(|n_{i_1}| + 2)^k \ldots (|n_{i_2}| + 2)^k$$
$$\le \sum (|n_{i_1}| + 2)^{k'} \ldots (|n_{i_2}| + 2)^{k'} \cdot$$
$$\sum_j \left\{ \left( \sum_{\ell\ell'} A(i_1, n_{i_1}, \ldots, i_{j-1}, n_{i_j}-1, i_\ell, \ell', j_j, n_{i_j}, \ldots, i_r, n_{i_r}) \cdot \right.\right.$$
$$\left.\left. (|\ell'|^6 + 2) \right) \left( \sum_{\ell,\ell'} \frac{B^2(i_\ell, \ell')}{|\ell'|^6 + 2} \right) \right\}$$
$$\le \sum (|n_{i_1}| + 2)^{k''} \ldots (|n_{i_r+1}| + 2)^{k''} \cdot A^2(i_1, n_{i_1}, \ldots, i_{r+1}, n_{i_{r+1}})) \cdot$$
$$\left( \sum_{\ell,\ell'} \frac{B^2(i_\ell, \ell')}{|\ell'|^6 + 2} \right)$$

for $k'' \gg k' \gg k$.

We conclude that $i_{X_\infty}$ is continuous from $L^{2p}(k'')$ into $L^p(k)$. ∎

# References

[Bis84]   J. M. Bismut. *Large deviations and the Malliavin calculus Progress in Math. 45.* Birkhaǔser, 1984.

[Bis85] J. M. Bismut. Index theorems and equivariant cohomology of the loop space. *C. M. P.*, 98:213–237, 1985.

[Bis86] J. M. Bismut. Localisation formula, superconnections and the index theorem for families. *C. M. P.*, 103:127–166, 1986.

[GJP90] E. Getzler, J. D. S. Jones, and S. Petrack. differential forms on a loop space and the cyclic bar complex. *Topology*, 30:339–373, 1990.

[HK74] R. Hoegh-Krohn. Relativistic quantum mechanics in 2 dimensional space time. *C. M. P.*, 38:195–224, 1974.

[HKPS93] T. Hida, H.H. Kuo, J. Potthoff, and L. Streit. *White noise: an infinite dimensional calculus*. Kluwer, 1993.

[JL] J. D. S. Jones and R. Leandre. A stochastic approach to the Dirac operator over free loop spaces. In preparation.

[JL91] J. D. S. Jones and R. Leandre. $l^p$ chen forms over loop spaces. In *Stochastic Analysis*, pages 104–162. Cambridge University Press, 1991.

[JP90] J. Jones and S. Petrack. The pinced point theorems in equivariant cohomology. *Trans. Amer. Math. Soci.*, 1:35–49, 1990.

[Leaa] R. Leandre. Brownian cohomology of an homogeneous manifold. In preparation.

[Leab] R. Leandre. Brownian motion over a KHahler manifold and elliptic genera of level N. To be published in the proceedings of the conference probability and physic, ed. Streit, L.

[Leac] R. Leandre. Cohomologie de Bismut-Nualart-Pardoux et cohomologie de Hochschild entière. Preprint.

R. Leandre
dept de Mathematiques, University de Nancy I, BP 239. 54506 Vandoeuvre.
Les Nancy, France
Mathematics Institute, University of Warwick, Coventry CV4 7AL, U.K.

# Linear Skorohod stochastic differential equations on Poisson space

Nicolas Privault

*Equipe d'Analyse et Probabilités, Université d'Evry-Val d'Essonne*
*Boulevard des Coquibus, 91025 Evry Cedex, France*

### Abstract

We study the absolute continuity of transformations defined by anticipative flows on Poisson space, and show that the process of densities associated to those transformations allows to solve anticipative linear stochastic differential equations on the Poisson space.

*Mathematics Subject Classification: 60H05, 60H07, 60J75.*

# 1   Introduction

Linear Skorohod stochastic differential equations have been studied on the Wiener space, cf. [3], using the anticipative stochastic calculus developed in [6]. It has been shown in particular that the solutions of such equations are associated to the density induced by absolutely continuous transformations defined by flows on the Wiener space. Such absolute continuity results have been extended in [11]. Our goal here is to investigate the Poisson space case. The anticipative stochastic calculus on Poisson space, cf. [5], [9], permits to introduce anticipative stochastic differential equations by means of an extension of the compensated Poisson stochastic integral, also called the Skorohod integral. We study the absolute continuity of some anticipative flows on Poisson space and show that their associated densities allow to solve Skorohod stochastic differential equations. Let us describe the Poisson space interpretation that we are working with, cf. [9]. Let $B$ be a space of sequences with a probability measure $P$ such that the coordinate functionals

$$\tau_k : B \longrightarrow \mathbf{R} \quad k \in \mathbf{N},$$

are independent identically distributed exponential random variables. The space $B$ is endowed with the norm $\| x \|_B = \sup_{n \in \mathbf{N}} | x_n | /(n+1)$ such that $P$ is defined on the Borel $\sigma$-algebra of $B$. Let $T_k = \sum_{i=0}^{i=k-1} \tau_i$, $k \geq 0$, denote the $k$-th jump time of the Poisson process $(N_t)$ defined as $N_t = \sum_{k \geq 0} 1_{[T_k, \infty[}(t)$, $t \in \mathbf{R}_+$. Denote by $(e_k)_{k \in \mathbf{N}}$ the canonical basis of the space of square-summable sequences $H = l^2(\mathbf{N})$. We define an operator $i$ that turns any discrete time stochastic process $u = (u_k)_{k \in \mathbf{N}}$ into a continuous time process $i(u)$ by $i_t(u) = u_{N_{t^-}}$, or

$$i_t(u) = \sum_{k \geq 0} u_k 1_{]T_k, T_{k+1}]}(t), \quad t \in \mathbf{R}_+. \tag{1}$$

The flow that we will consider is the family $(\mathcal{T}_t)_{t \in [0,1]}$ of transformations $\mathcal{T}_t : B \to B$, defined by

$$\mathcal{T}_t(\omega) = \omega + \left( \int_0^t i_s(e_k)(\mathcal{T}_s(\omega)) \sigma_s(\mathcal{T}_s(\omega)) ds \right)_{k \geq 0},$$

where $\sigma$ is a process satisfying some boundedness conditions. If the transformation $\mathcal{T}_t$, $t \in [0,1]$, is absolutely continuous, then the process of densities $\left( \frac{d(\mathcal{T}_t^{-1})_* P}{dP} \right)_{t \in [0,1]}$ solves the anticipative stochastic differential equation

$$X_t = 1 + \int_0^t \sigma_s(\omega) X_s \delta \tilde{N}_s,$$

where $\int_0^t u_s \delta \tilde{N}_s = \tilde{\delta}\left( u 1_{[0,t]} \right)$ is the Skorohod integral of $u 1_{[0,t]}$ on the Poisson space, as defined in [1], [5], [9]. This integral is an extension to anticipative integrands of the stochastic integral with respect to the compensated Poisson process. It is the adjoint of a derivation operator defined by shifting the Poisson process jump times, and has in particular the property of being an integral with zero expectation. As a consequence, we will be able to solve the anticipative stochastic differential equation

$$X_t = X_0 + \int_0^t \sigma_s X_s \delta \tilde{N}_s + \int_0^t b_s X_s ds \quad t \in [0,1], \tag{2}$$

where $X_0$ and $b$ are bounded random variables. In case the processes $b$ and $\sigma$ are predictable, the equation defining the inverse $(\mathcal{A}_t)_{t \in [0,1]}$ of $(\mathcal{T}_t)_{t \in [0,1]}$ becomes

$$\mathcal{A}_t(\omega) = \omega + \left( \int_0^t i_s(e_k) \sigma_s(\omega) ds \right)_{k \geq 0},$$

and we retrieve a classical result, cf. for instance [2].

We proceed as follows. In Sect. 2 the definitions and main results of the anticipative stochastic calculus on the Poisson space as introduced in [5], [9] are recalled. Sect. 3 is devoted to the definition of the flow $(\mathcal{T}_t)_{t \in [0,1]}$ of anticipative transformations of the Poisson process trajectories and to the study of its absolute continuity. Those results are applied in Sect. 4, where the solution of the linear Skorohod stochastic differential equation (2) is given.

## 2 Anticipative stochastic calculus on the Poisson space

Let $\mathcal{S}$ denote the set of functionals of the form

$$F = f(\tau_0, \ldots, \tau_n),$$

with $n \in \mathbf{N}$ and $f \in C_c^\infty(\mathbf{R}_+^{n+1})$. We define a gradient operator $D : L^2(B) \to L^2(B) \otimes H$ by

$$DF = (\partial_k f(\tau_0, \ldots, \tau_n))_{k \in \mathbf{N}}, \quad F \in \mathcal{S}. \tag{3}$$

We also define $\tilde{D} : L^2(B) \to L^2(B) \otimes L^2(\mathbf{R}_+)$ as

$$\tilde{D}F = -i \circ DF, \quad F \in \mathcal{S}.$$

The operators $\tilde{D}$ and $D$ are closable, cf. [9]. Denote by $\mathbf{D}_{1,2}$ the domain of the closed extension of $D$. Let $\tilde{\delta}$ be the adjoint of $\tilde{D}$, which is also closable and can be extended to a closed operator

$$\tilde{\delta} : L^1(B \times [0,1]) \longrightarrow L^1(B),$$

of domain $Dom(\tilde{\delta})$. Let $\mathcal{V}$ denote the class of processes of the form

$$v = \sum_{i=1}^{i=n} 1_{\Delta_i} f_i(\tau_0, \ldots, \tau_n),$$

where $f_i \in C_c^\infty(\mathbf{R}_+^{n+1})$, $1 \le i \le n$, and $\Delta_1, \ldots, \Delta_n \subset [0,1]$. We have the following formula, cf. [5], [9]:

$$\tilde{\delta}(v) = \int_0^\infty v(s) d(N_s - s) - \int_0^\infty \tilde{D}_s v(s) ds, \quad v \in \mathcal{V}. \tag{4}$$

The interpretation of $\tilde{\delta}$ as an extension of the stochastic integral with respect to the compensated Poisson process comes from the following proposition, cf. [5], [9].

**Proposition 1** *Let $v \in L^2(B) \otimes L^2(\mathbf{R}_+)$ be predictable with respect to the filtration generated by the Poisson process $(N_t)$. We have*

$$\tilde{\delta}(v) = \int_0^\infty v(s)d(N_s - s).$$

Denote by $\mathbf{D}_{1,\infty}$ the subset of $D_{1,2}$ made of the random variables $F$ for which

$$\| F \|_{L^\infty(B)} + \| |DF|_H \|_{L^\infty(B)}$$

is bounded and let $L_{1,\infty} = L^2([0,1], \mathbf{D}_{1,\infty})$, $L_{1,2} = L^2([0,1], D_{1,2})$. If $\mathcal{T} : B \longrightarrow B$ is a measurable mapping, we denote by $\mathcal{T}_* P$ the image measure of $P$ by $\mathcal{T}$, and say that $\mathcal{T}$ is absolutely continuous if $\mathcal{T}_* P$ is absolutely continuous with respect to $P$. A flow $(\phi_{s,t})_{0 \le s < t \le 1}$ of transformations of $B$ is said to be absolutely continuous if $\phi_{s,t}$ is absolutely continuous, $0 \le s < t \le 1$. We end this section with four propositions which will be useful in the sequel. Their statements and proofs are adapted from [4]. Proofs are given in the appendix.

**Proposition 2** *Let $F \in \mathbf{D}_{1,2}$. For any $\varepsilon > 0$, there is a sequence $(F_n)_{n \in \mathbf{N}} \subset \mathcal{S}$ that converges to $F$ in $\mathbf{D}_{1,2}$ and such that*

1. *ess inf $F < F_n <$ ess sup $F$, $n \in \mathbf{N}$.*

2. *$\| |DF_n|_H \|_\infty \le \| |DF|_H \|_\infty + \varepsilon$, $n \in \mathbf{N}$.*

We obtain in the same way the following result.

**Proposition 3** *Let $\sigma \in L_{1,\infty}$ with $\sigma > -1$ a.s. and $\int_0^1 \| \frac{1}{1+\sigma_r} \|_\infty^2 dr < \infty$. For any $\varepsilon > 0$, there is a sequence $(\sigma^n)_{n \in \mathbf{N}} \subset \mathcal{V}$ that converges to $\sigma$ in $L_{1,2}$ and such that for $n \in \mathbf{N}$,*

1. *$\sigma^n > -1$.*

2. *$\int_0^1 |\sigma_s^n|_\infty^2 ds \le \int_0^1 |\sigma_s|_\infty^2 ds$.*

3. *$\left(\int_0^1 \| |D\sigma_s^n|_H \|_\infty^2 ds\right)^{1/2} \le \varepsilon + \left(\int_0^1 \| |D\sigma_s|_H \|_\infty^2 ds\right)^{1/2}$*

4. *$\int_0^1 \| \frac{1}{1+\sigma_r^n} \|_\infty^2 \le \int_0^1 \| \frac{1}{1+\sigma_r} \|_\infty^2 dr$.*

5. *$\| \sigma^n \|_{L^\infty(B \times [0,1])} \le \| \sigma \|_{L^\infty(B \times [0,1])}$.*

6. *$\| D\sigma^n \|_{L^\infty(B \times [0,1] \times \mathbf{N})} \le \varepsilon + \| D\sigma \|_{L^\infty(B \times [0,1] \times \mathbf{N})}$.*

*If $\sigma$ has continuous trajectories a.s., then $(\sigma_{T_k}^n)_{n \in \mathbf{N}}$ converges in $L^2(B)$ to $\sigma_{T_k}$, $k \ge 1$.*

**Proposition 4** *Let $T^1, T^2$ be two absolutely continuous transformations, respectively defined by*

$$T^1(\omega) = \omega + \left( \int_0^1 i_s(e_k)\sigma_s^1(\omega)ds \right)_{k\in\mathbb{N}}$$

*and*

$$T^2(\omega) = \omega + \left( \int_0^1 i_s(e_k)\sigma_s^2(\omega)ds \right)_{k\in\mathbb{N}},$$

$\omega \in B$, *with* $\sigma^1, \sigma^2 \in L^2(B \times [0,1])$. *Let* $F \in D_{1,\infty}$. *We have*

$$| F \circ T^1(\omega) - F \circ T^2(\omega) | \leq ||| DF |_H||_\infty| \sigma^1(\omega) - \sigma^2(\omega) |_{L^2([0,1])}.$$

*If* $F \in S$, *then*

$$| F(\omega) - F(\omega + h) | \leq ||| DF |_H||_\infty|| h \|_H \quad h \in H, \ \omega \in B.$$

**Proposition 5** *Let* $(T^n)_{n\in\mathbb{N}}$ *be a sequence of absolutely continuous transformations with*

$$T^n\omega = \omega + \left( \int_0^\infty i_s(e_k)\sigma_s^n(\omega)ds \right)_{k\in\mathbb{N}},$$

*defined by a sequence* $(\sigma^n)_{n\in\mathbb{N}}$ *of processes that converges in* $L^2(B) \otimes L^2([0,1])$ *to a process* $\sigma$, *such that the sequence of densities* $(L^n)_{n\in\mathbb{N}} = \left(\frac{dT_*^nP}{dP}\right)_{n\in\mathbb{N}}$ *is uniformly integrable. If* $(F_n)_{n\in\mathbb{N}}$ *converges to* $F$ *in* $L^2(B)$, *then* $(F_n \circ T^n)_{n\in\mathbb{N}}$ *converges to* $F \circ T$ *in probability, where* $T$ *is defined by*

$$T\omega = \omega + \left( \int_0^\infty i_s(e_k)\sigma_s(\omega)ds \right)_{k\in\mathbb{N}}.$$

*Moreover,* $T$ *is absolutely continuous.*

## 3 Absolute continuity of anticipative flows

**Proposition 6** *Let* $\sigma \in \mathcal{V}$. *The equation*

$$T_t\omega = \omega - \left( \int_0^t i_s(e_k)(T_s\omega)\sigma_s(T_s\omega)ds \right)_{k\geq 0}, \quad t \in [0,1], \qquad (5)$$

*has a unique solution which is invertible. For* $s,t \in [0,1]$, $s < t$, *let* $A_t = T_t^{-1}$ *and* $\phi_{s,t} = T_s \circ A_t$, $s \leq t$. *Then* $\phi_{s,t}$ *satisfies to*

$$\phi_{s,t}\omega = \omega + \left( \int_s^t i_r(e_k)(\omega)\sigma_r(\phi_{r,t}\omega)dr \right)_{k\in\mathbb{N}} \quad \omega \in B, \ 0 \leq s < t \leq 1. \ (6)$$

*Let* $\psi_{s,t} = T_t \circ A_s$, $s \leq t$. *We have*

$$\psi_{s,t}\omega = \omega - \left( \int_s^t i_r(e_k)(\psi_{s,r}\omega)\sigma_r(\psi_{s,r}\omega)dr \right)_{k\in\mathbb{N}} \quad \omega \in B, \ 0 \leq s < t \leq 1.$$

*Proof.* The equations (5) and (6) can be solved as differential equations in finite dimension since $\sigma_r$ is Lipschitz, $r \in [0,1]$, cf. Prop. 4. Denote by $\mathcal{A}_t$ and $\mathcal{T}_t$ the solutions of (5) and (6) for $s = 0$. It remains to show that $\phi_{s,t} \circ \mathcal{T}_t = \mathcal{T}_s$, $s \leq t$. Let us show that for $r \leq t$, $i_r(e_k)(\mathcal{T}_t\omega) = i_r(e_k)(\mathcal{T}_r\omega)$, $k \in \mathbf{N}$. For $s \leq t$, we notice from (1) that $T_k(\mathcal{T}_s\omega) \leq s \Rightarrow T_k(\mathcal{T}_s\omega) = T_k(\mathcal{T}_t\omega)$, and $T_k(\mathcal{T}_s) \geq s \Rightarrow T_k(\mathcal{T}_t) \geq s$. Hence

$$T_k(\mathcal{T}_s\omega) \leq s \leq T_{k+1}(\mathcal{T}_s\omega) \iff T_k(\mathcal{T}_t\omega) \leq s \leq T_{k+1}(\mathcal{T}_s\omega)$$
$$\iff T_k(\mathcal{T}_t\omega) \leq s \leq T_{k+1}(\mathcal{T}_t\omega).$$

This gives

$$\phi_{s,t}(\mathcal{T}_t\omega) = \mathcal{T}_t\omega + \left( \int_s^t i_r(e_k)(\mathcal{T}_t\omega)\sigma(r,\mathcal{T}_r\omega)dr \right)_{k \in \mathbf{N}} = \mathcal{T}_s\omega,$$

which implies (5). Finally, $i_r(e_k)(\psi_{s,t}) = i_r(e_k)(\psi_{s,r})$, $0 \leq s < r < t \leq 1$, and

$$\omega = \phi_{s,t} \circ \psi_{s,t}\omega = \psi_{s,t}\omega + \left( \int_s^t i_r(e_k)(\psi_{s,t}\omega)\sigma_r(\phi_{r,t} \circ \psi_{s,t}\omega)dr \right)_{k \in \mathbf{N}}$$
$$= \psi_{s,t} + \left( \int_s^t i_r(e_k)(\psi_{s,r}\omega)\sigma_r(\psi_{s,r}\omega)dr \right)_{k \in \mathbf{N}}.$$

$\square$

**Theorem 1** *For $\sigma \in L_{1,\infty}$ with $\sigma > -1$ a.s., and $\int_0^1 \| \frac{1}{1+\sigma_r} \|_\infty dr < \infty$, Eq. (5) has a unique absolutely continuous solution which is invertible and whose inverse flow $\{\psi_{s,t} : 0 \leq s \leq t \leq 1\}$ satisfies to (6). Assume that $\sigma$ has continuous trajectories, a.s. Then*

$$L_{s,t} = \frac{d(\phi_{s,t})_* P}{dP} \tag{7}$$
$$= \exp\left( -\int_s^t \tilde{D}_r\sigma_r(\phi_{r,t})dr - \int_s^t \sigma_r(\phi_{r,t})dr \right) \prod_{s \leq T_k \leq t} (1 + \sigma_{T_k}(\phi_{T_k,t})),$$

$0 \leq s \leq t \leq 1$.

*Remark.* $\tilde{D}_r\sigma_r(\phi_{r,t})$ is here interpreted as $i_r(D\sigma_r(\phi_{r,t}))$.
*Proof.* We start by assuming that $\sigma \in \mathcal{V}$ and depends only on $\tau_0, \ldots, \tau_n$ for some $n \in \mathbf{N}$.

**Lemma 1** *If $\sigma \in \mathcal{V}$, we have with $\phi$ given by Prop. 6:*

$$\det(D\phi_{s,t}) = \exp\left( -\int_s^t \tilde{D}_r\sigma_r(\phi_{r,t})dr \right) \prod_{s \leq T_k \leq t} (1 + \sigma_{T_k}(\phi_{T_k,t})) \quad 0 \leq s \leq t \leq 1.$$

*Proof.* We have that $\phi_{s,t}(l)$ is differentiable since it is expressed with the solution of a differential equation with $C^\infty$ coefficients, and

$$
\begin{aligned}
D_k\phi_{s,t}(l) \\
= \; & 1_{\{k=l\}} + \sigma_{T_{l+1}}(\phi_{T_{l+1},t})1_{\{k\leq l\}}1_{\{s<T_{l+1}<t\}} - \sigma_{T_l}(\phi_{T_l,t})1_{\{k<l\}}1_{\{s<T_l<t\}} \\
& + \int_s^t \sum_{i=0}^{i=n} i_r(e_l)D_k\phi_{r,t}(i)D_i\sigma_r(\phi_{r,t})dr \quad k,l \in \mathbf{N}.
\end{aligned}
$$

Letting $U_{s,t} = (D_k\phi_{s,t}(l))_{0\leq k,l\leq n}$, this gives the following differential equation in the space of $(n+1)\times(n+1)$ matrices:

$$
U_{s,t} = A_{s,t} + \int_s^t U_{r,t}B_{r,t}dr, \tag{8}
$$

where

$$
A_{s,t}(k,l) = 1_{\{k=l\}} + \sigma_{T_{l+1}}(\phi_{T_{l+1},t})1_{\{k\leq l\}}1_{\{s<T_{l+1}<t\}} - \sigma_{T_l}(\phi_{T_l,t})1_{\{k<l\}}1_{\{s<T_l<t\}}
$$

$0 \leq k,l \leq n$, and $B_{s,t} = (i_s(e_l)D_k\sigma_s(\phi_{s,t}))_{0\leq k,l\leq n}$. Solving this differential equation in $s \in [0,t]$ for fixed $\omega$ on the intervals $]T_l, T_{l+1}[\cap[0,t]$, $k \in \mathbf{N}$, we get

$$
\begin{aligned}
(D_k\phi_{s,t}(l))_{0\leq k,l\leq n} \\
= \; & \exp\left(\int_s^{t\wedge T_{N_s+1}} B_{r,t}dr\right) \prod_{s<T_l<t}\left(\exp\left(\int_{s\wedge T_l}^{t\wedge T_{l+1}} B_{r,t}dr\right) + C_l\right),
\end{aligned} \tag{9}
$$

$0 \leq s \leq t \leq 1$, where $C_l$, $l \geq 1$, is a matrix such that $C_l(l,l) = \sigma_{T_l}(\phi_{T_l,t})$ and $C_l(i,j) = 0$ if $i \neq l$ or $j > i$. Since $B_{r,t}(i,j) = 0$ if $r < T_i$, $j = 0,\ldots,n$, we have

$$
\begin{aligned}
\det\left(\exp\left(\int_{s\wedge T_l}^{t\wedge T_{l+1}} B_{r,t}dr\right) + C_l\right) \\
= \; & (1 + \sigma_{T_l}(\phi_{T_l,t}))\det\left(\exp\left(\int_{s\wedge T_l}^{t\wedge T_{l+1}} B_{r,t}dr\right)\right).
\end{aligned} \tag{10}
$$

Hence

$$
\det(U_{s,t}) = \exp\left(\int_s^t \mathrm{trace}(B_{r,t})dr\right) \prod_{s<T_k<t,\ k\leq n}(1 + \sigma_{T_k}(\phi_{T_k,t})).
$$

Noticing that for $k > n$,

$$
\begin{aligned}
D_k\phi_{s,t}(l) \\
= \; & 1_{\{k=l\}} + \sigma_{T_{l+1}}(\phi_{T_{l+1},t})1_{\{k\leq l\}}1_{\{s<T_{l+1}<t\}} - \sigma_{T_l}(\phi_{T_l,t})1_{\{k<l\}}1_{\{s<T_l<t\}}
\end{aligned}
$$

and $\mathrm{trace}(B_{r,t}) = \sum_{k=0}^{k=n} D_k\sigma_r(\phi_{r,t})i_r(e_k) = \tilde{D}_r\sigma_r(\phi_{r,t})$, we obtain

$$
\det(D\phi_{s,t}) = \exp\left(-\int_s^t \tilde{D}_r\sigma_r(\phi_{r,t})dr\right) \prod_{s\leq T_k\leq t}(1 + \sigma_{T_k}(\phi_{T_k,t})).
$$

Define for $k \in \mathbf{N}$ $\pi_k : B \longrightarrow H$ by $\pi_k(w) = (1_{\{k \leq n\}} \tau_k)_{k \in \mathbf{N}}$. Let $\Phi_{s,t} = \phi_{s,t} - I_B$, $0 \leq s \leq t \leq 1$, and $F_k = \pi_k \Phi_{s,t}$ for $k \geq n$. The mapping $I_B - F_k$ is a diffeomorphism of $B^+ = \{\omega \in B : \omega_k \geq 0, \ k \in \mathbf{N}\}$, and we have for $f \in C_b^+(B)$, from the finite dimensional Jacobi theorem:

$$
\begin{aligned}
E[f] &= E\left[ f(I_B + F_k) \mid \det(I_H + DF_k) \mid \exp\left( - \sum_{i=0}^{i=k} F_k(i) \right) \right] \\
&= E[f(I_B + F_k) \mid \Lambda_k \mid], \quad k \geq n,
\end{aligned}
$$

with from (9):

$$
\Lambda_k = \exp\left( - \int_s^t \tilde{D}_r \sigma_r(\phi_{r,t}) dr - \int_s^t \sigma_r(\phi_{r,t}) dr \right) \prod_{s \leq T_i \leq t, \ i \leq k} (1 + \sigma_{T_i}(\phi_{T_i,t})).
$$

Now,

$$
\begin{aligned}
E\left[ \Lambda_k \mid \log \Lambda_k \mid \right] \\
&= E\left[ \mid \log \Lambda_k \circ (I_B + F_k)^{-1} \mid \right] \\
&\leq E\left[ \sum_{i=1}^{i=k} \parallel \sigma_v \parallel_\infty \mid_{v=T_i} + \parallel \frac{1}{1+\sigma_r} \parallel_\infty \mid_{r=T_i} \right] + \int_0^1 \parallel\parallel D\sigma_r \mid_H \parallel_\infty dr \\
&\leq \int_0^1 \parallel \sigma_r \parallel_\infty^2 dr + \int_0^1 \parallel \frac{1}{1+\sigma_r} \parallel_\infty^2 dr + \int_0^1 \parallel\parallel D\sigma_r \mid_H \parallel_\infty^2 dr, \quad k \geq n.
\end{aligned}
$$

Hence by uniform integrability of $(\Lambda_k)_{k \in \mathbf{N}}$, we obtain

$$
E[f] = E[f \circ \phi_{s,t} L_{s,t}]
$$

for $f \in C_b^+(B)$. We now return to the case of a general $\sigma$. From Prop. 3, we can choose a sequence $(\sigma^n)_{n \in \mathbf{N}} \subset \mathcal{V}$ that converges to $\sigma$ in $L_{1,2}$, with $\sigma_n > -1$, $n \in \mathbf{N}$. The sequence $(\sigma^n)_{n \in \mathbf{N}}$ defines a sequence of transformations $\left( \phi_{s,t}^n \right)_{n \in \mathbf{N}}$, $\left( \psi_{s,t}^n \right)_{n \in \mathbf{N}}$ and density functions $(L_{s,t}^n)_{n \in \mathbf{N}}$. The uniform integrability of the sequence $(L_{s,t}^n)_{n \in \mathbf{N}}$ is shown as above:

$$
\begin{aligned}
E\left[ L_{s,t}^n \mid \log L_{s,t}^n \mid \right] &= E\left[ \mid \log L_{s,t}^n (\psi_{s,t}^n) \mid \right] \\
&\leq E\left[ \sum_{k \geq 1} \mid \sigma_{T_k}^n(\psi_{T_k,t}^n) \mid + \mid \frac{1}{1 + \sigma_{T_k}^n(\psi_{T_k,t}^n)} \mid \right. \\
&\quad \left. + \int_s^t \mid \tilde{D}_r \sigma_r^n(\psi_{s,r}^n) \mid dr + \int_s^t \mid \sigma_r^n(\psi_{s,r}^n) \mid dr \right]
\end{aligned}
$$

$$
\leq E\left[\sum_{k\geq 1} \| \sigma_v^n \|_\infty|_{v=T_k} + \| \frac{1}{1+\sigma_v^n} \|_\infty^2|_{v=T_k}\right.
$$

$$
\left. + \int_0^t | \tilde{D}_r\sigma_r^n(\psi_{s,r}^n) | \, dr\right] + \int_0^t \| \sigma_r^n \|_\infty^2 \, dr
$$

$$
\leq 2\int_0^1 \| \sigma_r \|_\infty^2 \, dr + \int_0^1 \| \frac{1}{1+\sigma_r} \|_\infty^2 \, dr + \int_0^1 \| | D\sigma_r |_H \|_\infty^2 \, dr + \varepsilon.
$$

where $\varepsilon$ does not depend on $n$. Let $\Phi_{s,t}^n = \phi_{s,t}^n - I_B$, and let us show that $(\Phi_{r,t}^n)_{n\in\mathbf{N}}$ converges in $L^2(B) \otimes H$. We have

$$
E\left[| \phi_{s,t}^n - \phi_{s,t}^m |_H^2\right]
$$

$$
\leq E\left[\int_s^t | \sigma_r^n(r,\phi_{r,t}^n) - \sigma_r^m(r,\phi_{r,t}^m) |^2 \, dr\right]
$$

$$
\leq 2E\left[\int_s^t | \sigma_r^n - \sigma_r^m |^2 L_{r,t}^n dr + \int_s^t | \sigma_r^m(\phi_{r,t}^n) - \sigma_r^m(\phi_{r,t}^m) |^2 \, dr\right]
$$

$$
\leq 2E\left[\int_s^t | \sigma_r^n - \sigma_r^m |^2 L_{r,t}^n dr\right.
$$

$$
\left. + \int_s^t (\| | D\sigma_r |_H \|_\infty +1)^2 \int_s^r | \sigma_u^n(\phi_{u,r}^n) - \sigma_u^m(\phi_{u,r}^m) |^2 \, du dr\right]
$$

$$
\leq 2E\left[\int_s^t | \sigma_r^n - \sigma_r^m |^2 L_{r,t}^n dr\right] \exp\left(\int_s^t (\| | D\sigma_r |_H \|_\infty +1)^2 \, dr\right)
$$

$n, m \in \mathbf{N}$, $0 \leq s \leq t \leq 1$, by the Gronwall lemma and Prop. 4. This converges to 0 by uniform integrability. Denote by $\phi_{s,t}$ the limit of $(\phi_{s,t}^n)_{n\in\mathbf{N}}$. From Prop. 5, the sequence $(\sigma_r^n(\phi_{r,t}^n))_{n\in\mathbf{N}}$ converges to $\sigma_r(\phi_{r,t})$ in $L^2(B)$, for $r \in [0,1]$, hence by boundedness of $\sigma$ the limit $\phi_{s,t}$ solves Eq. (5). Moreover, $\phi_{s,t}$ is absolutely continuous from Prop. 5 and is the only absolutely continuous solution from Prop. 4. We can now show that $\left(\tilde{D}.\sigma^n(\phi_{.,t}^n)\right)_{n\in\mathbf{N}}$ converges to $\tilde{D}.\sigma.(\phi_{.,t})$ in $L^2(B) \otimes L^2([0,t])$:

$$
E\left[\int_0^t | \tilde{D}_r\sigma_r^n(\phi_{r,t}^n) - \tilde{D}_r\sigma(\phi_{r,t}) |^2 \, dr\right]
$$

$$
\leq 2E\left[\int_0^t \left(| D_r\sigma_r^n(\phi_{r,t}^n) - D_r\sigma_r(\phi_{r,t}^n) |_H^2 + | D_r\sigma_r(\phi_{r,t}^n) - D_r\sigma(\phi_{r,t}) |_H^2\right) dr\right]
$$

$$
\leq 2E\left[\int_0^t | D(\sigma_r^n - \sigma_r) |_H^2 L_{r,t}^n dr + \int_0^t | D\sigma_r(\phi_{r,t}^n) - D\sigma_r(\phi_{r,t}) |_H^2 \, dr\right],
$$

which converges to 0 as $n$ goes to infinity since $| D\sigma_r(\phi_{r,t}^n) |_H \leq \| | D\sigma_r |_H \|_\infty$, $r \in [0,1]$. We also have that $(\sigma_{T_k}^n(\phi_{T_k,t}^n))_{n\in\mathbf{N}}$ converges to $\sigma_{T_k}(\phi_{T_k,t})$ in $L^2(B)$, $k \in \mathbf{N}$, from Prop. 3. Hence by uniform integrability and convergence in probability of $\left\{L_{s,t}^n \; : \; n \in \mathbf{N}\right\}$ to $L_{s,t}$, we have for $f \in \mathcal{C}_b^+(B)$:

$$
E[f] = \lim_{n\to\infty} E\left[f(\phi_{s,t}^n)L_{s,t}^n\right] = E[f(\phi_{s,t})L_{s,t}].
$$

Since $1 + \sigma > 0$ a.s., it is not difficult to see that $\phi_{s,t}$ is bijective and that its inverse $\psi_{s,t}$ satisfies (6).

□

*Remark.* The expression of the density can be written in a form which is closer to its expression on Wiener space, cf. [4], [11], i.e.

$$
\begin{aligned}
L_{s,t} &= \frac{d(\phi_{s,t})_* P}{dP} \\
&= \exp\left(\int_s^t \tilde{D}_r(\sigma_r(\phi_{r,t}\omega))dr - \int_s^t \tilde{D}_r\sigma_r(\phi_{r,t}\omega)dr + \tilde{\delta}(1_{[0,t]}\sigma_{\cdot}(\phi_{\cdot,t}))\right) \\
&\quad \times \prod_{s \le T_k \le t} (1 + \sigma_{T_k}(\phi_{T_k,t})) \exp(-\sigma_{T_k}(\phi_{T_k,t})),
\end{aligned}
$$

using (4). From [10] and (4), we obtain the following formal expression for the Carleman-Fredholm determinant of $D\phi_{s,t}$:

$$
\begin{aligned}
\det_2(D\phi_{s,t}) &= \exp\left(\int_s^t \tilde{D}_r(\sigma_r(\phi_{r,t}\omega))dr - \int_s^t \tilde{D}_r\sigma_r(\phi_{r,t}\omega)dr\right) \\
&\quad \times \prod_{s \le T_k \le t} (1 + \sigma_{T_k}(\phi_{T_k,t})) \exp(-\sigma_{T_k}(\phi_{T_k,t}))
\end{aligned}
$$

**Lemma 2** *If $F \in S$ depends only on $\tau_0, \ldots, \tau_m$ and $\sigma \in \mathcal{V}$, then*

$$
\begin{aligned}
&\mid D(F(\mathcal{A}_t)) \mid_H \\
&\le 2(m+1) \parallel \sigma \parallel_{L^\infty(B \times [0,1])} \\
&\quad \times \left(1 + \int_0^1 \parallel\mid D\sigma_r \mid_H\parallel_\infty dr \exp\left(\int_0^1 \parallel\mid D\sigma_r \mid_H\parallel_\infty dr\right)\right) \mid DF \mid_H,
\end{aligned}
$$

$t \in [0,1]$.

*Proof.* We have from (8) and the Gronwall lemma, since $A_{0,t}(k,l) = 0$ if $k > l$:

$$
\begin{aligned}
&\mid D(F(\mathcal{A}_t)) \mid_H \\
&\le \left(\int_0^t \parallel\mid D\sigma_r \mid_H\parallel_\infty \mid DA_{0,r} \mid_{\mathbf{R}^{m+1} \otimes \mathbf{R}^{m+1}} dr \right. \\
&\quad \left. \times \exp\left(\int_0^t \parallel\mid D\sigma_r \mid_H\parallel_\infty dr\right) + \mid DA_{0,t} \mid_{\mathbf{R}^{m+1} \otimes \mathbf{R}^{m+1}}\right) \mid DF \mid_H \\
&\le 2(m+1) \parallel \sigma \parallel_{L^\infty(B \times [0,1])} \\
&\quad \times \left(1 + \int_0^1 \parallel\mid D\sigma_r \mid_H\parallel_\infty dr \exp\left(\int_0^1 \parallel\mid D\sigma_r \mid_H\parallel_\infty dr\right)\right) \mid DF \mid_H,
\end{aligned}
$$

□

# 4 Solution of a linear Skorohod equation

We need the following lemma.

**Lemma 3** *Let $F \in S$ and let $(T_t)_{t\in[0,1]}$ be the flow defined by $\sigma \in L_{1,2}$, $\sigma > -1$ a.s. We have*

$$\frac{d}{dt} F \circ T_t = \sigma_t(T_t) \left( \tilde{D}_t F \right) \circ T_t. \tag{11}$$

*If moreover $\sigma \in V$, then*

$$\frac{d}{dt} F \circ A_t = -\sigma_t \tilde{D}_t (F \circ A_t).$$

*Proof.* Eq. (11) comes from (3) and (5). We also have if $\sigma \in V$

$$
\begin{aligned}
0 &= \frac{d}{dt} F \circ A_t \circ T_t = \frac{d}{dt} (F \circ A_t) \circ T_t + \frac{d}{ds} F \circ A_t \circ T_s \mid_{s=t} \\
&= \frac{d}{dt} (F \circ A_t) \circ T_t + \sigma_t(T_t) \tilde{D}_t (F \circ A_t) \circ T_t.
\end{aligned}
$$

$\square$

**Theorem 2** *Let $\sigma \in L_{1,\infty}$ with continuous trajectories a.s., such that $\sigma > -1$ and $\int_0^1 \| \frac{1}{1+\sigma_r} \|_\infty < \infty$, $b \in L^2([0,1], L^\infty(B))$ and $\eta \in L^\infty(B)$. The anticipative stochastic differential equation*

$$X_t = \eta + \int_0^t \sigma_r X_r \delta \tilde{N}_r + \int_0^t b_s X_s ds \quad t \in [0,1] \tag{12}$$

*has for solution*

$$
\begin{aligned}
X_t &= \eta(T_t^{-1}) \exp \left( -\int_0^t \tilde{D}_s \sigma_s(\phi_{s,t}\omega) ds - \int_0^t \sigma_s(\phi_{s,t}) ds + \int_0^t b_s(\phi_{s,t}) ds \right) \\
&\quad \prod_{0 \leq T_k \leq t} (1 + \sigma_{T_k}(\phi_{T_k,t})), \quad t \in [0,1].
\end{aligned}
$$

*If moreover $\| b \|_{L^\infty(B\times[0,1])}$, $\| \sigma \|_{L^\infty(B\times[0,1])}$, $\| D\sigma \|_{L^\infty(B\times[0,1]\times\mathbb{N})}$ are finite, then $X$ is the unique solution of (12) in $L^1(B \times [0,1])$.*

*Proof.* The proof is close to [3], [7]. We have $X \in L^1(B \times [0,1])$ by integrability of the density $L_{0,t}$. Let $G \in S$.

$$E \left[ \int_0^t \sigma_s X_s \tilde{D}_s G ds \right]$$

$$= E\left[\int_0^t \sigma_s(T_s)\eta \exp\left(\int_0^s b_r(T_r)dr\right)\tilde{D}_sG(T_s)ds\right]$$

$$= E\left[\int_0^t \eta \exp\left(\int_0^s b_r(T_r)dr\right)\frac{d}{ds}G(T_s)ds\right]$$

$$= E\left[\exp\left(\int_0^t b_s(T_s)ds\right)G(T_t)\eta - \eta G\right.$$
$$\left. - \int_0^t \eta b_s(T_s)\exp\left(\int_0^s b_r(T_r)dr\right)G(T_s)ds\right]$$

$$= E\left[\eta(A_t)\exp\left(\int_0^t b_s(\phi_{s,t})ds\right)L_{0,t}G - \eta G\right.$$
$$\left. - \int_0^t \eta(A_s)b_s\exp\left(\int_0^s b_r(\phi_{r,s})ds\right)L_{0,s}Gds\right]$$

$$= E\left[\left(X_t - \eta - \int_0^t b_sX_sds\right)G\right],$$

and $X_t - \eta - \int_0^1 b_sX_sds \in L^1(B)$. Hence $\sigma X1_{[0,t]} \in Dom(\tilde{\delta})$, $t \in [0,1]$, and $(X_t)_{t\in[0,1]}$ is solution to Eq. (12). We now show the uniqueness of the solution in $L^1(B \times [0,1])$. Let $(\sigma^n)_{n\in\mathbb{N}}$ be a sequence given by Prop. 3, and let $(Y_t)_{t\in[0,1]}$ be the difference of two solutions, which satisfies

$$Y_t = \int_0^t b_sY_sds + \int_0^t \sigma_sY_s\delta\tilde{N}_s.$$

Let $F \in \mathcal{S}$.

$$E[Y_tF(A_t^n)] = E\left[\int_0^t \sigma_sY_s\tilde{D}_s\left(F(A_t^n)\right)ds + \int_0^t b_sY_sF(A_t^n)ds\right]$$

$$= E\left[\int_0^t \sigma_sY_s\tilde{D}_s\left(F(A_s^n) - \int_s^t \sigma_r^n\tilde{D}_r\left(F(A_r^n)\right)dr\right)ds\right.$$
$$\left. + \int_0^t b_sY_s\left(F(A_s^n) - \int_s^t \sigma_r^n\tilde{D}_r\left(F(A_r^n)\right)dr\right)ds\right].$$

We have for $u \in \mathcal{V}$

$$E\left[\int_0^t \sigma_sY_s\tilde{D}_s\int_s^t u_rdrds\right] = E\left[\int_0^t \int_0^r \sigma_sY_s\tilde{D}_su_rdsdr\right]$$
$$= E\left[\int_0^t \int_0^r \sigma_sY_s\delta\tilde{N}_su_rdr\right].$$

This relation can be extended by density to the process $u = \sigma^n\tilde{D}(F(A^n))$ since $\int_s^t \sigma_r^n\tilde{D}_r(F(A_r^n))dr = F(A_s^n) - F(A_t^n) \in D_{2,1}$ and gives

$$E\left[\int_0^t \sigma_sY_s\tilde{D}_s\int_s^t \sigma_r^n\tilde{D}_r(F(A_r^n))drds\right]$$

$$= E\left[\int_0^t \int_0^r \sigma_s Y_s \delta \tilde{N}_s \sigma_r^n \tilde{D}_r(F(\mathcal{A}_r^n))dr\right]$$

$$= E\left[\int_0^t \left(Y_r - \int_0^r b_s Y_s ds\right) \sigma_r^n \tilde{D}_r(F(\mathcal{A}_r^n))dr\right]$$

$$= E\left[\int_0^t Y_r \sigma_r^n \tilde{D}_r(F(\mathcal{A}_r^n))dr - \int_0^t b_s Y_s \int_s^t \sigma_r^n \tilde{D}_r(F(\mathcal{A}_r^n))drds\right].$$

Hence

$$E\left[Y_t F(\mathcal{A}_t^n)\right] = E\left[\int_0^t (\sigma_s - \sigma_s^n)Y_s \tilde{D}_s(F(\mathcal{A}_s^n))ds + \int_0^t b_s Y_s F(\mathcal{A}_s^n)ds\right].$$

From Lemma 2, $\mid D(F(\mathcal{A}_s^n)) \mid_H$ is uniformy bounded in $n$ and $\omega$, hence letting $n$ go to infinity we get

$$E\left[Y_t F(\mathcal{A}_t)\right] = E\left[\int_0^t b_s Y_s F(\mathcal{A}_s)ds\right].$$

Then

$$E\left[Y_t(\mathcal{T}_t)F\mathcal{L}_t\right] = E\left[\int_0^t \mathcal{L}_s b_s(\mathcal{T}_s)Y_s(\mathcal{T}_s)Fds\right],$$

with $\mathcal{L}_s = (L_{0,s}(\mathcal{T}_t))^{-1}$, which is satisfied by density for

$$F = sign(\mathcal{L}_t Y_t(\mathcal{T}_t)).$$

This gives

$$E\left[\mid Y_t \mid\right] \leq \int_0^t E\left[\mid Y_s \mid\right] ds$$

and $Y = 0$ by the Gronwall lemma. Consequently the solution is unique. □

*Remark.* If moreover the processes $\sigma$ and $b$ are $(\mathcal{F}_t)$-adapted and $\eta = 1$, then the solution coincides with the usual result, i.e.

$$X_t = \exp\left(\int_0^t b_s ds - \int_0^t \sigma_s ds\right) \prod_{0 \leq T_k \leq t} (1 + \sigma_{T_k}) \quad 0 \leq t \leq 1,$$

since $\phi_{s,t}(k) = \tau_k$ if $T_{k+1} < s$ and $\sigma_s$, $b_s$ depend on $\tau_k$ only if $T_{k+1} < s$, $s \in [0,1]$.

**Appendix.**

Let $\mathcal{F}_n$ denote the $\sigma$-algebra generated by $\tau_0, \ldots, \tau_n$, $n \in \mathbf{N}$.
*Proof of Prop. 2.* Let $F_n = (1 - \frac{1}{n})E[F \mid \mathcal{F}_n]$, $n \in \mathbf{N}$. We have *ess inf* $F <$ $F_n <$ *ess sup* $F$, $n \in \mathbf{N}$. If $(G_k)_{k \in \mathbf{N}} \subset \mathcal{S}$ converges to $F$ in $D_{2,1}$, then

$$\|| E[DG_k \mid \mathcal{F}_n] \mid_H\|_\infty \leq \| \left(\sum_{i=0}^{i=n} (D_i E[G_k \mid \mathcal{F}_n])^2\right)^{1/2} \|_\infty$$

$$\leq \ \left\| \left( \sum_{i=0}^{\infty} (E[D_i G_k \mid \mathcal{F}_n])^2 \right)^{1/2} \right\|_{\infty}$$

$$\leq \ \left\| \left( \sum_{i=0}^{\infty} E[(D_i G_k)^2 \mid \mathcal{F}_n] \right)^{1/2} \right\|_{\infty} \leq \| \mid DG_k \mid_H \|_{\infty} \ .$$

This gives $\| \mid DF_n \mid_H \|_{\infty} \leq \| \mid DF \mid_H \|_{\infty}$. We also have the convergence of $(F_k)_{k \in \mathbf{N}}$ to $F$ in $\mathbf{D}_{1,2}$. Hence it suffices to prove the result for $F \in \mathbf{D}_{1,2}$ of the form $F = f(\tau_0, \ldots, \tau_n)$. Assume first that $f$ has a compact support in $\mathbf{R}_+^{n+1}$. Let $\Psi \in \mathcal{C}_c^{\infty}(\mathbf{R}_+^{n+1})$ with $\int_{\mathbf{R}_+^{n+1}} \Psi(x) dx = 1$, $\Psi \geq 0$, and let $f_k(y) = \frac{1}{k^{n+1}} \int_{\mathbf{R}_+^{n+1}} \Psi(kx) f(y + x) dx$, $k > 0$, $y \in \mathbf{R}_+^{n+1}$. With $F_k = f_k(\tau_0, \ldots, \tau_n)$, we still have $ess \ inf \ F \leq F_k \leq ess \ sup \ F$, $k \in \mathbf{N}$, and

$$\| \mid DF_k \mid_H \|_{\infty} \leq \| \mid DF \mid_H \|_{\infty} \ .$$

If $f$ does not have a compact support, let $\Phi \in \mathcal{C}_c^{\infty}(\mathbf{R}^n)$ such that $\Phi(x) = 1$ for $\mid x \mid < 1$ and $0 \leq \Phi \leq 1$ on $\mathbf{R}^n$. Let $F_k = E[F \mid \mathcal{F}_n] \Phi(\tau_0/k, \ldots, \tau_m/k)$. Then $(F_k)_{k \in \mathbf{N}}$ converges to $F$ in $\mathbf{D}_{1,2}$ and

$$\| \mid DF_k \mid_H \|_{\infty} \ = \ \left\| \frac{1}{k} E[F \mid \mathcal{F}_n] D\Phi + \phi E[DF \mid \mathcal{F}_n] \mid_H \right\|_{\infty}$$

$$\leq \ \| \Phi \mid DF \mid_H \|_{\infty} + \frac{1}{k} \| F_k \|_{\infty} \| \mid D\Phi \mid_H \|_{\infty}$$

$$\leq \ \| \mid DF \mid_H \|_{\infty} + \frac{1}{k} \| F \|_{\infty} \sup \sum_{i=0}^{i=n} (\partial_i \Phi)^2$$

$$\leq \ \| \mid DF \mid_H \|_{\infty} + \varepsilon$$

for $k$ great enough. $\qquad\qquad\qquad\qquad\qquad\qquad\qquad\qquad\qquad\qquad\qquad\quad \square$

*Proof of Prop. 3.* For $\pi = \{\Delta_1, \ldots, \Delta_n\}$ a partition of $[0, 1]$, let

$$\sigma^{\pi} = \sum_{i=1}^{i=n} 1_{\Delta_i} \int_{\Delta_i} \sigma_r dr / \mid \Delta_i \mid \ .$$

Let $(\pi_n)_{n \in \mathbf{N}}$ be a sequence of partitions of $[0, 1]$, mutually increasing with $\max_{1 \leq i \leq n} \mid \Delta_i^n \mid$ converging to $0$ as $n$ goes to infinity. We have that $(\sigma^{\pi_n})_{n \in \mathbf{N}}$ converges to $\sigma$ in $L_{1,2}$ with

$$\int_0^1 \| \sigma_s^{\pi_n} \|_{\infty}^2 \, ds \leq \int_0^1 \| \sigma_s \|_{\infty}^2 \, ds,$$

$$\int_0^1 \left\| \frac{1}{1 + \sigma_s^{\pi_n}} \right\|_{\infty}^2 \, ds \leq \int_0^1 \left\| \frac{1}{1 + \sigma_s} \right\|_{\infty}^2 \, ds,$$

and

$$\int_0^1 \| \mid D\sigma_s^{\pi_n} \mid_H \|_{\infty}^2 \, ds \leq \int_0^1 \| \mid D\sigma_s \mid_H \|_{\infty}^2 \, ds.$$

We can apply Prop. 2 to $\frac{1}{|\Delta_i|} \int_{\Delta_i} \sigma_s ds$, $1 \leq i \leq n$.

□

*Proof of Prop. 4.* Assume $F = f(\tau_0, \ldots, \tau_n)$.

$$| F \circ T^1(\omega) - F \circ T^2(\omega) |$$

$$= \left| f\left( \int_0^{T_1} \sigma_s^1(\omega)ds, \ldots, \int_{T_n}^{T_{n+1}} \sigma_s^1(\omega)ds \right) \right.$$

$$\left. - f\left( \int_0^{T_1} \sigma_s^2(\omega)ds, \ldots, \int_{T_n}^{T_{n+1}} \sigma_s^2(\omega)ds \right) \right|$$

$$\leq \ ||| \, DF \, |_H ||_\infty \left( \sum_{i=0}^{i=n} \left( \int_{T_i}^{T_{i+1}} \sigma_s^1(\omega)ds - \int_{T_i}^{T_{i+1}} \sigma_s^2(\omega)ds \right)^2 \right)^{1/2}$$

$$\leq \ ||| \, DF \, |_H ||_\infty | \, \sigma^1(\omega) - \sigma^2(\omega) \, |_{L^2([0,1])}, \quad \omega \in B.$$

The same argument holds for the second part. If $F \in D_{1,\infty}$, then there is a sequence $(F_n)_{n \in \mathbb{N}} \subset \mathcal{S}$ that converges to $F$ in $D_{1,2}$ and

$$||| \, DF \, |_H ||_\infty \leq ||| \, DF \, |_H ||_\infty + \varepsilon.$$

Since $T^1$ and $T^2$ are absolutely continuous, $P(| \, F_n \circ T^1 - F \circ T^1 \, | \geq \delta)$ goes to 0 as $n$ goes to $\infty$, for any $\delta > 0$. The same is true for $T^2$. This gives

$$| \, F \circ T^2 - F \circ T^2 \, | \leq (||| \, DF \, |_H ||_\infty + \varepsilon) \, | \, \sigma^1 - \sigma^2 \, |_{L^2([0,1])},$$

where $\varepsilon$ is arbitrary.

□

*Proof of Prop. 5.* For any $\varepsilon > 0$, there is $M_\varepsilon > 0$ such that

$$\sup_{n \in \mathbb{N}} E\left[ L^n 1_{\{L^n > M_\varepsilon\}} \right] \leq \varepsilon/2.$$

For any $\delta > 0$, there is $n_0 \in \mathbb{N}$ such that for $n \geq n_0$,

$$P(| \, F(T^n) - F_n(T^n) \, | \geq \delta) \ = \ E\left[ 1_{\{|F - F_n| \geq \delta\}} L^n \right]$$

$$\leq \ E\left[ 1_{\{L^2 > M_\varepsilon\}} L^n \right] + M_\varepsilon P(| \, F - F_n \, | \geq \delta)$$

$$\leq \ \varepsilon/2 + M_\varepsilon P(| \, F - F_n \, | \geq \delta) \leq \varepsilon.$$

Let $(G^n)_{n \in \mathbb{N}} \subset \mathcal{S}$ be a sequence that converges to $F$ in $L^2(B)$. The density $L$ of $T$ is the weak limit of $(L^n)_{n \in \mathbb{N}}$. For any $\varepsilon, \delta > 0$, there is $k_0 \in \mathbb{N}$ such that

$$P(| \, F \circ T^n - G^{k_0} \circ T^n \, | \geq \delta) + P(| \, F \circ T - G^{k_0} \circ T \, | \geq \delta)$$

$$\leq \ E\left[ 1_{\{|F - G^{k_0}| \geq \delta\}} (L^n + L) \right]$$

$$\leq \ \varepsilon + 2M_\varepsilon \leq 2\varepsilon$$

for any $n \in \mathbf{N}$. We also have

$$P(| \; G^{k_0} \circ \mathcal{T} - G^{k_0} \circ \mathcal{T}^n \; | \geq \delta)$$

$$\leq \frac{1}{\delta} \; ||| \; DG^{k_0} \; |_H ||_\infty | \; \sigma - \sigma^n \; |_{L^2([0,1])} \leq \varepsilon$$

for $n$ great enough, from Prop. 4. Hence there is $n_0 \in \mathbf{N}$ such that for $n \geq n_0$,

$$P(| \; F \circ \mathcal{T} - F \circ \mathcal{T}^n \; | \; 3 \geq \delta) \leq 3\varepsilon. \qquad \square$$

# References

[1] N. Bouleau and F. Hirsch. *Dirichlet Forms and Analysis on Wiener Space*. de Gruyter, Berlin/New York, 1991.

[2] P. Brémaud. *Point Processes and Queues. Martingale Dynamics*. Springer-Verlag, Berlin/New-York, 1981.

[3] R. Buckdahn. Linear Skorohod stochastic differential equations. *Probability Theory and Related Fields*, 90:223–240, 1991.

[4] R. Buckdahn. Anticipative Girsanov transformations. *Probability Theory and Related Fields*, 91:211–238, 1992.

[5] E. Carlen and E. Pardoux. Differential calculus and integration by parts on Poisson space. In *Stochastics, Algebra and Analysis in Classical and Quantum Dynamics*, pages 63–73. Kluwer, 1990.

[6] D. Nualart and E. Pardoux. Stochastic calculus with anticipative integrands. *Probability Theory and Related Fields*, 78:535–582, 1988.

[7] E. Pardoux. Applications of stochastic calculus to stochastic differential equations. In H. Korezlioglu and A.S. Üstünel, editors, *Stochastic Analysis and Related Topics II*, volume 1444 of *Lecture Notes in Mathematics*, Berlin/New-York, 1988. Springer-Verlag.

[8] N. Privault. Décompositions chaotiques sur l'espace de Poisson et applications. *Comptes Rendus de l'Académie des Sciences de Paris, Série I*, 317:385–388, 1993.

[9] N. Privault. Chaotic and variational calculus in discrete and continuous time for the Poisson process. *Stochastics and Stochastics Reports*, 51:83–109, 1994.

[10] N. Privault. Girsanov theorem for anticipative shifts on Poisson space. *Probability Theory and Related Fields*, 103, 1995.

[11] A.S. Üstünel and M. Zakai. Transformation of Wiener measure under anticipative flows. *Probability Theory and Related Fields*, 93:91–136, 1992.

# Diffusion approximation for elliptic stochastic differential equations

by

Samy Tindel

## Abstract

In this note we prove a diffusion approximation result for an elliptic SPDE with and additive white noise. The proof is based on the continuity property of the map that sends the solution of the linear problem into the general solution. We also get a large deviation result as a direct corollary of the continuity property.

**Key words and phrases:** Stochastic partial differential equations, diffusion approximation, large deviations.

**AMS Classification Number :** 60H15.

## 1 Introduction

Let us consider a random field $F$ on $R_+^2$ defined by

$$F(s,t) = \sum_{k=1}^{\infty} \sum_{\ell=1}^{\infty} Z_{k,\ell} \; \mathbb{1}_{[k-1,k)\times[\ell-1,\ell)} \, (s,t)\,, \quad (s,t) \in R_+^2$$

where $\{Z_{k,\ell}; \, k \geq 1, \, \ell \geq 1\}$ is a family of independent identically distributed random variables of zero mean and unit variance. In this paper, we are interested in the behaviour of the solution $U^\varepsilon$ of the following elliptic stochastic equation on the domain $D = (0,1)^2$:

$$\begin{cases} -\Delta U^\varepsilon(x) + f(x, U^\varepsilon(x)) &= \frac{1}{\varepsilon} F^\varepsilon(x), \quad x \in D \\ U^\varepsilon|_{\partial D} &= 0 \end{cases}$$

with $F^\varepsilon(x) = F(x/\varepsilon)$. In particular, we will show a diffusion approximation type theorem, i.e. the distribution of $\{U^\varepsilon(x); x \in \overline{D}\}$ converges weakly as $\varepsilon \searrow 0$, on the Banach space of continuous function on $\overline{D}$ which vanish on $\partial D$, towards the unique solution $U$ of the white noise driven stochastic elliptic equation

$$\begin{cases} -\Delta U(x) + f(x, U(x)) &= \dot{W}(x), \quad x \in D \\ U_{|\partial D} &= 0. \end{cases}$$

Such a problem has been studied for hyperbolic equations by Carmona and Fouque in [3]. In a similar way as mentioned in their introduction, we can remark that the corresponding one-parameter problem is easily treated, even in the non linear case (i.e. $f \not\equiv 0$), since the solution $U$ of

$$\begin{cases} -\frac{d^2 U}{dt^2}(t) + f(t, U(t)) &= \dot{W}(t), \quad t \in (0,1) \\ U(0) = U(1) &= 0 \end{cases}$$

is a continuous functional of the Brownian motion $\{W_t; t \in [0,1]\}$. Indeed, in this case, the solution $U_0$ of the linear problem ($f \equiv 0$) can be written

$$U_0(t) = t \int_0^1 W_s \, ds - \int_0^t W_s \, ds$$

(cf Nualart-Pardoux [7]), and the solution $U$ of the nonlinear problem is a continuous function of $U_0$, as will be shown in Section 2. The continuity of $U_0$ as a function of $W$ is no longer true in the case of a two-parameters process, because the Poisson kernel has singularities in $D$.

Nevertheless, our proof will be based on the continuity of the function that maps the solution $U_0$ of the linear problem into $U$. We show this continuity in the general case in Section 2. We then prove the diffusion approximation for $U_0$, which is done in Section 3.

Note that the continuity property will also lead us very easily to a large deviation principle for the white noise driven quasi-linear stochastic elliptic differential equation.

# 2  Preliminary results on elliptic equations

We consider here a complete probability space $(\Omega, \mathcal{F}, P)$, and an open bounded subset $D$ of $R^k$, with $k = 1, 2, 3$, which is regular in the potential theory sense. Let $\{W(B); B \in \mathcal{B}(D)\}$ be a Gaussian family of random variables defined in $(\Omega, \mathcal{F}, P)$, such that $E[W(B)] = 0$ and

$$E[W(A)\,W(B)] = \lambda(A \cap B)$$

where $\lambda$ is the Lebesgue measure on $R^k$. The formal derivative of $W$ with respect to $\lambda$ is called white noise on $D$, i.e. for any $A \in \mathcal{B}(D)$,

$$W(A) = \int_D \dot{W}(x)\lambda(dx)$$

Our aim is to study a continuity property of a nonlinear stochastic elliptic equation of the type

$$\begin{cases} -\Delta U(x) + f(x, U(x)) & = \dot{W}(x), \quad x \in D \\ U_{|\partial D} & = 0 \end{cases} \qquad (2.1)$$

where $f : D \times R \longrightarrow R$ is a measurable function. Moreover, we assume that:

(H) the function $f$ is locally bounded, and continuous and nondecreasing as a function of the second variable.

We will denote by $f(U)$ the function $f(U)(x) = f(x, U(x))$, by $\| \cdot \|_\infty$ the supremum norm on $D$, and by $C_k^\infty(D)$, the set of infinitely differentiable functions on $D$ with compact support included in $D$.

Note that equation (2.1) is formal. The weak form would be: an a.s. continuous random field $U = \{U(x); x \in D\}$ is said to satisfy the weak form of (2.1) if for any $\phi \in C_k^\infty(D)$,

$$-\int_D U(x)\,\Delta\phi(x)\,dx + \int_D f(U)(x)\,\phi(x)\,dx = \int \phi(x)\,W(dx) \qquad (2.2)$$

where the last term is a stochastic integral with respect to $W$.

It is shown in Buckdahn–Pardoux [2] that under hypothesis (H), equation (2.2) has a unique solution, and it is equivalent to its integral form. In order to give that second form, let us first introduce some notation. The kernel $G_D$ will denote the fundamental solution of the elliptic partial differential equation

$$\begin{cases} -\Delta g(x) & = \varphi(x), \quad x \in D \\ g_{|\partial D} & = 0. \end{cases} \qquad (2.3)$$

That is, for any $\varphi \in L^2(D)$, the unique solution of (2.3) can be written

$$g(x) = \int_D G_D(x,y)\,\varphi(y)\,dy = (G_D\varphi)(x).$$

Donati–Martin proves in [5] that

$$\sup_{x \in D} \int_D G_D^2(x,y)dy = K^2 < \infty.$$

We then define the Gaussian random field $\{U_0(x); x \in \overline{D}\}$ by

$$U_0(x) = \int_D G_D(x,y)\,W(dy)$$

which is the unique solution of the linear version of equation (2.1):

$$\begin{cases} -\Delta U_0(x) & = \dot{W}(x) \quad x \in D \\ U_0|\partial D & = 0. \end{cases}$$

Buckdahn and Pardoux show in [2] that $U_0$ is $\alpha_k$–Hölder continuous for $k = 1, 2, 3$, with $\alpha_1 = 1$, $\alpha_2 < 1$ and $\alpha_3 < 3/8$. Moreover, equation (2.2) is equivalent to

$$U(x) + \int_D G_D(x,y)\,f(U)(y)\,dy = U_0(x) \tag{2.4}$$

which is called the integral form of (2.1). This last relation can be seen as a functional equation: set $B = \{\xi \in C(\overline{D}); \xi|\partial D = 0\}$ and define $T$ by

$$\begin{aligned} T: \quad B &\longrightarrow B \\ \xi &\longmapsto \xi + G_D\left(f(\xi)\right). \end{aligned}$$

Equation (2.4) can be written $T(U(\omega)) = U_0(\omega)$. Assuming $(H)$, it can be proved that $T$ is bijective, which leads to the uniqueness of the solution of (2.1). We now can state the main result of this section: the function that maps $U_0$ into $U$ is continuous.

**Theorem 2.1** *If $f$ verifies $(H)$, the function $T^{-1} : B \longrightarrow B$ defined above is continuous for $\|\cdot\|_\infty$.*

*Proof:*

*Step 1:* Suppose $f$ is continuous, nondecreasing and bounded.

Let $\{\xi_n; n \in \mathbf{N}\} \subset B$ be a sequence converging to a given $\xi$. We set $\overline{\xi}_n = T^{-1}\xi_n$ and $\overline{\xi} = T^{-1}\xi$. It is sufficient to show that $\overline{\xi}_n$ converges to $\overline{\xi}$. For any $n \in \mathbf{N}$, $\overline{\xi}_n$ satisfies, for $x \in \overline{D}$,

$$\overline{\xi}_n(x) + \int_D G_D(x, y)\, f(\overline{\xi}_n)(y)\, dy = \xi_n(x). \tag{2.5}$$

Let us first prove that $\{\overline{\xi}_n; n \in \mathbf{N}\}$ is relatively compact in $B$. We call $V_D$ the volume of $D$, and set $M = \|f\|_\infty$. For any $x \in \overline{D}$, we then have, by Schwarz inequality

$$\left| \int_D G_D(x, y)\, f(\overline{\xi}_n)(y)\, dy \right| \leq \left( \int_D G_D^2(x, y)\, dy \right)^{1/2} \cdot V_D^{1/2} \cdot M$$
$$\leq K\, M\, V_D^{1/2}.$$

Moreover, for any $x, z \in D$, using the ideas of the proof of Lemma 2.1 in Buckdahn–Pardoux [2], we get, for $k = 1, 2, 3$,

$$\sup_n \left| \int_D \left( G_D(x, y) - G_D(z, y) \right) f(\overline{\xi}_n)(y)\, dy \right|$$
$$\leq M\, V_D^{1/2} \Big[ \int_D |G_D(x, y) - G_D(z, y)|^2\, dy \Big]^{1/2}$$
$$\leq M\, V_D^{1/2}\, C_k\, |x - z|^{\alpha_k/2}.$$

From Ascoli–Azrela theorem, we can deduce that $\{G_D\, f(\overline{\xi}_n); n \in \mathbf{N}\}$ is relatively compact. By convergence of $\xi_n$, the family $\{\xi_n; n \in \mathbf{N}\}$ is also relatively compact, and from (2.5), we then get that $\{\overline{\xi}_n; n \in \mathbf{N}\}$ is relatively compact.

Let us now consider a converging subsequence $\{\overline{\xi}_{n_k}; k \in \mathbf{N}\}$ of $\overline{\xi}_n$, and set $\widehat{\xi} = \lim_k \overline{\xi}_{n_k}$. From continuity and boundedness of $f$, we have $\lim_k G_D\, f(\overline{\xi}_{n_k}) = G_D\, f(\widehat{\xi})$, and thus, passing to the limit in (2.5), $\widehat{\xi}$ verifies, for $x \in \overline{D}$

$$\widehat{\xi}(x) + \int_D G_D(x, y)\, f(\widehat{\xi}(y))\, dy = \xi(x).$$

By uniqueness of the solution of (2.4), we have $\widehat{\xi} = \overline{\xi}$. Consequently, any converging subsequence $(\overline{\xi}_{n_k})$ converges to $\overline{\xi}$, which shows the continuity in this case.

*Step 2:* Suppose $f$ satisfies $(H)$ and $f \geq \alpha$ for $\alpha \in R$.

We will use here again the method developed by Buckdahn and Pardoux in [2] Theorem 2.5. Denote by $T^{-1}$ the mapping corresponding to $f$. For

each $n \in \mathbf{N}$, we set $f_n = f \wedge n$, and consider the associated mapping $T_n^{-1}$, which is continuous and bijective from $B$ to $B$. Then for any $\xi \in B$ we call $\tilde{\xi}_n$ the function $T_n^{-1}\xi$ for $n \in \mathbf{N}$. The function $\tilde{\xi}_0$ is continuous on $\overline{D}$. The function $f$ being locally bounded, there exists $M(\xi) < \infty$ such that $M(\xi) = \sup \{f(\tilde{\xi}_0(x)); x \in \overline{D}\}$.
We denote by $\Omega_n$ the set

$$\Omega_n = \{\xi \in B; M(\xi) < n\}.$$

From what we have just seen,

$$\bigcup_{n \in \mathbf{N}} \Omega_n = B. \tag{2.6}$$

Moreover, from Buckdahn–Pardoux [2] Lemma 2.6 and Theorem 2.5, for any $\xi \in B$ the sequence $\{\tilde{\xi}_m; m \in \mathbf{N}\}$ is nonincreasing and converges to $\tilde{\xi} = T^{-1}\xi$. The function $\tilde{\xi}_m$ satisfies

$$\tilde{\xi}_m(x) + \int_D G_D(x,y) \left[(f \wedge m)(\tilde{\xi}_m)\right](y) \, dy = \xi(x).$$

Thus, if $\xi \in \Omega_n$, for each $m \geq n$, since $f(\tilde{\xi}_m) \leq n \leq m$, $\tilde{\xi}_m$ satisfies

$$\tilde{\xi}_m(x) + \int_D G_D(x,y) f(\tilde{\xi}_m)(y) \, dy = \xi(x)$$

which shows, by uniqueness of the solution of equation (2.4), that $\tilde{\xi}_m = T_m^{-1}\xi = \tilde{\xi}$. Hence, on $\Omega_n$, we have $T^{-1} = T_n^{-1}$. By relation (2.6), we get that $T^{-1}$ is continuous.

*Step 3:* Suppose $f$ satisfies $(H)$.

The proof is completed for $f$ unbounded by considering the sequence $\{f_n = f \vee (-n); n \in \mathbf{N}\}$ and applying the same arguments as in step 2. $\square$

The continuity property showed above leads us easily to a large deviations principle for the family

$$\begin{cases} -\Delta U^\varepsilon(x) + f(U^\varepsilon)(x) &= \sqrt{\varepsilon} \, \dot{W}(x) \quad x \in D \\ U^\varepsilon|_{\partial D} &= 0 \end{cases}$$

where $\varepsilon > 0$. We will call $\mu_\varepsilon$ the law on $B$ of the process $\{U^\varepsilon(x); x \in \overline{D}\}$, $H$ the set of square integrable functions on $D$, $\langle ., . \rangle$ and $\| . \|_H$ the scalar product and norm on $H$.

**Theorem 2.2** *If $f$ verifies (H), $\{\mu_\varepsilon; \varepsilon > 0\}$ satisfies a large deviations principle governed by the rate function*

$$\begin{cases} I(X) & = \dfrac{1}{2}\|\Delta X + f(X)\|_H & \text{if} \quad \Delta X \in H \\ & = \infty & \text{if} \quad X \in B, \ \Delta X \notin H \end{cases}$$

*Proof:* $T^{-1}$ being continuous, from Deuschel–Stroock [4] p. 37 Lemma 2.14, it is sufficient to prove the result for the family $\{U_0^\varepsilon; \varepsilon > 0\}$ defined by

$$\begin{cases} -\Delta U_0^\varepsilon(x) & = \sqrt{\varepsilon}\ \dot{W}(x) \\ U_0^\varepsilon|\partial D & = 0. \end{cases}$$

We call $\nu^\varepsilon$ the law of $\{U_0^\varepsilon(x); x \in D\}$, and $\nu$ the law of $\{U_0(x); x \in D\}$, both with support in $B$ a.s. We also define the map $i : H \longrightarrow B$ by

$$i(h) = \int_D G_D(\cdot, y)\, h(y)\, dy.$$

Donati–Martin proves in [5] that $(B, H, i, \nu)$ is an abstract Wiener space. Moreover, it is easily shown that $X \in i(H)$ if and only if $\Delta X \in H$ and if $X \in i(H)$, $i^{-1}(X) = \Delta X$. From Deuschel–Stroock [4] p. 88 Theorem 3.4.12, we then get directly that $\{\nu_\varepsilon; \varepsilon > 0\}$ satisfies a large deviation principle governed by the rate function

$$\begin{cases} J(X) & = \dfrac{1}{2}\|\Delta X\|_H & \text{if} \quad \Delta X \in H \\ \\ & = \infty & \text{if} \quad X \in B, \ \Delta X \notin H. \end{cases}$$

The proof is then easily completed. $\qquad\qquad\qquad\qquad\qquad\qquad\qquad\square$

## 3  Diffusion approximation

We will establish a diffusion approximation result analog to that of Carmona–Fouque [3]. We choose $D = (0,1)^2$, and a family $\{Z_{k,\ell}; k \geq 1, \ell \geq 1\}$ of independent identically distributed random variables of zero mean and unit variance satisfying: there exists $\alpha > 0$ such that

$$Z_{k,\ell} \in L^{4+\alpha} \qquad \forall k, \ell \geq 1$$

We then set

$$F(s,t) = \sum_{k=1}^{\infty} \sum_{\ell=1}^{\infty} Z_{k,\ell}\ \mathbb{1}_{[k-1,k)\times[\ell-1,\ell)}\,(s,t), \quad (s,t) \in [0,\infty)^2,$$

and are interested in the limiting behaviour when $\varepsilon \searrow 0$ of the solution $U^\varepsilon$ of the equation

$$\begin{cases} -\Delta U^\varepsilon(x) + f(U^\varepsilon)(x) & = \frac{1}{\varepsilon} F^\varepsilon(x), \quad x \in D \\ \\ U^\varepsilon|\partial D & = 0 \end{cases}$$

where $F^\varepsilon(x) = F(x/\varepsilon)$. The main result of this section is the following:

**Theorem 3.1** *Suppose that $f$ verifies (H). Then the distribution of $\{U^\varepsilon(x);$ $x \in \overline{D}\}$ converges weakly as $\varepsilon \searrow 0$, on the Banach space $B$, towards the distribution of $U$.*

Here again, we will first prove the result for $f \equiv 0$, i.e. establish that for the family of equations

$$\begin{cases} -\Delta U_0^\varepsilon(x) & = \frac{1}{\varepsilon} F^\varepsilon(x), \quad x \in D \\ \\ U_0^\varepsilon|\partial D & = 0 \end{cases}$$

we have $U_0^\varepsilon \longrightarrow U_0$ in law on $B$ as $\varepsilon \searrow 0$. Note that in this case, we have

$$U_0^\varepsilon(x) = \frac{1}{\varepsilon} \int_D G_D(x, y) \, F^\varepsilon(x) \, dx \,, \qquad x \in \overline{D}.$$

Let us first prove the

**Lemma 3.2** *The family $\{U_0^\varepsilon; \varepsilon > 0\} \subset B$ defined above is tight.*

*Proof:* We will show that for a $p \in (1, \infty)$, there exists a $q \in (1, \infty)$ and a constant $C$ such that, for any $\varepsilon > 0$ and any $x, v \in \overline{D}$

$$E\big[\,|U_0^\varepsilon(x) - U_0^\varepsilon(v)|^{2p}\big] \le C \,|x - v|^q.$$

For each $y \in \overline{D}$, we note $y = (z, w)$ with $z, w \in [0, 1]$ and set

$$H_D(z, w) = G_D(x, z, w) - G_D(v, z, w).$$

We also set, for $k, \ell \le \left[\frac{1}{\varepsilon}\right] + 1$,

$$A_{k\ell} = ((k-1)\,\varepsilon, \, k\,\varepsilon) \times ((\ell-1)\,\varepsilon, \ell\,\varepsilon).$$

Using the expression of $F^\varepsilon$, we get, if $Z_{k,\ell} \in L^{2p}$,

$$E[\,|U_0^\varepsilon(x) - U_0^\varepsilon(v)|^{2p}\,] = E\Big\{\Big|\sum_{k,\ell \leq [\frac{1}{\varepsilon}]+1} \frac{1}{\varepsilon}\, Z_{k\ell} \int_{A_{k\ell}} H_D(z,w)\,dz\,dw\Big|^{2p}\Big\}$$

and by Burkholder inequality for discrete two–parameters martingales, this last expectation is bounded from above by

$$\frac{C_p}{\varepsilon^{2p}}\, E\Big\{\Big|\sum_{k,\ell \leq [\frac{1}{\varepsilon}]+1} Z_{k\ell}^2 \Big(\int_{A_{k\ell}} H_D(z,w)\,dz\,dw\Big)^2\Big|^p\Big\}$$

$$\leq\ C_p E\Big\{\Big|\sum_{k,\ell \leq [\frac{1}{\varepsilon}]+1} Z_{k\ell}^2 \int_{A_{k\ell}} H_D^2(z,w)\,dz\,dw\Big|^p\Big\}.$$

Thus

$$\big\|(U_0^\varepsilon(x) - U_0^\varepsilon(v))^2\big\|_p \leq C_p^{1/p}\Big\|\sum_{k,\ell \leq [\frac{1}{\varepsilon}]+1} Z_{k\ell}^2 \int_{A_{k\ell}} H_D^2(z,w)\,dz\,dw\Big\|_p$$

$$\leq C_p^{1/p} \sum_{k,\ell \leq [\frac{1}{\varepsilon}]+1} \Big\|Z_{k\ell}^2 \int_{A_{k\ell}} H_D^2(z,w)\,dz\,dw\Big\|_p$$

$$= C_p^{1/p}\, \|\,Z_{11}^2\,\|_p\, \Big(\int_D H_D^2(y)\,dy\Big)$$

and

$$E[\,|U_0^\varepsilon(x) - U_0^\varepsilon(v)|^{2p}\,] \leq C_p\, E[Z_{11}^{2p}]\, \Big(\int_D |G_D(x,y) - G_D(v,y)|^2\,dy\Big)^p.$$

From Buckdahn–Pardoux [2], in dimension 2, for any $\zeta \in (0,1)$, there exists a constant $C_\zeta'$ such that

$$E[\,|U_0^\varepsilon(x) - U_0^\varepsilon(v)|^{2p}\,] \leq C_p\, C_\zeta'\, E[Z_{11}^{2p}]\, |x - v|^{(1-\zeta)p}.$$

Choosing $p > 2\cdot(1+\eta)/(1-\zeta)$ for $\eta > 0$, we get the desired result. Note that $p$ can be chosen in an interval $(2, 2 + \alpha/2)$ arbitrarely small. Consequently, the property $Z_{k,\ell} \in L^{4+\alpha}$ is sufficient. $\quad\square$

The family $\{U_0^\varepsilon;\ \varepsilon > 0\}$ being relatively compact, we can now give properties of the possible limit laws. To this purpose, we will need the following lemma, whose proof is given by J. Galambos in [6].

**Lemma 3.3** *Let $X_{jn}$ be a double indexed sequence of random variables where $j \in \{1, \ldots, k(n)\}$ and $\lim_n k(n) = +\infty$. For fixed $n$, the finite number $k(n)$ of random variables $X_{jn}$ are independent. We set*

$$S_{(n)} = \sum_{j=1}^{k(n)} X_{jn}.$$

*Assume that $S_{(n)}$ has a nondegenerate limit distribution $G(x)$. Then $G(x)$ is normal and $X_{jn}$ satisfies the uniform assymptotic negligibility condition $(P(\max_j |X_{jn}| > \varepsilon) \longrightarrow 0, \forall \varepsilon)$ if and only if $\max_j |X_{jn}| \longrightarrow 0$ in probability.*

**Lemma 3.4** *Let $\{\varepsilon_n; n \in \mathbf{N}\}$ be a sequence of positive real numbers converging to 0, such that $U_0^{\varepsilon_n}$ converges in distribution to a function $\widehat{U} \in B$. Let be $\phi_1, \ldots, \phi_k \in C_k^\infty(D)$ no null with disjoint support. Then $\langle \Delta\phi_1, \widehat{U} \rangle, \ldots, \langle \Delta\phi_k, \widehat{U} \rangle$ are Gaussian centered independent random variables with variances $\|\phi_1\|_H^2, \ldots, \|\phi_k\|_H^2$.*

*Proof:*

*Step 1:* Independence.

For any $n \in \mathbf{N}$ and $p = 1, \ldots, k$

$$\langle \Delta\phi_p, U_0^{\varepsilon_n} \rangle = \langle \phi_p, \Delta U_0^{\varepsilon_n} \rangle = \frac{1}{\varepsilon_n} \int_D \phi_p(x)\, F^{\varepsilon_n}(x)\, dx.$$

Let us introduce some notation: we call $D_p$ the support of $\phi_p$, and set

$$B_{k,\ell}^{p,n} = \Big( [(k-1)\varepsilon_n, k\varepsilon_n) \times [(\ell-1)\varepsilon_n, \ell\varepsilon_n) \Big) \cap D_p$$

$$G_{p,n} = \{(k,\ell) \in \mathbf{N}^2;\ B_{k,\ell}^{p,n} \neq \emptyset\}.$$

We have

$$\langle \Delta\phi_p, U_0^{\varepsilon_n} \rangle = \frac{1}{\varepsilon_n} \sum_{(k,\ell) \in G_{p,n}} Z_{k,\ell} \Big( \int_{B_{k,\ell}^{p,n}} \phi_p(x)\, dx \Big).$$

For $n$ large enough, the supports of the $\phi_p$ being disjoint, the sets $G_{p,n}$ are disjoint too. From independence of the random variables $Z_{k,\ell}$, we thus get that there exists an integer $N_0$ such that $\langle \Delta\phi_1, U_0^{\varepsilon_n} \rangle, \ldots \langle \Delta\phi_k, U^{\varepsilon_n} \rangle$ is an independent family for any $n \geq N_0$. Consequently, for each $(\alpha_1, \ldots, \alpha_k) \in R^k$ and $n \geq N_0$,

$$E\left\{ \exp\left[ i \sum_{p=1}^{k} \alpha_p \langle \Delta\phi_p, U_0^{\varepsilon n} \rangle \right] \right\} = \prod_{p=1}^{k} E\left\{ \exp[i\alpha_p \langle \Delta\phi_p, U_0^{\varepsilon n} \rangle] \right\}$$

and by distribution convergence of $U_0^{\varepsilon n}$ to $\widehat{U}$,

$$E\left\{ \exp\left[ i \sum_{p=1}^{k} \alpha_p \langle \Delta\phi_p, \widehat{U} \rangle \right] \right\} = \prod_{p=1}^{k} E\left\{ \exp\left[ i\alpha_p \langle \Delta\phi_p, \widehat{U} \rangle \right] \right\}$$

which proves independence of the random variables $\langle \Delta\phi_1, \widehat{U} \rangle, \ldots, \langle \Delta\phi_k, \widehat{U} \rangle$.

*Step 2:* Variance calculus.

Let be $\phi \in C_k^\infty(D)$. From convergence in distribution and the boundedness of the moments of $\langle \Delta\phi, \widehat{U} \rangle$ uniformly in $n$ proved as in Lemma 3.2, we know that

$$\mathrm{Var}(\langle \Delta\phi, \widehat{U} \rangle) = \lim_n \mathrm{Var}(\langle \Delta\phi, U_0^{\varepsilon n} \rangle) = \lim_n \mathrm{Var}(\langle \phi, \Delta U_0^{\varepsilon n} \rangle).$$

Using the expression of $F^{\varepsilon n}$ and the definition of $A_{k,\ell}$, we get, for each $n \in \mathbf{N}$

$$\langle \Delta\phi, U_0^{\varepsilon n} \rangle = \sum_{k,\ell \leq [\frac{1}{\varepsilon n}]+1} \left( \frac{1}{\varepsilon_n} \int_{A_{k,\ell}} \phi(x)\, dx \right) Z_{k,\ell} \qquad (3.1)$$

and by independence of the random variables $Z_{k,\ell}$, together with the fact that $\mathrm{Var}(Z_{k,\ell}) = 1$,

$$E[\langle \Delta\phi, U_0^{\varepsilon n} \rangle^2] = \sum_{k,\ell \leq [\frac{1}{\varepsilon n}]+1} \frac{1}{\varepsilon_n^2} \left( \int_{A_{k,\ell}} \phi(x)\, dx \right)^2.$$

We set $x_{k,\ell} = ((k-1)\varepsilon_n, (\ell-1)\varepsilon_n)$ for $k, \ell \leq [\frac{1}{\varepsilon n}]+1$. On $A_{k,\ell}$, since $\phi$ is infinitely differentiable, we have, as $n \to \infty$,

$$\phi(x) = \phi(x_{k,\ell}) + \nabla\phi \cdot (x - x_{k,\ell}) + o(\|x - x_{k,\ell}\|)$$

and by boundedness of $\nabla\phi$ on $\overline{D}$

$$\phi(x) = \phi(x_{k,\ell}) + O(\varepsilon_n).$$

Series developement arguments and Riemann's approximations thus leads to

$$\sum_{k,\ell \leq [\frac{1}{\varepsilon_n}]+1} \frac{1}{\varepsilon_n^2} \left( \int_{A_{k,\ell}} \phi(x)\, dx \right)^2 = \sum_{k,\ell \leq [\frac{1}{\varepsilon_n}]+1} (\varepsilon_n^2\, \phi^2(x_{k,\ell}) + o(\varepsilon_n^2))$$

$$= \sum_{k,\ell \leq [\frac{1}{\varepsilon_n}]+1} \varepsilon_n^2\, \phi^2(x_{k,\ell}) + o(1)$$

$$= \int_D \phi^2(x)\, dx + o(1).$$

Consequently

$$\lim_n \mathrm{Var}\,(\langle \Delta\phi, U_0^{\varepsilon_n} \rangle) = \|\phi\|_H^2.$$

*Step 3:* Normality.

Pick again $\phi \in C_k^\infty(D)$, with $\phi \neq 0$. Relation (3.1) shows that $\langle \Delta\phi, U_0^{\varepsilon_n} \rangle$ has the form $S_{(n)}$ of Lemma 3.3 with $j = (k,\ell)$, $k(n) = [\frac{1}{\varepsilon_n}]+1$ and

$$X_{jn} = \frac{1}{\varepsilon_n} \left( \int_{A_{k,\ell}} \phi(x)\, dx \right) Z_{k,\ell}.$$

Moreover, $\langle \Delta\phi, U_0^{\varepsilon_n} \rangle$ converges in distribution to $\langle \Delta\phi, \widehat{U} \rangle$. From step 2, $\mathrm{Var}\,(\langle \Delta\phi, \widehat{U} \rangle) = \|\phi\|_H^2$, and if $\phi \neq 0$, it is nondegenerate. Finally

$$\left| \left( \frac{1}{\varepsilon_n} \int_{A_{k,\ell}} \phi(x)\, dx \right) Z_{k,\ell} \right| \leq |Z_{k,\ell}|\, \|\phi\|_\infty\, \varepsilon_n. \tag{3.2}$$

If $Z_{k,\ell} \in L^2(\Omega)$ for any $k, \ell \geq 1$, by a direct application of central limit theorem, we have

$$\lim_n \sum_{k,\ell \leq [\frac{1}{\varepsilon_n}]+1} \varepsilon_n\, Z_{k,\ell} \overset{(d)}{=} N(0,1).$$

By the equivalence theorem established in Lemma 3.3, we thus get

$$\lim_n \max_{k,\ell \leq [\frac{1}{\varepsilon_n}]+1} |\varepsilon_n\, Z_{k,\ell}| \overset{(P)}{=} 0$$

and from (3.2),

$$\lim_n \max_{k,\ell \leq [\frac{1}{\varepsilon_n}]+1} \left| \left( \frac{1}{\varepsilon_n} \int_{A_{k,\ell}} \phi(x)\, dx \right) Z_{k,\ell} \right| \overset{(P)}{=} 0,$$

which shows that the sumands maximum converges towards 0 in probability. Hence, Lemma 3.3 can be applied here, and $\langle \Delta\phi, \widehat{U} \rangle$ is a Gaussian random variable. $\quad\square$

We can now establish the diffusion approximation result for $U_0$.

**Theorem 3.5** *The distribution of $\{U_0^\varepsilon(x); x \in \overline{D}\}$ converges weakly on the Banach space $B$ towards the distribution of $U_0$.*

*Proof:* By Lemma 3.2, $\{U_0^\varepsilon; \varepsilon > 0\}$ is tight. It remains to show that every subsequence $\{U_0^{\varepsilon_n}; n \in \mathbf{N}\}$ defined as in Lemma 3.4 converges in distribution to a random field $\widehat{U}$ such that $\widehat{U} \stackrel{(d)}{=} U_0$.
From Lemma 3.4 and using the fact that $C_k^\infty(D)$ is dense in $H$, the mapping

$$A(\phi) = \langle \Delta\phi, \widehat{U} \rangle, \quad \phi \in C_k^\infty(D)$$

can be extended to an isometry of $H$ into $L^2(\Omega)$. Then for any $h \in H$, $A(h)$ is still a Gaussian random variable with variance $\|h\|_H^2$. In particular, set for any $(s, t) \in \overline{D}$

$$\widehat{W}_{s,t} = A(\mathbb{1}_{[0,s] \times [0,t]}).$$

The isometry property implies $\widehat{W} = 0$ on the axes. Moreover, for any $(s, t) \in \overline{D}$, $\widehat{W}_{s,t}$ is a Gaussian random variable with variance $st$. Finally, for $z_1 = (s_1, t_1)$ and $z_2 = (s_2, t_2)$ in $\overline{D}$, let us denote by $\Delta_{[z_1, z_2]}(\widehat{W})$ the increment of $\widehat{W}$ on $[z_1, z_2]$, i.e.

$$\Delta_{[z_1,z_2]}(\widehat{W}) = \widehat{W}_{s_2,t_2} - \widehat{W}_{s_2,t_1} - \widehat{W}_{s_1,t_2} + \widehat{W}_{s_1,t_1}.$$

Let $R_1, \ldots, R_p$ be disjoints closed rectangles of the form $R_i = [s_i, t_i]$. Clearly the functions

$$\mathbb{1}_{R_1}, \ldots, \mathbb{1}_{R_p}$$

can be approximated in $H$ by functions of $C_k^\infty(D)$ with disjoint support, which shows the independence of

$$\Delta_{R_1}(\widehat{W}), \ldots, \Delta_{R_p}(\widehat{W}).$$

This implies that the random process $\widehat{W} = \{\widehat{W}_{s,t}; (s, t) \in [0, 1]^2\}$ is a two-parameters Brownian motion. This characterizes the distribution of $U_0$, and completes the proof. $\quad\square$

We know (see Billingsley [1]) that if $X^\varepsilon$ and $X$ are $E$-valued random variables ($E$ Polish space) such that distribution limit of $X^\varepsilon$ is $X$, and if $h$ is a continuous application from $E$ to $E$, then $h(X^\varepsilon)$ converges in distribution to $h(X)$. This result applies here with $E = B$, $X^\varepsilon = U^\varepsilon$ and $h = T^{-1}$ to deduce Theorem 3.1 from Theorem 3.5.

*Note:* The diffusion approximation result can be extended to $D = (0,1)^3$. The proof is the same as in dimension 2. However, since the Hölder coefficient of continuity for $U^0$ is no longer $< 1$ but $< 3/8$, we get a stronger integrability condition for the random variables $Z_{k,\ell}$, i.e. there exists $\alpha > 0$ such that

$$Z_{k,\ell} \in L^{16/3+\alpha} \qquad \forall\, k, \ell \geq 1.$$

# References

[1] P. Billingsley: Convergence of probability measures. New York. Wiley and Sons, 1979.

[2] R. Buckdahn and E. Pardoux: Monotonicity methods for white noise driven SPDEs. In: M. Pinski (ed.), Diffusion processes and related topics in Analysis, vol. I, pp. 219–233. Boston Basel Stuttgart. Birkhaüser 1990.

[3] R. Carmona and J.P. Fouque: A diffusion approximation result for two parameter processes. Probab. Theory Rel. Fields **98**, pp. 277–298, 1994.

[4] J.D. Deuschel and D. Stroock: An introduction to the theory of large deviations. New York. Academic Press, 1988.

[5] C. Donati–Martin: Quasi–linear elliptic stochastic partial differential equation. Markov property. Stochastics **41**, pp. 219-240, 1992.

[6] J. Galambos: Advanced probability theory. New York and Basel. Marcel Dekker, Inc. 1988.

[7] D. Nualart and E. Pardoux: Second order stochastic differential equations with Dirichlet boundary conditions. Stochastics **39**, pp. 1-24, 1991.

# Absolute Continuity of Monotone Shifts on the Wiener Space

Ali Süleyman ÜSTÜNEL and Moshe ZAKAI

## Abstract

In this work we study the absolute continuity of the image of the Wiener measure under the shifts of the form $I_W + u$, where $u$ is a random variable with values in the Cameron-Martin space $H$, with the property that the linear operator (on $H$) $I_H + \nabla u(w)$ has almost surely a positive spectrum.

## 1  Introduction

Consider the abstract Wiener space $(W, H, \mu)$, where $H$ is the Cameron-Martin space. Let $u$ be an $H$-valued random variable, $T$ will denote the shift induced by $u$, i. e., $T(w) = w + u(w)$ and $T^\star \mu$ is the measure defined by

$$T^\star \mu(A) = \mu(T^{-1}(A)).$$

The problem of absolute continuity of $T^\star \mu$ with respect to $\mu$ goes back to the work of Cameron-Martin and was considered by many authors (c. f . e. g. the references in [3] and [10]). Here we consider the problem for the class of monotone shifts which, roughly speaking, means that

$$\frac{d(T(w + th), h)}{dt}\Big|_{t=0}$$

is almost surely positive. Positivity conditions similar to this have been very useful in nonlinear functional analysis in determining existence and uniqueness of solutions to equations of the type $T(w) = a$ (cf. [2]). We intend to consider here its relevance to the absolute continuity problem. In the next section we present definitions and results in stochastic analysis which are needed later. The notion of positivity is defined in the third section and the results of absolute continuity of $T^*\mu$ with respect to $\mu$ are derived.

## 2 Notations and Preliminaries

$(W, H, \mu)$ denotes an abstract Wiener space, i.e., $H$ is a separable Hilbert space, identified with its continuous dual, $W$ is a Banach space into which $H$ is injected continuously and densely. $\mu$ is the canonical Gaussian measure on $W$ whose reproducing kernel Hilbert space is $H$ and we will call it as the Cameron-Martin space. In the case of classical Wiener space we have $W = C([0,1])$, $H = \{h : [0,1] \to I\!\!R : h(t) = \int_0^t \dot{h}(s)ds, \|h\|_H^2 = \int_0^1 |\dot{h}(s)|^2 ds\}$. Let $X$ be a separable Hilbert space and $a$ be an $X$-valued (smooth) polynomial on $W$:

$$a(w) = \sum_{i=1}^m \eta_i(<h_1, w>, \ldots, <h_n, w>)x_i,$$

with $x_i \in X$, $h_i \in W^*$ and $\eta_i \in C_b^\infty(I\!\!R^n)$. The Gross-Sobolev derivative of $a$ is defined as

$$\nabla a(w) = \sum_{i=1}^m \sum_{j=1}^n \partial_j \eta_i(<h_1, w>, \ldots, <h_n, w>)x_i \otimes h_j,$$

and $\nabla^k a(w)$ is defined recursively. Thanks to the Cameron-Martin theorem, all these operators are closable on all the $L^p$ spaces and the Sobolev spaces $I\!\!D_{p,k}(X), p > 1, k \in I\!\!N$ can be defined as the completion of $X$-valued smooth polynomials with respect to the norm:

$$\|a\|_{p,k} = \sum_{i=0}^k \|\nabla a\|_{L^p(\mu, X \otimes H^{\otimes i})}.$$

From the Meyer inequalities, it is known that the $(p, k)$-norm, defined above, is equivalent to the following norm

$$\|(I + L)^{k/2} a\|_{L^p(\mu; X)}$$

where $L$ is the Ornstein-Uhlenbeck operator on $W$ (cf.[1]) and we denote these two norms with the same notation. Since $I + L$ is an invertible operator, we can also define the norms for negative values of $k$ which describe the dual spaces of the positively indexed Sobolev spaces. We denote by $\mathbb{D}(X)$ the intersection of the Sobolev spaces $\{\mathbb{D}_{p,k}(X); p > 1, k \in Z\}$, equipped with the intersection (i.e., projective limit ) topology. The continuous dual of $\mathbb{D}(X)$ is denoted by $\mathbb{D}'(X)$ and in case $X = \mathbb{R}$ we write simply $\mathbb{D}_{p,k}, \mathbb{D}, \mathbb{D}'$ for $\mathbb{D}_{p,k}(\mathbb{R}), \mathbb{D}(\mathbb{R}), \mathbb{D}'(\mathbb{R})$ respectively. Consequently, for any $p > 1, k \in Z$, $\nabla : \mathbb{D}_{p,k}(X) \mapsto \mathbb{D}_{p,k-1}(X \tilde{\otimes} H)$ continuously, where $X \tilde{\otimes} H$ denotes the completed Hilbert-Schmidt tensor product of $X$ and $H$. Therefore $\delta = \nabla^*$ is a continuous operator from $\mathbb{D}_{p,k}(X \tilde{\otimes} H)$ into $\mathbb{D}_{p,k-1}(X)$ for any $p > 1, k \in Z$. We call $\delta$ the divergence operator on $W$. Let us recall that, in the case of classical Wiener space, $\delta$ coincides with the Ito stochastic integral on the adapted processes. Recall that, if $F$ is in $\mathbb{D}_{p,1}(H)$ for some $p > 1$, then almost surely, $\nabla F$ is an Hilbert-Schmidt operator on $H$ and if $F$ is an $H$-valued polynomial, then $\delta F$ can be written as

$$\delta F = \sum_{i=1}^{\infty} [(F, e_i)_H - (\nabla(F, e_i)_H, e_i)_H],$$

where $(e_i; i \in \mathbb{N})$ is any complete orthonormal basis in $H$.

An $X$-valued random variable $F$ is said to be in $\mathbb{D}_{p,1}^{loc}(X)$ for some $p > 1$ if there exists a sequence of measurable subsets $(W_n; n \in \mathbb{N})$ of $W$ and $(F_n; n \in \mathbb{N}) \subset D_{p,1}(X)$ such that $\cup_n W_n = W$ and $F = F_n$ on $W_n$ almost surely.

Let $K$ be a Hilbert-Schmidt operator on $H$ and denote by $(\lambda_i, i \in \mathbb{N})$ its eigenvalues according to their multiplicity. The Carleman-Fredholm determinant of $I_H + K$ is defined as

$$\det\nolimits_2(I + K) = \prod_{i=1}^{\infty}(1 + \lambda_i)e^{-\lambda_i}.$$

For $F \in \mathbb{D}_{p,1}^{loc}(H)$, we define

$$\Lambda_F(w) = \det\nolimits_2(I + \nabla F(w)) \exp(-\delta F(w) - \frac{1}{2}\|F(w)\|_H^2).$$

Let us denote by $P_t$ the Ornstein-Uhlenbeck semigroup on $W$, recall that $P_t$ is defined on the smooth polynomials on $W$ with the Mehler formula:

$$P_t a(w) = \int_W a(e^{-t}w + \sqrt{1 - e^{-2t}}z)\mu(dz),$$

(cf., [1]). We have the following result which plays an important role in the construction of the Radon-Nikodym density:

**Lemma 1** Let $\phi$ be a real valued Wiener functional such that $\nabla\phi \in L^\infty(\mu; H)$.

$$\mu\{\phi > c\} \leq E\left[\exp -\frac{(c - P_T\phi)^2}{2\|\nabla\phi\|_\infty^2(1 - e^{-2T})} 1_{\{c \geq P_T\phi\}}\right] + \mu\{P_T\phi > c\},$$

and

$$\mu\{-\phi > c\} \leq E\left[\exp -\frac{(c + P_T\phi)^2}{2\|\nabla\phi\|_\infty^2(1 - e^{-2T})} 1_{\{c \geq -P_T\phi\}}\right] + \mu\{-P_T\phi > c\},$$

for any $T > 0$, where $E$ denotes the expectation with respect to $\mu$.

As a corollary of this theorem, we can show:

**Corollary 1** With the above hypothesis, we have

$$E[\exp \lambda\phi^2] < \infty,$$

for any $\lambda < \frac{1}{2\|\nabla\phi\|_\infty^2}$.

Let $u$ be an $H$-valued random variable. It is said to be locally $H - C^1$ if there exists an almost surely strictly positive random variable $\rho$, such that the map $h \mapsto u(w + h)$ is $C^1$ on the open (in H) set $\{h \in H : |h| < \rho(w)\}$. In [10], we have proved the following result:

**Theorem 0** Suppose that $u$ is locally $H - C^1$, let $M$ denote the set on which $\det_2(I_H + \nabla u(w)) \neq 0$ and let $T$ be the map $I_W + u$. Then, for almost all $w \in W$, the cardinal of the set $T^{-1}\{w\} \cap M$ is countable and for any $F \in C_b^+(W)$, we have

$$E[F \circ T|\Lambda_u|] = E[F \cdot N(w, M)],$$

where $N(w, M)$ is the cardinal of the set $T^{-1}\{w\} \cap M$.

# 3    Monotone shifts

**Definition 1**    Let $u \in D_{p,1}(H)$ for some $p > 1$, the shift $T : W \to W$, defined by $T(w) = w + u(w)$ is called **monotone** if

$$((I_H + \nabla u(w))h, h))_H \geq 0,$$

almost surely for any $h \in H$. $T$ is called **strictly monotone** if, for any $h$ in $H$

$$((I_H + \nabla u(w))h, h))_H > 0,$$

$T$ is called **strongly monotone** if, for any $h$ in $H$,

$$((I_H + \nabla u(w))h, h))_H \geq \alpha |h|^2,$$

almost surely, for some $\alpha > 0$.

**Theorem 1**    Suppose that $T$ is a strongly monotone shift such that $\nabla u \in L^\infty(\mu, H \otimes H)$. Then we have

$$E[F \circ T | \Lambda_u|] = E[F],$$

for any $F \in C_b(W)$, where

$$\Lambda_u = \det{}_2(I + \nabla u) \exp\{-\delta u - \frac{1}{2}|u|^2\}.$$

Since $\det_2(I + \nabla u)$ is almost surely non-zero, the image of $\mu$ under $T$ is equivalent to $\mu$.

**Proof:**    Let $(h_i; i \in \mathbb{N})$ be a complete, orthonormal basis of $H$, for $n \in \mathbb{N}$, denote by $\pi_n$ the orthogonal projection of $H$ onto the closed subspace spanned by $\{h_1, \ldots, h_n\}$. Let $V_n$ be the sigma algebra generated by $\{\delta h_1, \ldots, \delta h_n\}$. Define $u_n$ by

$$u_n = E[\pi_n P_{1/n} u | V_n],$$

where $P_{1/n}$ is the Ornstein-Uhlenbeck semigroup at $t = 1/n$. Since we have

$$\nabla u_n = e^{-\frac{1}{n}} E[\pi_n \otimes \pi_n P_{1/n} \nabla u | V_n],$$

from the positivity of $P_{1/n}$, we have

$$(\nabla u_n h, h) \geq -(1-\alpha)e^{-\frac{1}{n}}|\pi_n h|^2.$$

Note that $u_n$ can be written as

$$u_n = \sum_{i=1}^{n} a_i(\delta h_1, \ldots, \delta h_n) h_i,$$

where $a_i$'s are smooth functions on $\mathbb{R}^n$. Let $a$ be the map defined by $a(x) = (a_1(x), \ldots, a_n(x))$. The monotonicity of $\nabla u_n$ implies that

$$((I_{\mathbb{R}^n} + \partial a(x))y, y)_{\mathbb{R}^n} \geq \alpha e^{-\frac{1}{n}}|y|^2_{\mathbb{R}^n}.$$

Therefore

$$\sup_{x \in \mathbb{R}^n} |(I_{\mathbb{R}^n} + \partial a(x))^{-1}| \leq \frac{e^{1/n}}{\alpha},$$

and a well-known theorem of the classical analysis implies that $I_{\mathbb{R}^n} + a$ is a global diffeomorphism of $\mathbb{R}^n$. Consequently, splitting the Wiener space as $W_1 \times W_2$, where $W_1$ is the image of $W$ under $\tilde{\pi}_n(w) = \sum_{i \leq n} \delta h_i(w) h_i$ and $W_2$ is the image of $W$ under $I_W - \tilde{\pi}_n$, we have from the Jacobi formula and the Fubini theorem:

$$E[F(w + u_n(w))|\Lambda_{u_n}|] =$$

$$\int_{W_2} \int_{\mathbb{R}^n} F(w_2 + \sum_{i=1}^{n} a_i(y)h_i)|\det{}_2(I_H + \sum_{i \leq n} \partial a_i(y)h_i)|.$$

$$\exp - \sum_{i \leq n}(a_i(y)y_i - \partial_i a_i(y)) - \frac{1}{2}|a(y)|^2 \mu_n(dy)\mu_2(dw_2) = E[F],$$

where $\mu_n$ is the Gauss measure on $\mathbb{R}^n$, $\mu_2$ is the image of $\mu$ under the map $I_W - \tilde{\pi}_n$ and we have used the identity

$$\det{}_2(I_H + \nabla u_n) = \det{}_2(I_{\mathbb{R}^n} + \partial a(\delta h_1, \ldots, \delta h_n)).$$

Let $S_n$ be the inverse of the shift $T_n = I_W + u_n$. It is easy to see that $S_n$ is of the form $I_W + v_n$, with $v_n = -u_n \circ S_n$. Hence, using the ordinary differential calculus, we have

$$\|\nabla v_n\| \leq 1 + \alpha$$

and
$$\|\nabla v_n\|_2 \le \alpha \|\nabla u\|_2,$$
where $\|\cdot\|$ denotes the operator norm and $\|\cdot\|_2$ denotes the Hilbert-Schmidt norm. Consequently, from Corollary 1, there exists some $\lambda > 0$, such that
$$\sup_n \left( E[e^{\lambda |v_n|^2} + e^{\lambda |u_n|^2}] \right) < \infty.$$
Let us denote by $L_n$ the density corresponding to $v_n$, i.e.,
$$L_n = \det_2(I + \nabla v_n) \exp -\delta v_n - \frac{1}{2}|v_n|^2.$$
We claim that
$$\sup_n \left( E[|\Lambda_n \log \Lambda_n| + |L_n \log L_n|] \right) < \infty.$$
Let us show this for $\Lambda_n$'s: in fact
$$
\begin{aligned}
E[|\Lambda_n \log \Lambda_n|] &= E[|\log \Lambda_n \circ S_n| \\
&\le E[|\log \det_2(I + \nabla u_n \circ S_n)| + |\delta u_n \circ S_n| + \frac{1}{2}|u_n \circ S_n|^2].
\end{aligned}
$$
We have (cf.[8])
$$\delta u_n \circ S_n = -\delta v_n - |v_n|^2 + \text{trace}[(\nabla u_n \circ S_n).\nabla v_n],$$
hence
$$
\begin{aligned}
E[|\delta u_n \circ S_n|] &\le \|\delta v_n\|_{L^2} + E[|v_n|^2] + E[\|\nabla u_n \circ S_n\|_2 \|\nabla v_n\|_2] \\
&\le \|v_n\|_{L^2} + \|v_n\|_{L^2}^2 + \|\nabla v_n\|_{L^2(\mu, H \otimes H)} + \alpha K^2 \\
&\le \|v_n\|_{L^2}(1 + \|v_n\|_{L^2}) + \alpha K(1 + K),
\end{aligned}
$$
where $K = \|\nabla u\|_{L^\infty(\mu, H \otimes H)}$. Since $\sup_n E[\exp \lambda |v_n|^2]$ is finite, the claim for $\Lambda_n$'s is proved. For $(L_n)$ it is the same procedure, hence the proof of this case is omitted.

It follows from the de la Vallé-Poussin lemma that $(\Lambda_n; n \in I\!N)$ is uniformly integrable, hence, for any $F \in C_b(W)$, we have
$$
\begin{aligned}
E[F \circ T |\Lambda|] &= \lim_n E[F \circ T_n |\Lambda_n|] \\
&= E[F].
\end{aligned}
$$
To complete the proof, it suffices to remark that the measure $f \mapsto E[f \circ T]$ is equivalent to the measure $f \mapsto E[f \circ T |\Lambda|]$ since $|\Lambda|$ is almost surely different than zero and the latter is equal to the Wiener measure. $\|QED$

**Proposition 1**   Let $u \in D_{p,1}$ be strongly monotone, then we have

$$E[F \circ T | \Lambda |] \le E[F],$$

for any $F \in C_b^+(W)$. In particular, the forward image of the restriction of $\mu$ to the set $M = \{w : \det_2(I + \nabla u(w)) \ne 0\}$ is absolutely continuous with respect to $\mu$.

**Proof:**   For the approximating sequence of shifts, constructed as in the proof of the theorem, we have again

$$E[F \circ T_n | \Lambda_n |] = E[F],$$

then the proof follows from the Fatou lemma.                              ‖QED

**Theorem 2**   Let $u \in D_{p,1}$ be monotone, then we have

$$E[F \circ T | \Lambda |] \le E[F],$$

for any $F \in C_b^+(W)$. In particular, the forward image of the restriction of $\mu$ to the set $M = \{w : \det_2(I + \nabla u(w)) \ne 0\}$ is absolutely continuous with respect to $\mu$.

**Proof:**   Let $u_n$ be defined as

$$u_n = (1 - \frac{1}{n})u.$$

Since $(\nabla u h, h) \ge -|h|^2$, we have $((I_H \nabla u_n)h, h) \ge \frac{1}{n}|h|^2$. Then the result follows from the Proposition 1 and the Fatou lemma.

‖QED

We have also the following elementary

**Corollary 2**   Suppose that a sequence $(u_n; n \in \mathbb{N})$ converges to $u$ in $D_{p,1}$ for some $p > 1$. If the shifts $I_W + u_n$ are monotone for each $n \in \mathbb{N}$ then the shift $T = I_W + u$ is also monotone. Consequently the conclusion of the Theorem 2 is valid.

**Theorem 3** Suppose that $u : W \rightarrow H$ is a random variable such that there exists an increasing sequence of measurable sets $(W_n; n \in I\!N)$ covering $W$ almost surely. Suppose also that for any $n \in I\!N$, there exists some $u_n$ whose shift is monotone and $u_n = u$ on $W_n$ almost surely. Then we have

$$E[F \circ T | \Lambda|] \leq E[F],$$

for any $F \in C_b^+(W)$, in particular the conclusion of the Theorem 2 is valid.

**Proof:** Let us note first that, because of the locality of the operators $\delta$ and $\nabla$ ([1]), $\delta u$ and $\nabla u$ are well-defined.

We have, for any $n \in I\!N$,

$$
\begin{aligned}
E[1_{W_n} F \circ T | \Lambda|] &= E[1_{W_n} F \circ T_n | \Lambda_n|] \\
&\leq E[F \circ T_n | \Lambda_n|] \\
&\leq E[F],
\end{aligned}
$$

where $T_n$ is the shift and $\Lambda_n$ is the density corresponding to $u_n$. Since $(W_n; n \in I\!N)$ is increasing to $W$, the proof follows. $\qquad \|QED$

**Definition 2** A shift satisfying the hypothesis of the Theorem 3 will be called locally monotone.

**Theorem 4** Suppose that $u$ is an $H$-valued random variable whose corresponding shift is locally monotone. Suppose that $u$ is also locally $H - C^1$. Then for almost all $w \in M = \{w : \det_2(I + \nabla u(w)) \neq 0\}$, the set $T^{-1}\{w\}$ has at most one element, hence it is injective on $M$.

**Proof:** From [10], we have

$$E[F \circ T | \Lambda|] = E[F \, N(w, M)],$$

where $N(w, M)$ is the cardinal of the set $T^{-1}\{w\} \cap M$. From Theorem 3, we have

$$E[F \circ T | \Lambda|] \leq E[F],$$

for any $F \in C_b^+(W)$, hence $N(w, M) \leq 1$, and this implies that $T$ is injective on $M$. $\qquad \|QED$

**Corollary 3**   Suppose that $u$ is locally $H - C^1$, its shift is locally monotone and that the set of nondegeneracy $M$ has full $\mu$-measure. Then $T$ is almost surely injective.

**Remark:**   Note that if $T = I_W + u$ is strongly monotone and locally $H - C^1$, then $W = M$ almost surely.

**Corollary 4**   Suppose that $T = I_W + u$ is strongly monotone, $u$ is locally $H - C^1$ and that $\nabla u$ is in $L^\infty(\mu, H \otimes H)$. Then $T$ is almost surely a bijection of $W$.

**Proof:**   From Theorem 1, we have

$$E[F \circ T|\Lambda|] = E[F],$$

since $u$ is locally $H - C^1$, we have also

$$E[F \circ T|\Lambda|] = E[F\, N(w, M)].$$

Hence $N(w, M) = N(w, W) = 1$ almost surely.     ‖$QED$

# References

[1] N. Bouleau and F. Hirsch: Dirichlet Forms and Analysis on Wiener Space. De Gruyter Studies in Math., Vol. 14, Berlin-New York, 1991.

[2] F. E. Browder: "The solvability of non-linear functional equations". Duke Math. Jour. Vol. 30, pp. 557-566 (1963).

[3] S. Kusuoka: "The nonlinear transformation of Gaussian measures on Banach space and its absolute continuity, I", J. Fac. Sci. Univ. Tokyo, Sect.IA, Math. 29, pp. 567–598 (1982).

[4] A. S. Üstünel: "Intégrabilité exponentielle de fonctionnelles de Wiener". CRAS, Paris, Série I, Vol. 315, p.279-282 (1992).

[5] A. S. Üstünel: "Exponential tightness of Wiener functionals". In Stochastic Analysis and Related Topics, Proceedings of the Fourth Oslo-Silivri Workshop, p. 265-275. Stochastic Monographs, Vol.8, Gordon and Breach, 1994.

[6] A. S. Üstünel: "Some exponential moment inequalities for the Wiener functionals". Preprint, to appear in Jour. Func. Analysis.

[7] A. S. Üstünel: Analysis on Wiener Space. To appear in Lecture Notes in Mathematics.

[8] A. S. Üstünel and M. Zakai: "Transformations of Wiener measure under anticipative flows". Proba. Theory Relat. Fields 93, p.91-136 (1992).

[9] A. S. Üstünel and M. Zakai:"Applications of the degree theorem to absolute continuity on Wiener space". Probab. Theory Relat. Fields, vol. 95, p. 509-520 (1993).

[10] A. S. Üstünel and M. Zakai:"Transformation of the Wiener measure under non-invertible shifts". Probab. Theory Relat. Fields 99, p. 485-500 (1994).

A. S. Üstünel : ENST, Dépt. Réseaux, 46, rue Barrault, 75013 Paris, France

M. Zakai : Technion, Israel Institute of Technology, Dept. Electrical Engineering, 32000 Haifa, Israel.

# Hilbert Space Methods Applied to Elliptic Stochastic Partial Differential Equations

## Gjermund Våge

Department of Mathematical Sciences
Norwegian Institute of Technology
University of Trondheim
N-7034 Trondheim, NORWAY
e-mail: vage@imf.unit.no

## 1  Introduction

Over the last couple of years there has been a growing interest in stochastic partial differential equations (SPDEs). Various methods have been used to study SPDEs. Here we apply white noise analysis to obtain abstract existence and uniqueness theorems. More specifically we combine the ideas of Kondratiev spaces with the variational formulation for elliptic partial differential equations to study elliptic SPDEs. Hyperbolic and parabolic SPDEs can be treated similarly (see [10]).

To illustrate our ideas we prove existence and uniqueness of solution, $u$, for

$$-\nabla \cdot (F \diamond \nabla u) \;=\; f \text{ for } x \in D \tag{1}$$

$$u|_{\partial D} \;=\; g|_{\partial D}, \tag{2}$$

where $F$, $f$, and $g$ are given stochastic processes. If $F$ is the Wick exponential of smoothed white noise, $F = \exp^\diamond W_{\phi_x}$, we obtain the pressure equation in a stochastic medium, first solved in [3]. This equation is a model for flow in an (stochastic) isotropic porous media where $F$ is the permeability.

In [6], the authors define the Kondratiev spaces $(\mathscr{S})^\rho := \cap_{k=0}^{\infty}(\mathscr{S})^{\rho,k}$ endowed with the projective limit topology and $(\mathscr{S})^{-\rho} := \cup_{k=0}^{\infty}(\mathscr{S})^{-\rho,-k}$ endowed with the inductive limit topology. We, however, want to apply the Lax-Milgram theorem to obtain existence and uniqueness theorems for equations as (1), (2). This requires studying the problem in a space with a Hilbert space structure that captures the behavior in the space variable. We therefore introduce a family of Hilbert spaces $(\mathscr{S})^{\rho,k,m}$, which are isomorphic to the tensor product between the Kondratiev Hilbert space $(\mathscr{S})^{\rho,k}$ and the Sobolev space $H^m(D)$. $(\mathscr{S})^{\rho,k,m}$ has the additional advantage that it is possible to prove regularity results for the solution in the space variable. This is not possible using the theory developed in [3], [4], [5], [7], [8], and [9].

The Lax-Milgram theorem requires bounds on the bilinear form associated with the differential operator. To verify these conditions we have to obtain bounds on the norm of the Wick product in terms of the norm of the operands. We thus obtain a family of generalized noises which generate positive continuous bilinear forms. This approach has the advantage that it is easy to verify existence and uniqueness of solution for a large class of elliptic SPDEs.

## 2  Basic Definitions and Theorems

Let $\mathscr{S} := \mathscr{S}(\mathbb{R}^d)$ denote the Schwartz functions on $\mathbb{R}^d$, endowed with the usual Fréchet topology, and $\mathscr{S}' := \mathscr{S}'(\mathbb{R}^d)$ denote its dual endowed with the weak-$*$ topology. Then the Bochner-Minlos theorem ensures the existence of a probability measure $\mu$ on the Borel sets of $\mathscr{S}'$, $\mathscr{B} := \mathscr{B}(\mathscr{S}')$, satisfying

$$\int_{\mathscr{S}'} e^{i\langle \omega, \phi \rangle} d\mu(\omega) = \exp(-\frac{1}{2}\|\phi\|^2) \text{ for every } \phi \in \mathscr{S},$$

where $\|\phi\|^2 = \int_{\mathbb{R}^d} \phi(x)^2 \, dx$. The triple $(\mathscr{S}', \mathscr{B}, \mu)$ is called the white noise probability space.

If we apply the Bochner-Minlos theorem to the inverse Fourier transform of the Fourier transform of a function $f \in C_0^\infty(\mathbb{R})$ we obtain the useful identity

$$E[f(\langle \cdot, \phi \rangle)] = \frac{1}{\sqrt{2\pi}\,\|\phi\|} \int_{\mathbb{R}} f(x) \exp(-\frac{x^2}{2\|\phi\|^2}) \, dx. \tag{3}$$

When $f(x) = x^2$ is approximated from below, on an increasing sequence of compact sets converging to $\mathbb{R}$, (3) yields the isometry

$$E[\langle \cdot, \phi \rangle^2] = \|\phi\|^2 \tag{4}$$

in the limit. Using (4) we can extend the action of $\omega$ to functions $\phi \in L^2(\mathbb{R}^d)$ by defining

$$\langle \omega, \phi \rangle := \lim_{n \to \infty} \langle \omega, \phi_n \rangle \ (\text{limit in } L^2(\mu)),$$

where $\{\phi_n\}_{n=1}^{\infty} \subset \mathscr{S}$ is some sequence of functions converging to $\phi$ in $L^2(\mathbb{R}^d)$.

Let

$$h_n(x) = (-1)^n e^{x^2/2} \frac{d^n}{dx^n} e^{-x^2/2}, \ n \in \mathbb{N}_0 := \{0, 1, \ldots\}$$

denote the Hermite polynomials and define the Hermite functions by

$$\xi_n(x) := \pi^{-1/4}((n-1)!)^{-1/2} e^{-x^2/2} h_{n-1}(\sqrt{2}\,x), \ n \in \mathbb{N} := \{1, 2, \ldots\}.$$

Throughout this paper $\{e_i\}_{i=1}^{\infty} \subset \mathscr{S}$ denotes some fixed orthonormal basis for $L^2(\mathbb{R}^d)$ which is derived by taking tensor products of the $\xi_n$. Define

$$H_\alpha(\omega) := \prod_{j=1}^{m} h_{\alpha_j}(\theta_j(\omega))$$

when $\alpha = (\alpha_1, \ldots, \alpha_m)$ and

$$\theta_j(\omega) := \langle \omega, e_j \rangle.$$

It is proved in [2] that

$$\{H_\alpha(\omega) : \alpha \in \mathbb{N}_0^m \text{ for some } m = 0, 1, \ldots\}$$

forms an orthogonal basis for $L^2(\mu) := L^2(\mathscr{S}', \mathscr{B}, \mu)$, with $E[H_\alpha H_\beta] = \delta_{\alpha,\beta}\alpha!$.

We now turn to define the Kondratiev Hilbert spaces. For $-1 \leq \rho \leq 1$ and $k \in \mathbb{R}$ we consider the inner product spaces $(\mathscr{S})^{\rho,k}$ defined by

$$(\mathscr{S})^{\rho,k} := \{f = \sum_\alpha f_\alpha H_\alpha : f_\alpha \in \mathbb{R} \text{ and } \|f\|_{\rho,k} < \infty\}$$

where $\| \cdot \|_{\rho,k}$ is the norm associated with the inner product

$$(f, g)_{\rho,k} := \sum_\alpha f_\alpha g_\alpha (\alpha!)^{1+\rho} (2\mathbb{N})^{\alpha k}. \tag{5}$$

In (5) we use the definitions $\alpha! = \prod_{j=1}^{m} \alpha_j!$ and

$$(2\mathbb{N})^\alpha := \prod_{j=1}^{m} (2^d \beta_1^{(j)} \beta_2^{(j)} \cdots \beta_d^{(j)})^{\alpha_j}$$

if $\alpha = (\alpha_1, \alpha_2, \ldots, \alpha_m)$. $\beta^{(j)} = (\beta_1^{(j)}, \beta_2^{(j)}, \ldots, \beta_d^{(j)})$ is multi-index number $j$ in the fixed ordering of all multi-indices $\beta = (\beta_1, \ldots, \beta_d)$ related to $\{e_j\}$ by

$$e_j = \xi_{\beta_1^{(j)}} \otimes \xi_{\beta_2^{(j)}} \otimes \cdots \otimes \xi_{\beta_d^{(j)}}.$$

It is not difficult to verify that $(\mathscr{S})^{\rho,k}$ equipped with the inner product (5) is a separable Hilbert space when $-1 \leq \rho \leq 1$ and $k \in \mathbb{R}$.

**Remark 1** *When $\rho \in [-1, 0)$ and $k < 0$ an element in $(\mathscr{S})^{\rho,k}$ is a formal sum, i.e., the sum $\sum_\alpha f_\alpha H_\alpha$ does not necessarily converge in $L^1(\mu)$.*

*The reason these spaces are not considered for $|\rho| > 1$ is that in this case it is not possible to define the S-transform (see [6]).*

From the definition of $(\mathscr{S})^{\rho,k}$ we see $(\mathscr{S})^{0,0} = L^2(\mu)$. Moreover for fixed $0 \leq \rho \leq 1$

$$(\mathscr{S})^{\rho,k} \subset (\mathscr{S})^{\rho,\ell} \subset (\mathscr{S})^{\rho,0} \subseteq (\mathscr{S})^{-\rho,0} \subset (\mathscr{S})^{-\rho,-\ell} \subset (\mathscr{S})^{-\rho,-k} \qquad (6)$$

for every $k > \ell > 0$, and $(\mathscr{S})^{-\rho,-k}$ is the dual of $(\mathscr{S})^{\rho,k}$ since the duality bracket is given by

$$\langle F, f \rangle_{\rho,k} := \sum_\alpha F_\alpha f_\alpha \alpha!,$$

for $F = \sum_\alpha F_\alpha H_\alpha \in (\mathscr{S})^{-\rho,-k}$ and $f = \sum_\alpha f_\alpha H_\alpha \in (\mathscr{S})^{\rho,k}$.

We now introduce a family of Hilbert spaces which turn out to be useful when stochastic partial differential equations are investigated. Fix an open set $D \subseteq \mathbb{R}^d$ and let $(\cdot, \cdot)_{m,D}$, or simply $(\cdot, \cdot)_m$ if $D$ is clear from the context, denote the usual inner product on the real Sobolev space $H^m(D)$ for $m = 0, 1, \ldots$. We then define the inner product

$$(f, g)_{\rho,k,m,D} := \sum_\alpha (f_\alpha, g_\alpha)_{m,D} (\alpha!)^{1+\rho} (2\mathbb{N})^{k\alpha} \qquad (7)$$

on the set of functions of the form

$$f(x) = \sum_\alpha f_\alpha(x) H_\alpha,$$

where $f_\alpha(x) \in H^m(D)$ for every multi-index $\alpha$.

**Definition 1** *Let $(\mathscr{S})^{\rho,k,m}(D)$ (resp. $(\mathscr{S})_0^{\rho,k,m}(D)$) denote the set of functions*

$$\{f(x) = \sum_\alpha f_\alpha(x) H_\alpha \ : \ f_\alpha \in H^m(D) \, \forall \alpha \ (resp. \ H_0^m(D) \, \forall \alpha), \ and$$

$$\|f\|_{\rho,k,m,D} := (f, f)_{\rho,k,m,D}^{1/2} < \infty\}$$

*equipped with the inner product (7). We write $(\mathscr{S})^{\rho,k,m}$, $(\mathscr{S})_0^{\rho,k,m}$, and $\|\cdot\|_{\rho,k,m}$ if $D$ is clear from the context.*

285

It is not difficult to prove that if $-1 \leq \rho \leq 1$ and $k \in \mathbb{R}$, then $(\mathscr{S})^{\rho,k,m} \cong (\mathscr{S})^{\rho,k} \otimes H^m(D)$ and $(\mathscr{S})_0^{\rho,k,m} \cong (\mathscr{S})^{\rho,k} \otimes H_0^m(D)$ for $m = 0, 1, \dots$.

For $m \in \mathbb{N}_0$ we recover the inclusions (6):

$$(\mathscr{S})^{\rho,k,m} \subset (\mathscr{S})^{\rho,\ell,m} \subset (\mathscr{S})^{\rho,0,m} \subseteq (\mathscr{S})^{-\rho,0,m} \subset (\mathscr{S})^{-\rho,-\ell,m} \subset (\mathscr{S})^{-\rho,-k,m}$$

when $\rho \in [0,1]$ and $k > \ell > 0$. Moreover, for fixed $\rho$ and $k$ we find that

$$(\mathscr{S})^{\rho,k,m} \subseteq (\mathscr{S})^{\rho,k,n}$$

when $m, n \in \mathbb{N}_0$ and $m \geq n$. Combining these results we obtain

$$(\mathscr{S})^{\rho_1,k_1,m} \subseteq (\mathscr{S})^{\rho_2,k_2,n}$$

if $\rho_1 \geq \rho_2$, $k_1 \geq k_2$, and $m \geq n$. $(\mathscr{S})^{\rho_1,k_1,m}$ is a proper subset of $(\mathscr{S})^{\rho_2,k_2,n}$ if and only if one of the inequalities are strict. Similar results are also valid for $(\mathscr{S})_0^{\rho,k,m}$.

So far we have constructed the Hilbert spaces $(\mathscr{S})^{\rho,k,m}$ and $(\mathscr{S})_0^{\rho,k,m}$ and considered basic relations between the spaces. We now turn to study properties and define operations on elements from them.

One of the reasons for introducing $(\mathscr{S})^{\rho,k,m}$ and $(\mathscr{S})_0^{\rho,k,m}$ for $m \geq 1$, is that elements from these spaces possess derivatives.

**Definition 2** *Let $D \subseteq \mathbb{R}^d$ be an open set and $m \in \mathbb{N}$. If $\beta \in \mathbb{N}_0^d$ and $|\beta| \leq m$ we define*

$$\frac{\partial^\beta}{\partial x^\beta} f(x) := \sum_\alpha \frac{\partial^\beta f_\alpha}{\partial x^\beta}(x) H_\alpha,$$

*for any $f(x) = \sum_\alpha f_\alpha(x) H_\alpha \in (\mathscr{S})^{\rho,k,m}$. We interpret $\partial^\beta f_\alpha / \partial x^\beta$ in the usual $L^2(D)$ sense and often write $\partial_x^\beta$ or $\partial^\beta$ for $\partial^\beta / \partial x^\beta$.*

With this definition

$$\partial_x^\beta : (\mathscr{S})^{\rho,k,m} \to (\mathscr{S})^{\rho,k,m-|\beta|}$$

is a continuous linear operator, if $|\beta| \leq m$.

**Example:** The smoothed white noise process is usually defined as

$$W(\phi, x, \omega) := \langle \omega, \phi_x \rangle,$$

where $\phi_x(\cdot) := \phi(\cdot - x)$ and $\phi \in \mathscr{S}$. For each $x \in \mathbb{R}^d$ we obtain from (4) that

$$\langle \omega, \phi_x \rangle = \langle \omega, \sum_{i=1}^\infty (\phi_x, e_i)_{0,\mathbb{R}^d} e_i \rangle = \sum_{i=1}^\infty (\phi_x, e_i)_{0,\mathbb{R}^d} H_{e_i}(\omega) \text{ (in } L^2(\mu)),$$

where $\varepsilon_i = (0, \ldots, 0, 1)$ denotes the multi-index whose only nonzero entry is a 1 in the $i$th position. Lebesgue's monotone convergence theorem yields

$$\|\langle \omega, \phi_x \rangle\|^2_{\rho,k,0} = \int_D \left[ \sum_{i=1}^{\infty} (\phi_x, e_i)^2_{0,\mathbf{R}^d} (2\mathrm{N})^{k\varepsilon_i} \right] dx.$$

Since $\sum_{i=1}^{\infty} (\phi_x, e_i)^2_{0,\mathbf{R}^d} = \|\phi\|^2_{0,\mathbf{R}^d} < \infty$ for every $x \in \mathbf{R}^d$, the sum on the right hand side can not converge for all $\phi$ unless $(2\mathrm{N})^{k\varepsilon_i}$ is uniformly bounded in $i$. Hence, if we define

$$W_{\phi_x} := \sum_{i=1}^{\infty} (\phi_x, e_i)_{0,\mathbf{R}^d} H_{\varepsilon_i},$$

then $W_{\phi_x} \in (\mathscr{S})^{\rho,k,0}$ for all $\phi \in \mathscr{S}$ when $-1 \leq \rho \leq 1$, $k \leq 0$, and $D$ is a bounded open set in $\mathbf{R}^d$. Moreover, $W_{\phi_x}(\omega) = W(\phi, x, \omega)$ in $(\mathscr{S})^{0,0,0}$.

More generally, viewing $(\phi_x, e_i)_{0,\mathbf{R}^d}$ as the convolution of two Schwartz functions we have

$$\partial_x^{\beta} (\phi_x, e_i)_{0,\mathbf{R}^d} = (\partial_x^{\beta} \phi_x, e_i)_{0,\mathbf{R}^d}$$

for every $\beta \in \mathbf{N}_0^d$, and Definition 2 implies that

$$\partial_x^{\beta} W_{\phi_x} = \sum_{i=1}^{\infty} (\partial_x^{\beta} \phi_x, e_i) H_{\varepsilon_i}.$$

Since $\partial^{\beta} \phi \in \mathscr{S}(\mathbf{R}^d) \subset L^2(\mathbf{R}^d)$, summation over all $|\beta| \leq m$ shows $W_{\phi_x} \in (\mathscr{S})^{\rho,k,m}$ for all $-1 \leq \rho \leq 1$, $k \leq 0$, $m \in \mathbf{N}_0$ and bounded open sets $D \subset \mathbf{R}^d$.

Let $\tilde{\phi}(z) = \phi(-z)$, then $\langle \omega, \phi_x \rangle = \omega * \tilde{\phi}(x)$. Since $\omega * \tilde{\phi}(x)$ is a $C^{\infty}$-function with at most polynomial growth in $x$, $W_{\phi_x}$ can not in general belong to $(\mathscr{S})^{\rho,k,0}$ for unbounded sets $D$. For the same reason $W_{\phi_x}$ does not generally belong to $(\mathscr{S})^{\rho,k,0}_0$. But by extending the proof of Theorem 2.1 in [12] to $\mathbf{R}^d$, it is possible to show that $\sum_{\alpha} (2\mathrm{N})^{k\alpha} < \infty$ if $k < -1 - \log_2 d$ (see [10]). Hence the results above can be extended to unbounded sets $D \subseteq \mathbf{R}^d$ by requiring $k < -1 - \log_2 d$. Note also that

$$\frac{\partial^{\beta}}{\partial x^{\beta}} W_{\phi_x}(\omega) = \frac{\partial^{\beta}}{\partial x^{\beta}} \langle \omega, \phi_x \rangle = \frac{\partial^{\beta}}{\partial x^{\beta}} (\omega * \tilde{\phi})(x) = \omega * (\partial^{\beta} \tilde{\phi})(x) = \langle \omega, \partial_x^{\beta} \phi_x \rangle$$

for every $\omega$ and $\phi$. Thus for $W_{\phi_x}$ our definition of differentiation agrees with the ordinary one.

A closely related process we use in examples, is the singular white noise

$$W_x := \sum_{i=1}^{\infty} e_i(x) H_{\varepsilon_i}.$$

Formally, $W_x$ can be thought of as $W_{\delta_x}$ where $\delta_x$ is Dirac's point mass at $x$. It is not difficult to show that $W_x \in (\mathscr{S})^{\rho,k,0}(D)$ for any open set $D \subseteq \mathbb{R}^d$ when $-1 \leq \rho \leq 1$, and $k < -1 - \log_2 d$.

We would also like to compute the expectation of elements in $(\mathscr{S})^{\rho,k,m}$ for general $\rho$, $k$, and $m \in \mathbb{N}_0$. Inspired by [3] we observe that if $f = \sum_\alpha f_\alpha H_\alpha \in (\mathscr{S})^{0,0,0}$, then

$$E[f] = E[f \cdot 1] = \sum_\alpha f_\alpha E[H_\alpha \cdot H_{(0,0,\ldots)}] = f_{(0,0,\ldots)} \in L^2(D),$$

using

$$E[H_\alpha H_\beta] = \begin{cases} 0 & \text{if } \alpha \neq \beta \\ \alpha! & \text{if } \alpha = \beta \end{cases} .$$

It is therefore reasonable to define:

**Definition 3** *The* (generalized) expectation *of* $f = \sum_\alpha f_\alpha H_\alpha \in (\mathscr{S})^{\rho,k,m}$ *(resp.* $(\mathscr{S})_0^{\rho,k,m}$*) is the deterministic function*

$$E[f] := f_{(0,\ldots,0)}(x) : D \to \mathbb{C},$$

*which belongs to* $H^m(D)$ *(resp.* $H_0^m(D)$*.)*

We use the term generalized expectation since an $f \in (\mathscr{S})^{\rho,k,m}$ is not necessarily integrable with respect to the measure $\mu$.

Before we can proceed further we need the Wick product.

**Definition 4** *If* $f = \sum_\alpha f_\alpha H_\alpha$ *and* $g = \sum_\alpha g_\alpha H_\alpha$ *are two formal series, we define their* Wick product, $f \diamond g$, *to be the formal series*

$$f \diamond g := \sum_{\alpha,\beta} f_\alpha g_\beta H_{\alpha+\beta} = \sum_\gamma \Big( \sum_{\alpha+\beta=\gamma} f_\alpha g_\beta \Big) H_\gamma.$$

Recall that $L^2(\mu)$ is not closed under Wick multiplication. To see this let, for example, $f = H_{\varepsilon_1}$, then $g \mapsto f \diamond g$ is a densely defined unbounded linear operator on $L^2(\mu)$. If $f, g \in (\mathscr{S})^{0,0,0}$ we have the additional problem that $f_\alpha g_\beta$ need not belong to $L^2(D)$. To provide conditions on $f$ such that $g \mapsto f \diamond g$ gives a continuous linear operator on $(\mathscr{S})^{-1,k,0}$ we introduce the Banach space $\mathscr{F}_\ell$. For open $D \subseteq \mathbb{R}^d$ and $\ell \in \mathbb{R}$ we define

$$\mathscr{F}_\ell(D) := \{ f(x) = \sum_\alpha f_\alpha(x) H_\alpha \ :$$

$$f_\alpha(x) \text{ is measurable on } D \text{ for every } \alpha \text{ and}$$

$$\|f\|_{\ell,*} := \operatorname*{ess\,sup}_{x \in D} \Big( \sum_\alpha |f_\alpha(x)| (2\mathbb{N})^{\ell\alpha} \Big) < \infty \}.$$

We suppress the set, $D$, in the notation whenever it is clear from the context.

**Proposition 1** *Let $D$ be an open subset of $\mathbb{R}^d$ and $\ell \in \mathbb{R}$. Then $f \in \mathscr{F}_\ell$ defines a continuous linear operator on $(\mathscr{S})^{-1,k,0}$ by $g \mapsto f \diamond g$ when $k \leq 2\ell$. Moreover*

$$\|f \diamond g\|_{-1,k,0} \leq \|f\|_{k/2,*}\|g\|_{-1,k,0} \leq \|f\|_{\ell,*}\|g\|_{-1,k,0} \text{ for } g \in (\mathscr{S})^{-1,k,0}.$$

*Proof:* See [10].

The corresponding result for ordinary Kondratiev spaces, $(\mathscr{S})^{-1,k}$, with no $x$ dependence becomes a corollary.

**Corollary 1** *If $f = \sum_\alpha f_\alpha H_\alpha$ is such that $s := \sum_\alpha |f_\alpha|(2\mathbb{N})^{k\alpha/2} < \infty$, then $g \mapsto f \diamond g$ defines a continuous linear operator on $(\mathscr{S})^{-1,k}$ with norm less than or equal to $s$. Moreover, if $\ell > k+1+\log_2 d$ for some real $\ell$, then any $f \in (\mathscr{S})^{-1,\ell}$ defines a continuous linear operator on $(\mathscr{S})^{-1,k}$ by $g \mapsto f \diamond g$ such that*

$$\|f \diamond g\|_{-1,k} \leq \left(\sum_\alpha (2\mathbb{N})^{(k-\ell)\alpha}\right)^{1/2}\|f\|_{-1,\ell}\|g\|_{-1,k} < \infty \text{ for } g \in (\mathscr{S})^{-1,k}.$$

Corollary 1 generalizes results as Corollary 4.22 in [2]. It seems impossible to extend the Corollary to Kondratiev spaces with $\rho > -1$ since there is no constant $K$ such that $(m+n)! \leq Km!n!$ for all $m, n \in \mathbb{N}$.

$W_{\phi_x}$ belongs to $\mathscr{F}_\ell$ for arbitrary open $D \subseteq \mathbb{R}^d$ and $\ell < -(1+\log_2 d)/2$. To see this note that by Schwarz' inequality

$$\|W_{\phi_x}\|_{\ell,*} = \operatorname{ess\,sup}_{x \in D} \sum_{i=1}^\infty |(\phi_x, e_i)_{0,\mathbb{R}^d}|(2\mathbb{N})^{\ell\varepsilon_i} \leq \|\phi\|_{0,\mathbb{R}^d}\left(\sum_\alpha (2\mathbb{N})^{2\ell\alpha}\right)^{1/2},$$

which is finite if $2\ell < -1 - \log_2 d$. Using 22.14.17 in [1] to show $|e_i(x)| \leq (2\pi^{-1/4})^d$ for $x \in \mathbb{R}^d$ and $i \in \mathbb{N}$, it can be shown that $W_x \in \mathscr{F}_\ell$ if $\ell < -1 - \log_2 d$.

When we turn to consider the pressure equation in a stochastic medium it is important to be able to determine when a bilinear form $b(g_1, g_2) = (f \diamond g_1, g_2)_{-1,k,0}$ is strictly positive, that is, if there exists a $C > 0$ such that $b(g,g) \geq C\|g\|^2_{-1,k,0}$ for every $g \in (\mathscr{S})^{-1,k,0}$. Let

$$\mathscr{P}_\ell(D) := \{f \in \mathscr{F}_\ell(D) : \exists A > 0 \text{ such that}$$
$$(f_{(0,\ldots,0)}g, g)_{0,D} \geq A\|g\|^2_{0,D} \,\forall g \in L^2(D)\}.$$

Note that $f \in \mathscr{F}_\ell$ ensures $b(\cdot, \cdot)$ is continuous on $(\mathscr{S})^{-1,k,0}$ for $k \leq 2\ell$. The second condition is necessary, otherwise $b(\cdot, \cdot)$ would not be strictly positive on the subspace $\{g(x)H_{(0,\ldots,0)} : g \in L^2(D)\}$ of $(\mathscr{S})^{-1,k,0}$. The following proposition shows the conditions also are sufficient, provided $k$ is sufficiently small.

**Proposition 2** *Let $D \subseteq \mathbb{R}^d$ be open and $f \in \mathscr{P}_\ell$ for some real $\ell$. Then there exist constants $K = K(f) \leq 2\ell$ and $C = C(K, f) > 0$ such that*

$$(f \diamond g, g)_{-1,k,0} \geq C\|g\|^2_{-1,k,0} \text{ for every } g \in (\mathscr{S})^{-1,k,0},$$

*when $k < K$.*

*Proof:* See [10].

A useful result for what follows is:

**Proposition 3** *Let $D \subseteq \mathbb{R}^d$ be open and $\ell \in \mathbb{R}$.*

(i) *If $f, g \in \mathscr{F}_\ell$, then $\|f \diamond g\|_{\ell,*} \leq \|f\|_{\ell,*}\|g\|_{\ell,*}$.*

(ii) *Suppose $G(x) = \sum_{n=0}^{\infty} c_n x^n$ is analytic on the open interval $(-R, R)$ for some $R > 0$. If $f \in \mathscr{F}_\ell$ with $\|f\|_{\ell,*} < R$ then*

$$G^\diamond(f) := \sum_{n=0}^{\infty} c_n f^{\diamond n} \in \mathscr{F}_\ell$$

*and*

$$\|G^\diamond(f)\|_{\ell,*} \leq \sum_{n=0}^{\infty} |c_n| \|f\|^n_{\ell,*} < \infty.$$

(iii) *If $f \in \mathscr{F}_\ell$, then $\exp^\diamond f := \sum_{n=0}^{\infty} f^{\diamond n}/n! \in \mathscr{P}_\ell$.*

*Proof:* See [10].

**Remark 2** *Results similar to Proposition 2 and 3 are easily obtained for the ordinary Kondratiev spaces.*

By Proposition 3, $\exp^\diamond W_x \in \mathscr{P}_\ell \subset \mathscr{F}_\ell$ for arbitrary $\ell < -1 - \log_2 d$ and open $D \subseteq \mathbb{R}^d$. Therefore $g \mapsto \exp^\diamond W_x \diamond g$ is a continuous linear operator on $(\mathscr{S})^{\rho,k,0}$ if $\rho = -1$ and $k \leq 2\ell$. It is in fact possible to show that $\rho = -1$ also is a necessary condition for $g \mapsto \exp^\diamond W_x \diamond g$ to be a continuous linear transformation on $(\mathscr{S})^{\rho,k,0}$.

# 3  Stochastic Elliptic Operators

Recently an explicit solution to the smoothed pressure equation in a stochastic medium was found in [3]. We use this equation as an example to illustrate how our ideas can be applied to elliptic SPDEs in general.

The main problem is to determine a suitable Hilbert space on which the bilinear form associated with the equation satisfies the conditions in the Lax-Milgram theorem (consult for example [11]). When we have found a suitable Hilbert space, which depends on the problem, the Lax-Milgram theorem implies existence of a unique solution to the corresponding variational problem.

**Example:** Fix $F \in \mathscr{P}_\ell(D)$ for some real $\ell$ and consider

$$-\nabla \cdot (F \diamond \nabla u) = f \text{ in } D \times \mathscr{S}', \qquad (8)$$

$$u|_{\partial D} = g|_{\partial D}, \qquad (9)$$

where $D \subset \mathbb{R}^d$ is an open set of finite width, that is, $D$ lies between two parallel hyperplanes. We now intend to find a variational formulation in the Hilbert space $(\mathscr{S})^{\rho,k,1}$. Suppose a solution $u \in (\mathscr{S})^{\rho,k,1}$ of (8) for a suitable pair $(\rho, k)$ is known, then

$$(-\nabla \cdot (F \diamond \nabla u), v)_{\rho,k,0} = (f, v)_{\rho,k,0} \qquad (10)$$

for every $v \in (\mathscr{S})_0^{\rho,k,1}$. Using the definition of $(\cdot, \cdot)_{\rho,k,0}$ and integrating each term by parts, we obtain the bilinear form

$$b_{\rho,k}(u, v) := (F \diamond \nabla u, \nabla v)_{\rho,k,0}$$

for the left hand side of (10).

Given $f \in (\mathscr{S})^{\rho,k,0}$ and $g \in (\mathscr{S})^{\rho,k,1}$, the variational formulation of (8), (9) becomes: Find $u \in (\mathscr{S})^{\rho,k,1}$ such that

(i) $u - g \in (\mathscr{S})_0^{\rho,k,1}$, and

(ii) $b_{\rho,k}(u, v) = (f, v)_{\rho,k,0}$ for every $v \in (\mathscr{S})_0^{\rho,k,1}$.

For simplicity we interpret the boundary condition in the generalized sense, (i).

Existence and uniqueness of solution will follow from the Lax-Milgram theorem if $(\rho, k)$ is chosen such that $b_{\rho,k}(u, v)$ is continuous on $(\mathscr{S})^{\rho,k,1} \times (\mathscr{S})^{\rho,k,1}$ and strictly positive on $(\mathscr{S})_0^{\rho,k,1} \times (\mathscr{S})_0^{\rho,k,1}$. When homogeneous boundary data are considered, that is if $g = 0$, it suffices to assume $b_{\rho,k}(\cdot, \cdot)$ is continuous on $(\mathscr{S})_0^{\rho,k,1} \times (\mathscr{S})_0^{\rho,k,1}$.

Schwarz' inequality gives

$$|b_{\rho,k}(u, v)| = |(F \diamond \nabla u, \nabla v)_{\rho,k,0}| \le \|F \diamond \nabla u\|_{\rho,k,0} \|\nabla v\|_{\rho,k,0}.$$

From Proposition 1 and the inequality $\| \cdot \|_{\rho,k,0} \le \| \cdot \|_{\rho,k,1}$ we conclude

$$|b_{-1,k}(u, v)| \le \|F\|_{k/2,*} \|u\|_{-1,k,1} \|v\|_{-1,k,1},$$

for every $u, v \in (\mathscr{S})^{-1,k,1}$, when $k \leq 2\ell$ and $\rho = -1$. Thus $b_{-1,k}(\cdot, \cdot)$ is a continuous bilinear form on $(\mathscr{S})^{-1,k,1} \times (\mathscr{S})^{-1,k,1}$ if $k \leq 2\ell$.

By Proposition 2 there exist constants $K(F) \leq 2\ell$ and $C = C(K, F) > 0$ such that

$$b_{-1,k}(u, u) \geq C\|\nabla u\|_{-1,k,0}^2, \text{ for all } u \in (\mathscr{S})_0^{-1,k,1},$$

when $k < K(F)$. Since $D$ is assumed to have finite width, Poincaré's inequality implies

$$b_{-1,k}(u, u) \geq C\tilde{C}\|u\|_{-1,k,1}^2, \quad \tilde{C} = \tilde{C}(D) > 0,$$

for every $u \in (\mathscr{S})_0^{-1,k,1}$ when $k < K(F)$.

The following theorem is an almost immediate consequence of the example.

**Theorem 1** *Let $D$ be an open subset of $\mathbb{R}^d$ of finite width and suppose $F \in \mathscr{P}_\ell(D)$ for some $\ell$. Then there exists a constant $K(F) \leq 2\ell$ such that if $k < K(F)$, (8), (9) has a unique variational solution $u \in (\mathscr{S})^{-1,k,1}$, for every $f \in (\mathscr{S})^{-1,k,0}$ and $g \in (\mathscr{S})^{-1,k,1}$.*

*Moreover, $E[u]$ is the variational solution of the deterministic problem that results upon replacing $F$, $f$, and $g$ in (8), (9) with $E[F]$, $E[f]$, and $E[g]$, respectively.*

*Proof:* It only remains to prove the last statement. Suppose $u = \sum_\alpha u_\alpha H_\alpha$ is the variational solution of (8), (9) and let $F = \sum_\alpha F_\alpha H_\alpha$. Then $E[u]$ satisfies

$$\sum_\alpha (F_\alpha \cdot \nabla E[u], \nabla v_\alpha)_{0,D} (2\mathbb{N})^{k\alpha} = \sum_\alpha (f_\alpha, v_\alpha)_{0D} (2\mathbb{N})^{k\alpha}$$

for every $v = \sum_\alpha v_\alpha H_\alpha \in (\mathscr{S})_0^{-1,k,1}$. In particular it must be true if $v = w H_{(0,0,\ldots)}$ for arbitrary $w \in H_0^1(D)$. The boundary condition becomes

$$E[u] - E[g] = u_{(0,0,\ldots)} - g_{(0,0,\ldots)} \in H_0^1(D),$$

which shows $E[u]$ satisfies the deterministic variational problem specified in the theorem and concludes the proof. $\square$

Proposition 3 provides many choices for $F \in \mathscr{P}_\ell$ we could solve for in the example. The case when $F$ is chosen to be the smoothed exponential white noise, $F = \exp^\circ W_{\phi_x}$, is of special interest. An explicit solution of (8), (9) with this choice for $F$ was found in [3] for deterministic data $f$ and $g$. It is difficult to compare the solutions since differentiation is defined differently in [3]. In principle, however, the solutions agree. By this we mean that our solution may be thought of as obtained by: (i) applying the Hermite transform to the problem, (ii) solving the resulting (complex) deterministic equation, and (iii) using the inverse Hermite transform to obtain the solution. This is the same strategy as the authors use in [3]. Our approach has the advantage that it lets us solve for various choices of $F$ and consider stochastic right hand sides as well as stochastic boundary data. Moreover, it captures regularity in the $x$-variable. The main disadvantage is that we do not obtain an explicit expression for the solution.

# 4   Concluding Remarks

A consequence of our Hilbert space approach is that we can obtain stability results. To see this assume (8), (9) are solved with respect to two noises $F_1$ and $F_2 \in \mathscr{P}_\ell$. Let $f \in (\mathscr{S})^{-1,k,0}$ and $g \in (\mathscr{S})^{-1,k,1}$ be given for a suitably chosen $k < 2\ell$. Suppose $u_1$ and $u_2 \in (\mathscr{S})^{-1,k,1}$ solves (8), (9) with respect to the noises $F_1$ and $F_2$, respectively. Since $b_i(u,v) = (F_i \diamond \nabla u, \nabla v)_{-1,k,0}$ is strictly positive on $(\mathscr{S})_0^{-1,k,1} \times (\mathscr{S})_0^{-1,k,1}$ ($i = 1, 2$), there is a $C > 0$ such that

$$
\begin{aligned}
C\|u_2 - u_1\|_{-1,k,1}^2 &\leq b_2(u_2 - u_1, u_2 - u_1) \\
&= (f, u_2 - u_1)_{-1,k,0} - b_2(u_1, u_2 - u_1) \\
&\quad + (F_1 \diamond \nabla u_1, \nabla(u_2 - u_1))_{-1,k,0} - (f, u_2 - u_1)_{-1,k,0} \\
&\leq \|F_1 - F_2\|_{k/2,*}\|u_1\|_{-1,k,1}\|u_2 - u_1\|_{-1,k,1}.
\end{aligned}
$$

Hence $\|u_2 - u_1\|_{-1,k,1}$ is bounded by $\|F_1 - F_2\|_{k/2,*}$.

In particular, if we let $F_1 = \exp^\diamond W_x$ and $F_\phi = \exp^\diamond W_{\phi_x}$ for any $\phi \in \mathscr{S}(\mathbb{R}^d)$ with $\|\phi\|_{L^1(\mathbb{R}^d)} \leq K$. (This condition on the test functions ensures $\|F_\phi\|_{k/2,*}$ is uniformly bounded. Therefore $b_\phi(\cdot, \cdot)$ is strictly positive with a constant $C > 0$ which is independent of $\phi$.) Repeated applications of Proposition 3 shows

$$
\begin{aligned}
\|\exp^\diamond W_x - \exp^\diamond W_{\phi_x}\|_{k/2,*} &= \|\exp^\diamond W_x \diamond (1 - \exp^\diamond(W_{\phi_x} - W_x))\|_{k/2,*} \\
&\leq \|\exp^\diamond W_x\|_{k/2,*}\|1 - \exp^\diamond(W_{\phi_x} - W_x)\|_{k/2,*} \\
&\leq \exp\|W_x\|_{k/2,*}(1 - \exp\|W_{\phi_x} - W_x\|_{k/2,*}).
\end{aligned}
$$

Thus $\|u_{\phi^{(n)}} - u_1\|_{-1,k,1} \to 0$ as $n \to \infty$, where $\phi^{(n)}(x) = n^d \phi(nx)$ for $n \in \mathbb{N}$ and $\phi \in \mathscr{S}(\mathbb{R}^d)$ with $\int \phi \, dx = 1$ and $\|\phi\|_{L^1(\mathbb{R}^d)} \leq K$. This shows that the smoothed problem considered in [3] is a reasonable approximation to the singular problem where $F = \exp^\diamond W_x$.

In the previous section we interpreted the boundary condition in the generalized sense, by requiring $u - g \in (\mathscr{S})_0^{\rho,k,m}(D)$ for a given boundary data $g \in (\mathscr{S})^{\rho,k,m}(D)$. This was done for simplicity. We could, alternatively, define stochastic trace spaces

$$(\mathscr{S})^{\rho,k,m-1/2}(\partial D) \cong (\mathscr{S})^{\rho,k} \otimes H^{m-1/2}(\partial D),$$

for $m \in \mathbb{N}$ and bounded open sets $D \subset \mathbb{R}^d$. If $D$ is sufficiently regular (see [11]), the trace operator has an inverse

$$Z_m : H^{1/2}(\partial D) \times \cdots \times H^{m-1/2}(\partial D) \to H^m(D),$$

with a natural extension

$$\tilde{Z}_m : (\mathscr{S})^{\rho,k,1/2}(\partial D) \times \cdots \times (\mathscr{S})^{\rho,k,m-1/2}(\partial D) \to (\mathscr{S})^{\rho,k,m}(D).$$

Suppose Dirichlet boundary data is specified in the stochastic trace spaces, then $\tilde{Z}_m$ provides an element $g \in (\mathscr{S})^{\rho,k,m}(D)$. Since the kernel of the trace operator on $(\mathscr{S})^{\rho,k,m}(D)$ is $(\mathscr{S})_0^{\rho,k,m}(D)$ we return to the generalized interpretation used above.

As mentioned above the main advantage with the approach we have presented here, is that it with simplicity can be applied to other elliptic SPDEs. It is for instance straightforward to obtain results similar to Theorem 1 for the differential equations

$$-\Delta u + F \diamond u \ = \ f \text{ in } D \times \mathscr{S}', \tag{11}$$

$$u|_{\partial D} \ = \ g|_{\partial D}, \tag{12}$$

and

$$\Delta(F \diamond \Delta u) \ = \ f \text{ in } D \times \mathscr{S}', \tag{13}$$

$$u|_{\partial D} \ = \ g|_{\partial D}. \tag{14}$$

In the forthcoming paper [10], hyperbolic and parabolic problems are considered as well.

# References

[1] Milton Abramowitz and Irene A. Stegun, *Handbook of Mathematical Functions*, Dover Publications, New York, 1972.

[2] Takeyuki Hida, Hui-Hsiung Kuo, Jürgen Potthoff, and Ludwig Streit, *White Noise*, Kluwer Academic Publishers, Dordrecht, 1993.

[3] Helge Holden, Tom Lindstrøm, Bernt Øksendal, Jan Ubøe, and Tu-Sheng Zhang, *The Pressure Equation for Fluid Flow in a Stochastic Medium*, Potential Analysis, in print.

[4] _____ , *Stochastic Boundary Value Problems. A White Noise Functional Approach*, Probab. Theory Relat. Fields **95** (1993), 391–419.

[5] _____ , *The Burgers Equation with a Noisy Force and the Stochastic Heat Equation*, Comm. PDE **19 (1& 2)** (1994), 119 – 141.

[6] Yuri G. Kondratiev, Peter Leukert, and Ludwig Streit, *Wick Calculus in Gaussian Analysis*, Manuscript (1994).

[7] Tom Lindstrøm, Bernt Øksendal, and Jan Ubøe, *Stochastic Differential Equations Involving Positive Noise*, in M. Barlow and N. Bingham (eds): Stochastic Analysis, Cambridge University Press (1991), 261–303.

[8] _____ , *Stochastic Modelling of Fluid Flow in Porous Media*, in S. Chen & J. Young (eds): Control Theory, Stochastic Analysis and Applications, World Scientific (1991), 156–172.

[9] Bernt Øksendal and Tu-Sheng Zhang, *The Stochastic Volterra Equation*, D. Nualart and Sang Solé: Barcelona Seminar on Stochastic Analysis (Birkhäuser 1993), 168–202.

[10] Gjermund Våge, *Hilbert Space Methods for Stochastic Partial Differential Equations*, Manuscript (1995).

[11] Joseph Wloka, *Partial Differential Equations*, Cambridge University Press, Cambridge, 1987.

[12] Tu-Sheng Zhang, *Characterizations of White Noise Test Functions and Hida Distributions*, Stochastics **41** (1992), 71–87.

# Progress in Probability

*Editors*

Professor Thomas M. Liggett
Department of Mathematics
University of California
Los Angeles, CA 90024-1555

Professor Charles Newman
Courant Institute of
Mathematical Sciences
251 Mercer Street
New York, NY 10012

Professor Loren Pitt
Department of Mathematics
University of Virginia
Charlottesville, VA 22903-3199

*Progress in Probability* is designed for the publication of workshops, seminars and conference proceedings on all aspects of probability theory and stochastic processes, as well as their connections with and applications to other areas such as mathematical statistics and statistical physics. It acts as a companion series to *Probability and Its Applications,* a context for research level monographs and advanced graduate texts.

We encourage preparation of manuscripts in some form of TeX for delivery in camera-ready copy, which leads to rapid publications, or in electronic form for interfacing with laser printers or typesetters.

Proposals should be sent directly to the editors or to:
Birkhäuser Boston, 675 Massachusetts Avenue, Cambridge, MA 02139, U.S.A.